Broadband Last Mile

Access Technologies for Multimedia Communications

Signal Processing and Communications

1. Digital Signal Processing for Multimedia Systems, *edited by Keshab K. Parhi and Takao Nishitani*

2. Multimedia Systems, Standards, and Networks, *edited by Atul Puri and Tsuhan Chen*

3. Embedded Multiprocessors: Scheduling and Synchronization, *Sundararajan Sriram and Shuvra S. Bhattacharyya*

4. Signal Processing for Intelligent Sensor Systems, *David C. Swanson*

5. Compressed Video over Networks, *edited by Ming-Ting Sun and Amy R. Reibman*

6. Modulated Coding for Intersymbol Interference Channels, *Xiang-Gen Xia*

7. Digital Speech Processing, Synthesis, and Recognition: Second Edition, Revised and Expanded, *Sadaoki Furui*

8. Modern Digital Halftoning, *Daniel L. Lau and Gonzalo R. Arce*

9. Blind Equalization and Identification, *Zhi Ding and Ye (Geoffrey) Li*

10. Video Coding for Wireless Communication Systems, *King N. Ngan, Chi W. Yap, and Keng T. Tan*

11. Adaptive Digital Filters: Second Edition, Revised and Expanded, *Maurice G. Bellanger*

12. Design of Digital Video Coding Systems, *Jie Chen, Ut-Va Koc, and K. J. Ray Liu*

13. Programmable Digital Signal Processors: Architecture, Programming, and Applications, *edited by Yu Hen Hu*

14. Pattern Recognition and Image Preprocessing: Second Edition, Revised and Expanded, *Sing-Tze Bow*

15. Signal Processing for Magnetic Resonance Imaging and Spectroscopy, *edited by Hong Yan*

16. Satellite Communication Engineering, *Michael O. Kolawole*

17. Speech Processing: A Dynamic and Optimization-Oriented Approach, *Li Deng*

18. Multidimensional Discrete Unitary Transforms: Representation: Partitioning and Algorithms, *Artyom M. Grigoryan, Sos S. Agaian, S.S. Agaian*

19. High-Resolution and Robust Signal Processing, *Yingbo Hua, Alex B. Gershman, Qi Cheng*

20. Domain-Specific Processors: Systems, Architectures, Modeling, and Simulation, *Shuvra Bhattacharyya; Ed Deprettere; Jurgen Teich*

21. Watermarking Systems Engineering: Enabling Digital Assets Security and Other Applications, *Mauro Barni, Franco Bartolini*

22. Biosignal and Biomedical Image Processing: MATLAB-Based Applications, *John L. Semmlow*

23. Broadband Last Mile: Access Technologies for Multimedia Communications, *edited by Nikil Jayant*

Broadband Last Mile

Access Technologies for Multimedia Communications

edited by
Nikil Jayant

Georgia Institute of Technology
Atlanta, Georgia, U.S.A.

Taylor & Francis
Taylor & Francis Group

Boca Raton   London   New York   Singapore

A CRC title, part of the Taylor & Francis imprint, a member of the
Taylor & Francis Group, the academic division of T&F Informa plc.

Published in 2005 by
CRC Press
Taylor & Francis Group
6000 Broken Sound Parkway NW, Suite 300
Boca Raton, FL 33487-2742

International Standard Book Number-10: 0-8247-5886-2 (Hardcover)
International Standard Book Number-13: 978-0-8247-5886-8 (Hardcover)
Library of Congress Card Number 2004061833

Library of Congress Cataloging-in-Publication Data

Broadband last mile: access technologies for multimedia communications / edited by Nikil Jayant.
     p. cm. -- (Signal processing and communications)
  Includes bibliographical references and index.
  ISBN 0-8247-5886-2
  1. Broadband communication systems. I. Jayant, N. (Nikil) II. Series.

TK5103.4.B7645 2005
621.382--dc22   2004061833

Taylor & Francis Group
is the Academic Division of T&F Informa plc.

Visit the Taylor & Francis Web site at
http://www.taylorandfrancis.com

and the CRC Press Web site at
http://www.crcpress.com

Series Introduction

Over the past 50 years, digital signal processing has evolved as a major engineering discipline. The fields of signal processing have grown from the origin of fast Fourier transform and digital filter design to statistical spectral analysis and array processing; image, audio, and multimedia processing; and shaped developments in high-performance VLSI signal processor design. Indeed, few fields enjoy so many applications — signal processing is everywhere in our lives.

When one uses a cellular phone, the voice is compressed, coded, and modulated using signal processing techniques. As a cruise missile winds along hillsides searching for the target, the signal processor is busy processing the images taken along the way. When we watch a movie in HDTV, millions of audio and video data are sent to our homes and received with unbelievable fidelity. When scientists compare DNA samples, fast pattern recognition techniques are used. On and on, one can see the impact of signal processing in almost every engineering and scientific discipline.

Because of the immense importance of signal processing and the fast-growing demands of business and industry, this series on signal processing serves to report up-to-date developments and advances in the field. The topics of interest include but are not limited to:

1. Signal theory and analysis
2. Statistical signal processing
3. Speech and audio processing
4. Image and video processing

5. Multimedia signal processing and technology
6. Signal processing architectures and VLSI design
7. Signal processing for communications
8. Communication technologies and services

We hope this series will provide the interested audience with high-quality, state-of-the-art signal processing literature through research monographs, edited books, and rigorously written textbooks by experts in their fields.

Prologue

The idea for this book came out of the editor's involvement in a National Academies study on the same topic that resulted in an NRC report entitled *Broadband: Bringing Home the Bits* (www.nap.edu). That report provided one of the first integrated snapshots of the subject from a broad perspective that included technology, economics, and policy; the current monograph delves into in-depth treatments of access technologies and the applications that need and support them. The access part of the end-to-end broadband communication system is called the last mile (sometimes also referred to, in a user-centric fashion, as the first mile). This book is about the technologies needed to make sure that the last mile is not a weak link in the broadband chain. Written by experts spanning the academic as well as industrial segments in the field, these self-contained sections deal with topics that fall into the disciplines of communications, networking, computing, and signal processing.

The first chapter of the collection sets the stage by providing a multidimensional view of broadband as well as self-contained treatments of the broadband-enabling technologies of media compression and content distribution. These discussions are fundamental, implied in later chapters, and generally independent of the physical pipe that is the focus of the central part of the book. Explicitly addressed in this chapter is the topic of applications and application classes that are the reason for broadband services.

The second chapter places the last mile in perspective by relating it to the backbone network and the edge that separates that core from the access link. This chapter addresses the overarching issue of end-to-end networking and information routing while providing brief previews of different physical arrangements in the last mile.

The central core of the book contains contemporary views of broadband pipes in the classes of copper, cable, fiber, wireless, and satellite. These chapters are up-to-date treatments of technologies that will coexist in various ways, anchored by the asymptotic importance of optical communications for unprecedented bandwidth in the downlink as well as uplink, and wireless, which offers the important attribute of flexibility and mobility. Copper (DSL) and cable (HFC) technologies are already using an increasing degree of fiber as they approach the home; additionally, they are increasingly aware of the need and value of a wireless segment in the last meters.

Following the description of physical broadband media is a perspective on the increasingly important topic of network management, with notions that are largely, but not always, unspecific to the physical pipe. Concluding the collection is a second, closing-the-loop section on applications and broadband services.

This book provides a collocated treatment of the physical pipes and network architectures that make possible the rich and increasingly personalized applications in the brave new world of increasingly pervasive broadband. We trust that the collection proves to be interesting to researchers as well as practitioners in the field.

Nikil Jayant
Director, Georgia Tech Broadband Institute

Acknowledgments

The artwork on the cover is inspired by illustrations prepared under the auspices of the Georgia Tech Broadband Institute (www.broadband.gatech.edu) and its predecessor, the Broadband Telecommunications Center. For their roles in creating or popularizing these visual icons of broadband, and for permissions to use them in the cover design, I thank my colleagues Daniel Howard, Mary Ann Ingram, Nan Jokerst, and John Limb. Thanks as well to my colleague Rex Smith for his deftful arrangement of the visuals at short notice. Finally, I am grateful to Stefany Wilson, Kim Keeling, and Barbara Satterfield, for their invaluable contributions during the compilation of this handbook.

Nikil Jayant

About the Editor

Dr. Nikil Jayant joined the faculty of the Electrical and Computer Engineering Department at Georgia Tech in July 1998, as a Georgia Research Alliance Eminent Scholar, the John Pippin Chair in wireless systems, and the director of the Georgia Tech Wireless Institute. In April 1999, he created the position of and became the first director of the Georgia Tech Broadband Institute, with cross-campus responsibilities in research and industry partnership in broadband access, lifestyle computing, and ubiquitous multimedia. In October 2000, he was named executive director of the Georgia Centers for Advanced Telecommunications Technology (GCATT).

Earlier, in his career at Bell Laboratories, Dr. Jayant created and managed the Signal Processing Research Department, the Advanced Audio Technology Department, and the Multimedia Communications Research Laboratory. He also initiated several new ventures for AT&T and Lucent Technologies, including businesses in Internet multimedia, wireless communications, and digital audio broadcasting. His research has been in the field of digital coding and transmission of information signals. Dr. Jayant has published 140 papers and authored or coauthored five books, including an IEEE reprint book, *Waveform Quantization and Coding* (1976); a fundamental textbook, *Digital Coding of Waveforms* (Prentice Hall,1984), coauthored with Peter Noll; an edited book, *Signal Compression* (World Scientific, 1998); a National Academies Committee monograph, *Broadband: Bringing Home the Bits* (NRC Press, 2002); and the current book on broadband last mile technologies. He is also the author of 35 patents. Technologies created by Dr. Jayant's research and leadership span several aspects of audiovisual communications.

Dr. Jayant received his Ph.D. in electrical communication engineering from the Indian Institute of Science in Bangalore, India. As part of this doctoral program, he was a research associate at Stanford University for 1 year prior to joining Bell Labs in 1968. Dr. Jayant has received several honors, including the IEEE Browder J. Thompson Memorial Prize Award (for the best IEEE publications by an author under 30 years of age, 1974); the IEEE Donald G. Find Prize Paper Award (for the best tutorial in an IEEE publication, 1995); and the Lucent Patent Recognition Award (1997). In 1998, he was inducted into the New Jersey Inventors Hall of Fame. Dr. Jayant was the founding editor-in-chief of the *IEEE ASSP Magazine*. He is a fellow of the IEEE, a recipient of the IEEE Third Millennium Medal, and a member of the National Academy of Engineering.

Most recently, Dr. Jayant served as chairperson of the National Academies Committee on Broadband Last Mile Technologies. In parallel, he cofounded EGTechnology, an Atlanta-based startup company engaged in creating broadband platforms, with initial focus on software for advanced television. He is also the founder and president of MediaFlow, a consulting company. He has served on the advisory board of NTT-DoCoMo (U.S.A.) and is currently a scientific advisor to the Singapore Institute for Infocomm Research.

Contributors

Dharma P. Agrawal University of Cincinnati, Cincinnati, Ohio

John Apostolopoulos Hewlett-Packard Labs, Palo Alto, California

Benny Bing Georgia Institute of Technology, Atlanta, Georgia

G.K. Chang Georgia Institute of Technology, Atlanta, Georgia

Bruce Currivan Broadcom Corporation, Irvine, California

Carlos de M. Cordeiro University of Cincinnati, Cincinnati, Ohio

Brian Ford BellSouth, Atlanta, Georgia

Daniel Howard Quadrock Communications, Duluth, Georgia

Krista S. Jacobsen Texas Instruments, San Jose, California

Nikil Jayant Georgia Institute of Technology, Atlanta, Georgia

Thomas Kolze Broadcom Corporation, Irvine, California

Jonathan Min Broadcom Corporation, Irvine, California

Stephen E. Ralph Georgia Institute of Technology, Atlanta, Georgia

Henry Samueli   Broadcom Corporation, Irvine, California

Paul G. Steffes   Georgia Institute of Technology, Atlanta, Georgia

Jim Stratigos   Broadband Strategies LLC, Atlanta, Georgia

Mani Subramanian   Georgia Institute of Technology, Atlanta, Georgia

Contents

Chapter 1: Broadband in the Last Mile: Current and
Future Applications 1
John Apostolopoulos and Nikil Jayant

Chapter 2: The Last Mile, the Edge, and the Backbone 103
Benny Bing and G.K. Chang

Chapter 3: Last Mile Copper Access 159
Krista S. Jacobsen

Chapter 4: Last Mile HFC Access 251
*Daniel Howard, Bruce Currivan, Thomas Kolze,
Jonathan Min, and Henry Samueli*

Chapter 5: Optical Access: Networks and Technology 333
Brian Ford and Stephen E. Ralph

Chapter 6: Last Mile Wireless Access in Broadband and
Home Networks 387
Carlos de M. Cordeiro and Dharma P. Agrawal

Chapter 7: Satellite Technologies Serving as Last Mile
Solutions 495
Paul G. Steffes and Jim Stratigos

Chapter 8: Management of Last Mile Broadband Networks 523
Mani Subramanian

Chapter 9: Emerging Broadband Services Solutions 583
Mani Subramanian

Index 625

1

Broadband in the Last Mile: Current and Future Applications

JOHN APOSTOLOPOULOS AND NIKIL JAYANT

1.1 INTRODUCTION

In technical as well as nontechnical contexts, the term *broadband* (an adjective, used here as a noun) seems to imply an agile communication medium carrying rich information. The term *last mile*, used mostly in technical circles, connotes the access link that connects the information-rich Internet (and the World Wide Web) to the end user. The location of the user is fixed, quasi-stationary, or mobile. Although relevant to many earlier communication services, the notion of the last mile has become particularly common in the relatively newer context of its being a potential weak link in an otherwise high-speed communication network. It is the purpose of this book to describe communications technologies that support newer generations of information and entertainment services that depend on the notion of pervasive broadband. It is the aim of this chapter to point out the conceptual and quantitative connection between the access pipe and the application, and to describe application classes and core capabilities that the broadband capabilities in succeeding chapters will support.

A 2002 report from the National Research Council of the United States was one of the first authoritative treatments on the subject of broadband last mile, the topic of this book. The report, entitled *Broadband: Bringing Home the Bits,* looked at the subject from a comprehensive viewpoint that included technology timelines, economic considerations, and regulatory policy.[1] The report regarded "broadband" as a convergent platform capable of supporting a multitude of applications and services, and observed that "at best, broadband can be as big a revolution as Internet itself." It noted examples of significant broadband deployment worldwide and maintained that pervasive broadband access should be a national imperative in the U.S., keeping in mind the limited success of the 1996 Telecommunications Act in the country. The report went on to list significant findings and made several recommendations for stimulating broadband deployment, including pointed research in the area for industrial as well as academic laboratories. Among the salient technical observations of the report were

- An elastic definition of broadband
- The critical relationship among broadband, content, and applications
- The need to look at regulatory practices at the service layer, rather than at a technology level

Broadband Options. In the specific dimension of the physical broadband pipe, the NRC report eschewed a horse race view of alternate modalities (such as copper, fiber, wireless, and satellite) and instead predicted that, although wireless and optical accesses have obvious fundamental attractions (in terms of mobility and bandwidth affluence, respectively), the most realistic future scenario is one in which the alternate modalities will coexist in different locality-specific mixes. In fact, none of the modalities mentioned have reached the saturation point in terms of metrics, such as the bandwidth-range product, or tracked the well-known Moore's law in their evolution with time — not to mention the complementary opportunities at the network layer.

Also noteworthy is the fact that the advantageous optical and wireless modalities are currently less often deployed, for reasons such as cost, than the copper (DSL) and cable modalities. DSL is preparing to be a significant carrier of entertainment over copper lines known originally for (nearly universal) telephony. Likewise, the cable medium, originally deployed for entertainment services, is vying to be a serious carrier of (Internet protocol [IP]) telephony. The satellite modality has advantages in terms of geographical coverage and footprint, but is constrained by limitations on on-board power and inherent latency, particularly affecting the capability of interactive two-way services. The power-line channel has the advantage of ubiquity to and within the home, but services on this channel are currently limited by and large to low-rate data-monitoring applications.

The report serves as a natural point of departure for this collection of chapters on broadband access. Several of this book's chapters focus on the access modalities, dedicating a chapter for each of them. The purpose of this chapter is to provide some of the broad context pointed up in the NRC report, particularly, the application dimension. In this introductory chapter, we make liberal use of the NRC report, authored by a committee chaired by the editor of this collection. This chapter continues in Section 1.1.1 by providing two definitions of last mile broadband, as defined in the NRC report, and then briefly highlighting some of the applications of broadband.

We then introduce the notion of the broadband margin (BBM) for a network. A key element of the broadband margin is source coding or compression; media compression is arguably one of the most important enablers of broadband applications. Because no other chapter covers media compression, we devote an entire section to it. Therefore, Section 1.2 of this chapter continues by providing an overview of the science and art of media compression. A number of important broadband applications are examined in Section 1.3, including interactive applications such as voice over IP (VoIP), peer-to-peer (P2P) networks and file sharing, and media streaming. Applications

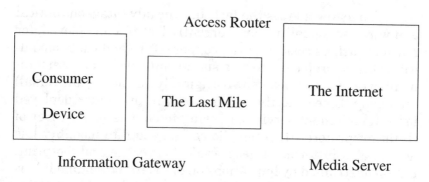

Figure 1.1 Simplified depiction of the access link: the last mile, sometimes called the first mile.

that involve video have the potential to become some of the most important uses of broadband; therefore, the problem of video delivery over broadband networks is examined in more depth as an illustrative example of a challenging broadband application.

1.1.1   Last Mile Broadband and Broadband Applications

1.1.1.1   Definition of Last Mile Broadband

The Last Mile. Figure 1.1 defines the access link that connects the end user to the Internet (the backbone or core network). This access link has been called the *last mile*. It has sometimes been referred to in a user-centric style as the *first mile*. Regardless of nomenclature, the simple way of characterizing the challenge is to say that this mile should not be the weakest link in the chain. The more difficult problem is to understand the many dimensions along which the strength of the link needs to be understood. It is also important to note that the landscape includes several technological functionalities besides the pipe and the application. Some of these are depicted in the smaller-font labels in Figure 1.1. Others, not shown in the figure, include the notion of the *penultimate mile* or the *second mile*. This is the mile that captures distribution granularity in the downlink and data rate aggregation in the uplink. As the problem of the last (or first) mile is solved, attention may well shift to the penultimate (or second) mile, as we seek to attain the ideal of end-to-end — and, ideally, user-steered — quality of service (QoS).

The Last Meter. For the purposes of this book and the technology challenges that it reflects, more important than the penultimate (or second) mile is the problem of the last meter. This notion is particularly meaningful in the context of the home or small office. Here, the access link includes the topologically and logically distinct segments to the home and within the home. Part of the challenge of pervasive wireless is to create a seamless unification of these two segments (supported and constrained by two classes of standards, NG-wireless standards for cellular wireless and IEEE 802.11 standards for wireless local area networks). Neither of these modalities has currently succeeded in making multimedia as ubiquitous as telephony or low bit rate data.

The Backbone Network. The box labeled "the Internet" in Figure 1.1 includes the notion of a backbone network. In a somewhat simplified view, this network has much higher capacity (arising mainly from the pervasive deployment of optical channels) compared to the last mile or the last meter. Assurance of QoS in the backbone in any rigorous sense is less clear, however. Yet, it is realistic to use the model in which the most serious broadband bottlenecks are attributed to the access part of the network, rather than its backbone. This is indeed the premise of this book. As the end-to-end network evolves and the last mile and meter get better and more heavily used, points of congestion can shift away from the access piece, back into the network. Furthermore, at any given time, the notion of end-to-end QoS, in any rigorous sense, is likely to remain an elusive ideal rather than a universal reality. Chapter 2 provides a description of the backbone network, especially as it relates to the last mile and meter, and the part in between — the edge of the network.

The Multiple Dimensions of Broadband. The most commonly held view of broadband focuses on connection speed in kilobits per second (kbps), megabits per second (Mbps), and gigabits per second (Gbps). A next level of characterization deals with the speed of the uplink and downlink and the extent of asymmetry. Still another level of understanding is the notion of "always-on," as opposed to dial-up connectivity. The NRC report lists the following distinct dimensions of broadband, including the attributes mentioned earlier, and supplements it with notions that include openness and multiservice capability. The dimensions are:

- Speed
- Latency and jitter
- Symmetry between uplink and downlink capacity
- Always-on
- Connectivity sharing and home networks
- Addressability
- Controls on applications and content
- Implications of network design/architecture

A Functional and Elastic Definition of Broadband. The broad spectrum of applications that broadband is expected to support indicates that certain convenient definitions such as "256 kbps each way" are not an adequate definition of broadband. An even more important point is that the mere association of any single number, however large, suggests a "broader band" in which efficiency is better in one way or another. On the flip side, if one were to associate some arbitrary large number for the "ultimate broadband link," the definition may well be overkill, especially regarding the quite modest thresholds that sometimes define the perception of broadband, or the tolerable latency for interactivity. Taking this thinking into account, the NRC report adopts the following two approaches for defining broadband:

- Local access link performance should not be the limiting factor in a user's capability for running today's applications.
- Broadband services should provide sufficient performance — and wide enough penetration of services reaching that performance level — to encourage the development of new applications.

1.1.1.2   Applications of Broadband

Pervasive broadband is needed for applications. A major weakness in current approaches to pervasive broadband is that they explore technology and the uses for the technology independently. The end result of such isolation is systems in which the uses and technology are incompatible. What is needed is a methodology that addresses application development consistent with state-of-the-art technological systems. Interdisciplinary interactions in the emerging research and business communities will allow applications research to build upon ongoing technological invention.

Although the applications of broadband are multifarious, it is useful to talk about the following classes:

- Faster general Internet access and general Internet applications
- Audio communications
- Video communications
- Telemetry
- New kinds of publishing
- Multiplexing applications demand in homes
- Communities and community networks

Security. Often implied, but not made explicit (see the preceding list as an example of this omission) is the attribute of security. It is important to note that information security (in general, and in broadband applications in particular) has several dimensions as well. Some of these are network security against computer system hacking and intrusion (especially in wireless); privacy in personal applications (which sometimes is antithetical to the needs of overall network security); and content security (for confidentiality of sender-to-receiver communication as well as protection against unintended parties engaging in piracy; protection against piracy has recently become very important because improvements in signal compression and broadband have overcome many technical barriers to piracy and distribution of pirated content). Network security is addressed as appropriate in different points in the book, but with some deliberation in Chapter 2.

Table 1.1   Seconds to Download Various Media Types for Different Access Speeds

Media	Typical file size (MB)	64 kbps	128 kbps	640 kbps	64 Mbps
Image	0.1	12.5	6.25	1.25	0.01
AudioSingle	1.9	237.5	118.75	23.75	0.24
AudioAlbum	34.6	4325.0	2162.50	432.50	4.33
VideoA	1000.0	125000.0	62500.00	12500.00	125.00

Table 1.2   Seconds to Download a 5-Minute Music Selection

Net capacity (kbps)	Low fidelity	High fidelity
50	480	4800
200	120	1200
800	30	300
1000	24	240
5000	4.8	48

Table 1.3   Broadband Service Capabilities and Application Classes

Capability	Application
Large downstream bandwidth	Streaming content, including video
Large upstream bandwidth	Home publishing
Always-on	Information appliances
Low latency	VoIP, interactive games

Core Capabilities. Basic metrics for data transfer and multimedia access are common to the preceding application classes and, perhaps, to applications yet to be invented, with the broad categories of download and real-time streaming in each case. Table 1.1 and Table 1.2 provide quantitative grounding for these core capabilities, and begin to relate application to access technology, at least in the primary dimension of (one-way) speed or bit rate. Table 1.3 provides a qualitative mapping between application class and a broader set of broadband dimensions, as previously discussed.

Illustrations of the Two Definitions of Broadband. Today, in browsing, navigating, file downloading, and games, yesterday's applications are running faster. This supports the first definition of broadband. Some new ideas have emerged, however, such as network storage, static image delivery, new publishing, P2P, local interest home hosting, and push content. The promise is that with increased penetration of broadband, new applications will follow. This illustrates the second definition of broadband. Video exemplifies the potential gap between the two definitions. Next-generation video incorporates the notion of everything on demand, which creates a nonlinear increase in the required capacity of the last mile because of the inherent switch from a broadcast paradigm to a switched paradigm.

1.1.1.3 The Role of Signal Compression: The Broadband Margin

The media file sizes and download times used in Table 1.1 and Table 1.2 are rates *after* signal compression, and the implied compression ratios are very significant, following advances in compression science, technology, and standards (Section 1.2). In fact, disruptive applications have often occurred when advances in signal compression (that caused a compact representation of the information, as measured in kilobits per second or megabits per second) intersected with advances in modulation (that caused a broadening of the access pipe, as measured in bits per second per hertz [bps/Hz]). To complete this perspective, advances in unequal error protection provide important further compaction in error-resilient data (which include error protection redundancy). At the same time — although this is a more recent trend — network coding further expands the effective bandwidth available to a user by countering network congestion over a given physical pipe bandwidth and modulation system.

Broadband margin (BBM) is defined as the ratio of total bit rate, $R_{last\text{-}mile}$, at the end of the last mile, as seen by the user, to the bit rate needed by an application, R_{app}. The BBM may also be defined with respect to the total rate of the first mile, $R_{first\text{-}mile}$, if the first mile is the bottleneck. The application rate R_{app} is a function of compression or source coding (SC) and error protection or channel coding (CC). Although source coding reduces the required rate to represent a signal, channel coding increases the total rate by introducing sophisticated redundancy for combating channel errors. The bit rate for the last mile, $R_{last\text{-}mile}$, is a function of pipe bandwidth (BW) and the modulation efficiency determined by the modulation coding (MC) in bits per second per hertz and the congestion-countering mechanisms of network coding (NC). The BBM is then expressed as:

$$BBM = \frac{R_{last\text{-}mile}\left(MC,NC\right)}{R_{app}\left(SC,CC\right)}$$

An idealized figure illustrating source coding, channel coding, network coding, and modulation coding in the context of the popular Internet protocol stack is illustrated in Figure 1.2. Although the preceding expression of the BBM and the idealized figure are conceptually useful, it is important to realize that they are simplifications. For example, the different types of coding are not independent and, in fact, are highly interrelated. In addition, some techniques may appear at multiple locations across the protocol stack. For example, channel coding may be performed at the application layer as well as at the physical layer. Furthermore, automatic repeat request (ARQ) or retransmission of a lost packet is often performed at a number of layers. For example, retransmissions may be performed at the MAC layer, where they provide fast, link-layer retransmits of lost packets, or at the transport or application layers, where they provide end-to-end retransmission of the lost packets.

Additionally, retransmits at the application layer are generally performed in a selective manner, based on the importance of each packet, and retransmits at the MAC or transport layer generally do not account for the specific content or importance of each particular packet. Also, the boundary

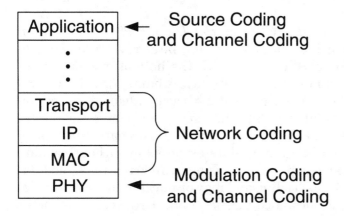

Figure 1.2 The five key layers of the popular IP protocol stack: application, transport, media access control (MAC), Internet protocol (IP), and physical layer (PHY). The locations where each of the four types of coding (source, channel, network, and modulation coding) is typically applied are indicated in this idealized figure.

between channel coding and network coding is quite fuzzy because both attempt to overcome or prevent losses. In the preceding discussion, the distinction was made based on whether feedback is used, e.g., channel coding techniques include block codes and convolutional codes, which do not require feedback, but network coding uses feedback to trigger retransmission of lost packets or adapt the sending rate, etc.

The bulk of this book examines advances in the domain of $R_{last\text{-}mile}$ as a function of the specific pipe, pipe bandwidth, modulation coding, etc. Network coding (NC) is mentioned when appropriate. Channel coding is not discussed in depth in this book because many excellent references on this topic already exist; however, it is useful to know that channel coding leads to a 10 to 50% increase or expansion of R_{app} compared to the case in which only source coding is applied. Advances in source coding, on the other hand, currently provide compression ratios on the order of 5 to 100 for audiovisual signals, with impact on a large set of application classes. This section continues by briefly describing the four types of coding:

modulation coding, channel coding, network coding, and source coding.

The Bandwidth (BW) of the Digital Pipe: Modulation Coding. When information-carrying bits traverse a physical (analog) channel, they do so with the help of modems (modulator–demodulator pairs) that convert binary data into waveforms that the analog channels can accept. Media compression reduces the number of bits that need to traverse the digital channel; the task of the modems is to prepare the analog channel to accept as many bits as possible without causing errors in their reception. This is often referred to as modulation or modulation coding and may or may not include channel coding. The performance metric is bits per second per hertz of physical channel bandwidth. This metric depends on channel characteristics such as signal-to-channel-noise ratio, signal-to-interference ratio, and intersymbol interference.[2–5] Therefore, this metric is also a function of physical distance traversed on the physical medium. In the kinds of broadband networks discussed in this book, the modem efficiency ranges typically from 0.5 to 5.0 bps/Hz.

For cable networks, the so-called 256 QAM cable modem system has a theoretical maximum efficiency of 8 bps/Hz. Practically, about 6 bps/Hz is more common, resulting in about 38 Mbps data rate over a 6-MHz cable channel, with very low probabilities of bit errors. In the fiber medium, because of the high speeds of operation, the emphasis is on more basic two-level modulation methods with a nominal 1-bps/Hz efficiency. However, the availability of inherently greater bandwidth in the optical fiber results in much higher data rates, in the realm of gigabits per second or higher. Expected and actual bit error rates are typically very low, on the order of one in a billion or lower, because of the association of the fiber medium with the very high quality backbone network. These trends continue as fiber is deployed in the last mile, although the case is stronger for higher modem efficiencies in this context, with recent signal processing technology helping in that quest. Modem efficiencies in DSL tend to span the intermediate range (between cable and fiber). Recent work on crosstalk mitigation is aimed at increasing the modem efficiency in DSL systems, as reflected ultimately in the (data rate) × (range) product.

Wireless and satellite media are characteristically power limited and, as a result, aggressive modulation methods are not common in these channels. Cell phone technology, for example, uses modem efficiencies on the order of 1 bps/Hz or lower. With the recent development of multiple antenna technology using MIMO (multiple input multiple output) algorithms, the stage has been set for a potentially dramatic jump in wireless modem efficiencies, with research demonstrations reaching up to 5 bps/Hz over short distances. For perspective, a 200-MHz band, if available, together with a 5-bps/Hz wireless modem, would provide the basis for gigabit wireless.

Network Coding (NC): The Efficient Use of a Shared Medium. A digital pipe provides a throughput determined by the physical layer system, i.e., the modem. Use of this throughput as well as the actual rate that an application sees depends on a number of factors, such as whether or not the medium is shared, the MAC efficiency, and the transport layer efficiency. In this discussion, the term "network coding" is used to describe the processing performed between the physical and application layers that affects the end-to-end communication. Network coding is a critical component of nearly all broadband systems and applications; it requires some overhead and, as discussed below, the actual rate that an end-user application sees is generally less than the throughput provided by the physical layer system. Three illustrative examples of network coding follow; note that they do not completely describe the range of functions performed by network coding.

Most wired and wireless access technologies are shared media that provide a number of desirable properties (e.g., statistical multiplexing gain); however, although efficiently utilizing the available bandwidth of the shared medium provides other important properties such as fairness, it is quite complicated. In a slotted system, all of the users are synchronized and each is given a time slot in which to transmit. On the other hand, systems that use contention-based channel access enable greater flexibility and allow asynchronous operation. Basically, if a central controller performs a resource allocation of the shared bandwidth, the efficiency can be very high.

However, when distributed processing is performed, such as that used in wired Ethernet and its wireless counterpart 802.11, the bandwidth utilization is generally much less. For example, the original 802.11 consists of two modes of media access control (MAC) operation for the access point (AP): (1) distributed coordination function (DCF) based on distributed contention for accessing the media; and (2) point coordination function (PCF) in which contention-free periods (based on polling) and contention periods (based on DCF) are provided for access to the medium. The centralized control provided by PCF mode leads to a more efficient use of the medium when the medium is heavily loaded (on the other hand, PCF is much more complex than DCF and thus was not implemented in most 802.11b systems). This behavior is also true for the new 802.11e MAC protocol designed to provide better QoS and higher channel efficiency over wireless LANs, in which the centralized hybrid coordination function (HCF [an extension of PCF in the original 802.11b]) provides about a 20% increase in channel utilization over its distributed counterpart when the channel is heavily loaded.[6]

The MAC design also has a significant effect on the effective throughput even when the channel is not shared with other users. For example, an 802.11b digital pipe has a channel rate of 11 Mbps; however, even when only a single user attempts a large file transfer, the contention-based MAC leads to a surprisingly low throughput of only about 5 to 6 Mbps. This is a result of the large overheads at the PHY and MAC layers due to the headers, acknowledgments (ACKs), and the timing strategy required to support the contention-based protocol. Furthermore, an application that transmits packets smaller than the large 1500-byte packets used in the large file transfer (mentioned earlier) would lead to even lower throughput over 802.11.

A third example of the efficiency of network coding, this time at the transport layer, is the notion of TCP "goodput" vs.

throughput. The popular transmission control protocol (TCP) is described in Section 1.3.5.2 and is a basic component of many applications running on the Internet. TCP goodput is the useful bit rate as it is provided to the application, as opposed to the network throughput. The difference between goodput and throughput is a function of the TCP overhead, potential retransmission of the same data, etc. Once again, the network coding results in the goodput, or useful rate provided to the application, being less than the throughput that the physical layer provides. Network coding can be thought of as providing a mapping between physical layer throughput and network goodput.

Channel Coding (CC): Overcoming Losses in the Network. The goal of channel coding is to overcome channel losses, such as bit or burst errors on a wireless link or packet loss on a wired link. Two general classes of channel coding approaches are based on the use of: (1) retransmission and (2) forward error correction (FEC). In retransmission-based approaches, the receiver notifies the sender of any lost information, and the sender retransmits this information. This approach is simple and efficient in the sense that available bandwidth is only used to resend lost packets (the overhead is zero if no packets are lost), and it straightforwardly adapts to changing channel conditions. As a result of these benefits, retransmission is widely used; for example, it forms the backbone for many Internet applications because it provides the error-recovery mechanism for the ubiquitous TCP protocol. However, retransmission-based approaches require a back channel, which may not be available in certain applications such as broadcast or multicast. In addition, persistent losses may lead to very large delays in delivery. Therefore, although retransmission-based approaches provide a simple and practical solution for some applications, they are not always applicable.

The goal of FEC is to add specialized redundancy that can be used to recover from errors. For example, to overcome packet losses in a packet network one typically uses block codes (e.g., Reed Solomon or Tornado codes) that take K data packets and output N packets, where $N - K$ of the packets are redundant packets. For certain codes, as long as any K of the N packets is correctly received, the original data can be recovered. On the other hand, the added redundancy increases the required bit rate by a factor of N/K. A second popular class of FEC codes is convolutional codes.

FEC provides a number of advantages and disadvantages. Compared to the use of retransmissions, FEC does not require a back channel and may provide lower delay because it does not depend on the round-trip time of retransmits. Disadvantages of FEC include the overhead for FEC, even when no losses occur, and possible latency associated with reconstruction of lost packets. Most importantly, FEC-based approaches are designed to overcome a predetermined amount of loss and are quite effective *if* they are appropriately matched to the channel. If the losses are less than a threshold, then the transmitted data can be perfectly recovered from the received, lossy data. However, if the losses are greater than the threshold, then only a portion of the data can be recovered and, depending on the type of FEC used, the data may be completely lost. Unfortunately, the loss characteristics for packet networks are often unknown and time varying. Therefore, the FEC may be poorly matched to the channel, thus making it ineffective (too little FEC) or inefficient (too much FEC).

Rate-compatible codes help address this mismatch problem by providing integrated designs of codes that support several possible rates (ratio of message bits to total bits after error protection overhead). An example is RCPC (rate-compatible punctured convolutional codes).[7] Recent work extends the rate-compatibility feature to a particular example of block codes called low-density parity check codes (LDPC).[67]

Unequal Error Protection (UEP). Unlike unqualified data, compressed bitstreams of audiovisual information exhibit bit-specific sensitivity to channel errors. For example, because header bits of various kinds are typically important, systems are highly sensitive or vulnerable to errors in these header bits. At the next level, bits signifying pitch information in speech and motion information in video can be quite sensitive to channel errors; bits signifying a low-order prediction coefficient in speech or a high-frequency discrete cosine transform (DCT) coefficient in video may be relatively less sensitive.

Unequal error protection addresses the preceding characteristics of nonuniform sensitivity to errors by appropriately distributing the error protection to minimize the vulnerability to errors while also minimizing the average (or total) overhead. This is in contrast to equal error protection (EEP) schemes that distribute the protection uniformly over all bits, irrespective of their importance. In an illustrative three-level UEP design typical of cell phone standards,[5] a very small fraction, $f1$, of highly error-sensitive bits will have significant error protection overhead; a fairly large fraction, $f3$, will have very little or zero error protection and the intermediate fraction will have an appropriate intermediate level of channel error protection. The end result is an average overhead that is typically much smaller than in the overdesigned case in which all the bits have, say, the same degree of error protection as the small fraction, $f1$. The resultant saving in overall bit rate can be typically in the range of 1.2 to 1.5.

Source Coding: Media Compression. Media compression is the science and art of representing an information source with a compact digital representation, for economies in transmission and storage. As such, compression is a fundamental enabler of multimedia services, including those on broadband networks. Very few broadband networks are able to handle uncompressed information in practical multiuser scenarios. For example, a single raw (uncompressed) high-definition television (HDTV) video signal requires (720 × 1280 pixels/frame) (60 frames/sec) (24 b/pixel) = 1.3 Gb/sec. Even a single raw HDTV signal would be too large to fit through many of today's broadband pipes.

Through the use of compression, the HDTV video signal is compressed to under 20 Mb/sec while preserving very high perceptual quality.[8] This factor of about 70 compression means that the single HDTV video signal can fit through a pipe that is 1/70 the size required by the raw HDTV signal or, alternatively, 70 HDTV signals can be sent through a pipe that would normally carry only a single raw HDTV signal. Clearly, media compression has a huge effect on enabling media applications, even for fat broadband pipes that provide tens of megabits of bit rate. Media compression is discussed in more detail in Section 1.2.

1.1.1.4 Managing the BBM

The modulation coding of a broadband system provides a physical layer bandwidth to the higher layers. As described before, the network coding supports sharing of the bandwidth among the multiple users that may want to use it. In addition, each user may have multiple flows to transport over that bandwidth. For example, a single user may simultaneously have a data flow, a video flow, and an audio flow to transport over the available bandwidth. Network coding addresses this, as well as potential different QoS requirements for each of the different flows.

Contrast between Data and Media Delivery. Another important consideration is the fundamental distinction between data delivery and media delivery. Generally, with data delivery all of the bits of data must be delivered and thus all of the bits of data are of equal importance. In contrast, the various bits of a compressed media stream are generally of different importance. Furthermore, the coded media can be somewhat tolerant of losses; portions of the coded media may be lost (or, as we will see in Section 1.3.5, may arrive late) without rendering the entire media useless. Also, data delivery typically does not have a strict delivery constraint. On the other hand, media packets often have strict delivery deadlines, especially for conversational applications such as VoIP and videophone, or interactive applications such as computer games. Media packets that arrive after their respective delivery deadlines may be useless — with an effect equivalent to that if they had been lost.

1.1.1.5 Outline of Remainder of Chapter

Many of the later chapters in this book focus on the last mile, discussing modulation coding and, to a lesser extent, channel coding and network coding. However, no other chapter discusses source coding — in particular, media compression.

Because media compression is a fundamental enabler in broadband applications that involve media and media applications are likely to be some of the most important applications used over broadband networks in the future, we therefore devote an entire section to it. Specifically, Section 1.2 provides a more detailed overview of the science and practice of media compression, covering basic principles as well as the most important media compression standards.

In Section 1.3, we examine a number of important applications of broadband networks. We begin by providing a taxonomy of the different media application and network operating conditions in Section 1.3.1 and the latency spectrum for these applications in Section 1.3.2. Interactive two-way communications such as VoIP are examined in Section 1.4.3. Peer-to-peer (P2P) networks and file sharing are discussed in Section 1.3.4 and media streaming in Section 1.3.5. An important theme of these applications is media delivery over broadband networks, and we focus on real-time and streaming applications, the network challenges that arise, and methods to overcome these challenges. Many of the practical solutions to these challenges involve careful codesign of the media compression, network coding, channel coding, and sometimes also the modulation coding.

The important problem of media streaming of encrypted content and secure adaptation of that content (without decryption) is highlighted in Section 1.3.6. Our discussion of media delivery considers operations that can be performed at the end hosts (the sender and receiver) as well as the important problem of network infrastructure design — how to design the infrastructure to better support media delivery over broadband networks. In particular, we discuss the important network protocols for streaming media over IP networks in Section 1.3.7 and provide an overview of the design of emerging streaming media content delivery networks (SM-CDNs) in Section 1.3.8. A number of potential future application areas and trends are discussed in Section 1.3.9. This chapter concludes in Section 1.4 with a summary and an overview of the subsequent chapters.

Now, before discussing the important broadband applications, many of which involve media, we first continue with a more detailed overview of media compression.

1.2 MEDIA COMPRESSION

Media for communications and entertainment have been important for many decades. Initially media were captured and transmitted in analog form, including broadcast television and radio. In the last two decades, the emergence of digital integrated circuits and computers and the advancement of compression, communication, and storage algorithms led to a revolution in the compression, communication, and storage of media. Media compression enabled a variety of applications, including:

- Video storage on DVDs and video-CDs
- Video broadcast over digital cable, satellite, and terrestrial (over-the-air) digital television (DTV)
- Video conferencing and videophone over circuit-switched networks
- Audio storage on portable devices with solid-state storage
- Digital cameras for still images as well as video
- Video streaming over packet networks such as the Internet
- The beginning of media delivery over broadband wired and wireless networks

Clearly, one of the fundamental enablers of these applications was media compression, which enables media signals to be compressed by factors of 5 to 100 (depending on specific media and application) while still solving the application goals.

This section provides a very brief overview of media compression and the most commonly used media compression standards; the following section discusses how the compressed media are sent over a network. Limited space precludes a detailed discussion; however, we highlight some of the important principles and practices of current and emerging media compression algorithms and standards especially relevant for media communication and delivery over broadband networks (additional details are available in the references).

An important motivation for this discussion is that the standards (such as the JPEG image coding,[23] MPEG video coding,[32-34] and MP3 and AAC audio coding standards) and the most popular proprietary solutions (e.g., Microsoft Windows Media[9] and RealNetworks[10]) are based on the same basic principles and practices; therefore, by understanding them, one can gain a basic understanding for standard as well as proprietary media compression systems. Additional goals of this section are to (1) describe the different media compression standards; (2) describe which of these standards are most relevant for broadband in the future; and (3) identify what these standards actually specify. Therefore, this section continues in Section 1.2.1 by providing a short overview of the principles and practice of media compression, followed in Section 1.2.2 by a brief overview of the most popular and practically important media compression standards. The standards' actual specifications are examined in Section 1.2.3.

This section concludes with a short discussion of how to evaluate the quality of different compression algorithms and how to adapt media to the available bit rate on the network and to conceal the effect of channel losses. Detailed overviews of media compression are given in References 11 through 14. Recent media compression standards are discussed in References 15 through 21 and Reference 24; additional references on specific topics of media compression are identified throughout this section.

1.2.1 Principles and Practice of Media Compression

Compression is the operation of taking a digitized signal and representing it with a smaller number of bits; it is usually classified as lossless or lossy. In lossless compression, the goal is to represent the signal with a smaller number of bits while preserving the ability to recover the original signal perfectly from its compressed representation. In lossy compression, the goal is to represent the signal with a smaller number of bits so that the signal can be reconstructed with a fidelity or accuracy that depends on the compressed bit rate. Typically, text and electronic documents are compressed in a lossless manner and media are typically compressed in a lossy manner. Media are compressed in a lossy manner for two reasons: (1) lossy compression can enable much higher compression rates; and (2) generally one does not require a perfect reconstruction of the original media signal, so it is sufficient (and practically more beneficial) to have a reconstructed signal perceptually similar (but not necessarily bitwise identical) to the original.

The Fundamental Techniques of Compression. Although a large number of compression algorithms have evolved for multimedia signals, they all depend on a handful of techniques that will be mentioned in the sections to follow. In turn, these techniques perform no more than two fundamental tasks: *removal of (statistical) redundancy* and *reduction of (perceptual) irrelevancy.* The fundamental operation of signal prediction performs the first function of removing redundancy, regardless of the signal-specific methods of performing prediction (in time and/or frequency domains). Prediction and its frequency-domain dual, transform coding, are well-understood staple methods for compressing speech, audio, image, and video signals. The even more fundamental function of quantization performs the function of minimizing irrelevancy, and relatively recent and sophisticated methods for perceptually tuned quantization provide the framework for high-quality audio compression. The final operation of binary encoding is a reversible operation that takes the quantized parameters and exploits their nonuniform statistics to pack them efficiently into a bitstream for transmission.

As described previously, the removal of statistical redundancy and reduction of perceptual irrelevancy are the two fundamental goals of media compression. Typical media signals exhibit a significant amount of (statistical) similarities or redundancies across the signal. For example, consecutive frames in a video sequence exhibit temporal redundancy because they typically contain the same objects, perhaps undergoing some movement between frames. Within a single frame, spatial redundancy occurs because the amplitudes of nearby pixels are often correlated. Similarly, the red, green, and blue color components of a given pixel are often correlated.

The second goal of media compression is to reduce the irrelevancy in the signal — that is, only to code features perceptually important to the listener or viewer and not to waste valuable bits on information that is not perceptually important or is irrelevant. Identifying and reducing the statistical redundancy in a media signal is relatively straightforward, as suggested by examples of statistical redundancy described earlier. However, identifying what is perceptually relevant and what is not requires accurate modeling of human perception, which is very difficult. This not only is highly complex for a given individual, but also varies from individual to individual. Therefore irrelevancy is difficult to exploit.

Dimensions of Performance. The four dimensions of performance in compression are signal quality and compressed bit rate, communication delay, and complexity (in encoding and decoding), as shown in Figure 1.3. As speed and memory

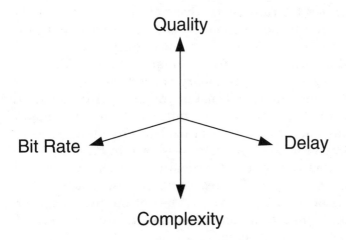

Figure 1.3 The four dimensions of compression performance: quality, bit rate, delay, and complexity. The design of a compression system typically involves trade-offs across these four dimensions. (From Reference 11. Reprinted with permission.)

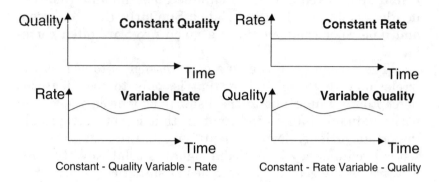

Figure 1.4 Example of constant-quality variable-rate (CQVR) coding on the left and variable-quality constant-rate (VQCR) coding on the right.

capabilities increase, the role of the complexity dimension tends to diminish while the other three dimensions remain fundamental. Of greatest importance is the quality–bit rate trade-off. The typical application requirement is that the compression provides the desired reconstructed quality at a practical (acceptable) bit rate. In the case of audiovisual signals that are inherently nonstationary, one can talk about two approaches to compression: constant-quality variable-rate (CQVR) and constant-rate variable-quality (CRVQ). Because of the time-varying or nonstationary characteristics of typical media signals, to achieve a constant quality across all time requires a variable bit rate across time. On the other hand, to achieve a constant bit rate across time leads to a time-varying quality. Examples of this behavior are shown in Figure 1.4.

This section continues by providing a brief overview of speech, audio, image, and video compression, which is followed in Section 1.2.2 by a brief review of the most popular and practically important compression standards for each of these different types of media.

Speech Compression. High-quality telephony uses a signal bandwidth of 3.2 kHz and commercial digital telephony uses a 64-kbps format called PCM. In some instances, the speech is compressed, with very little loss of quality, to 32 kbps by using adaptive differential pulse code modulation (ADPCM) in which the quantized information is an appropriate differential signal rather than the original speech sample. The differential signal is the result of a simple form of redundancy reduction.

Lower bit rate representations of speech use more efficient forms of redundancy removal using the formalism of LPC (linear predictive coding) with a typical prediction order of ten (i.e., each speech sample is predicted on the basis of the past ten samples). To increase the efficiency of the prediction process in the so-called voiced portions of speech, a pitch predictor is also employed to remove periodic (or long-term) redundancy in the speech signal. These methods can provide high-quality representations of speech at 16 kbps.

Extending the range of high quality to the realm of cellular and practical Internet telephony has depended on the inclusion of two additional functionalities in the speech coder. One is the removal of perceptual irrelevancy by means of frequency weighting of the reconstruction noise spectrum so that different frequency regions are coded with varying fidelity, to take into account the perceptual seriousness of the respective distortion components. In other words, overcoding of a given component can be perceptually irrelevant and therefore wasteful of overall coding resources. The second methodology, called analysis by synthesis, is based on the notion of considering a number of alternate encoding paradigms for each block of speech about 10 msec in length and picking the one that is best based on a perceptually weighted reconstruction error criterion. Although this methodology leads to a more complex encoder, it provides the key step towards realizing high-quality speech at bit rates as low as 4 kbps. To complete the picture, speech coding at bit rates on the order of 2 kbps is currently possible in terms of preserving intelligibility in the speech, but not from the point of view of quality, naturalness, or speaker recognition.

Wideband speech (with a bandwidth of, say, 7 to 8 kHz) is a step beyond telephone quality and is closer to the quality associated with AM radio. Digital representation of wideband speech using a simple waveform coder such as ADPCM but with variable bit assignment for the low- and high-frequency subbands (0 to 4 kHz and 4 to 8 kHz, respectively) requires a bit rate of 64 kbps. As with telephone quality speech, lower rates are possible with more complex algorithms, but these are not parts of commercial standards.

Audio Compression. Unlike the case of speech coding, in which powerful models for speech production are exploited for speech coding, for general audio, no corresponding models can adequately represent the rich range of audio signals produced in the world. Therefore, audio coding is usually tackled as a waveform coding problem in which the audio waveform is to be efficiently represented with the smallest number of bits. On the other hand, research in psychoacoustics has been quite successful in terms of creating models that identify which distortions are perceptible and which are not — that is, identify what is perceptually relevant or irrelevant. The application of these perceptual audio models to audio coding is key to the design of high-quality audio coders, which are often referred to as *perceptual audio coders* to stress the fact that the audio coders are designed to maximize the perceptual quality of the audio, as opposed to optimized for other metrics such as minimizing the mean-squared error.

A generic audio coder begins by taking the audio signal and passing it through an analysis filter bank, which decomposes the audio into different frequency subbands. The subband coefficients are then quantized and coded into a bitstream for transmission. The key point of a perceptual audio coder is that the quantization is driven by the auditory models in order to minimize the perceived distortion. Specifically, the original audio signal is examined in time and frequency domains (the frequency domain can be the output of the analysis filter bank, or another frequency transformation, such as a Fourier transform, that may be more appropriate for the perceptual models) in order to identify what is referred to as a masking threshold, below which the distortion is not perceptible. Note that this threshold is a function of the original audio signal and varies with time and frequency. The quantization is adapted to shape the distortion in time and frequency so that it is masked by the input signal, e.g., the coarseness/fineness of the quantization is adapted for each frequency component.

Similarly, the analysis filter bank is also adapted based on the perceptual models. For example, for stationary audio segments, it is important to have a long window for analysis, but for audio transients such as sharp attacks, it is important to have a short analysis window in order to prevent the infamous pre-echo artifacts. For further details on perceptual audio coding, see Jayant et al.[11] and Johnston et al.[22]

The preceding discussion focused on single-channel audio coding. Stereo and other forms of multichannel audio are quite pervasive, especially for entertainment applications. Although the multiple audio channels in these applications may be independently coded using separate audio coders, significant compression gains may be achieved by jointly coding the multiple channels. For example, the two audio channels in stereo are highly correlated, and joint coding across the two channels can lead to improved compression. Once again, for multiple-channel audio coding, as for single-channel audio coding, the perceptual effects of the coding must be carefully considered to ensure that the resulting distortion is perceptually minimized.

Image Compression. Image compression algorithms are designed to exploit the spatial and color redundancy that exists in a single still image. Neighboring pixels in an image are often highly similar, and natural images often have most of their energies concentrated in the low frequencies. The well-known JPEG image compression algorithm exploits these features by partitioning an image into 8 × 8 pixel blocks and computing the 2-D discrete cosine transform (DCT) for each block. The motivation for splitting an image into small blocks is that the pixels within a small block are generally more similar to each other than the pixels within a larger block. The DCT compacts most of the signal energy in the block into only a small fraction of the DCT coefficients in which this small fraction of the coefficients is sufficient to reconstruct an accurate version of the image.

Each 8 × 8 block of DCT coefficients is then quantized and processed using a number of techniques known as zigzag scanning, run-length coding, and Huffman coding to produce a compressed bitstream.[23] In the case of a color image, a color space conversion is first applied to convert the RGB image into a luminance/chrominance color space in which the different human visual perception for the luminance (intensity) and chrominance characteristics of the image can be better exploited. The quantization of the DCT coefficients is adapted as a function of a number of attributes, such as spatial frequency and luminance/chrominance, in order to minimize the perceptual distortion.

Video Compression. A video sequence consists of a sequence of video frames or images. Each frame may be coded as a separate image — for example, by independently applying JPEG-like coding to each frame. However, because neighboring video frames are typically very similar, much higher compression can be achieved by exploiting the similarity between frames. Currently, the most effective approach to exploit the similarity between frames is by coding a given frame by (1) first predicting it based on a previously coded frame; and then (2) coding the error in this prediction. Consecutive video frames typically contain the same imagery, although possibly at different spatial locations because of motion. Therefore, to improve predictability, it is important to estimate the motion between the frames and then to form an appropriate prediction that compensates for the motion.

The process of estimating the motion between frames is known as *motion estimation* (ME), and the process of forming a prediction while compensating for the relative motion between two frames is referred to as *motion-compensated prediction*. Block-based ME and MC prediction is currently the most popular form of ME and MC prediction: the current frame to be coded is partitioned into 16×16–pixel blocks; for each block, a prediction is formed by finding the best matching block in the previously coded reference frame. The relative motion for the best matching block is referred to as the *motion vector*.

The three basic common types of coded frames are:

- Intracoded frames, or I-frames: the frames are coded independently of all other frames

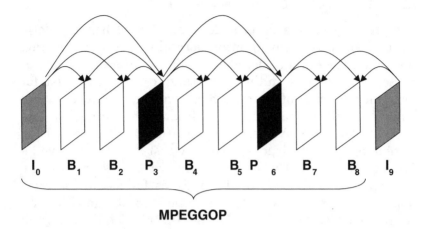

Figure 1.5 MPEG group of pictures (GOP) illustrating the three different types of coded frames within a GOP: I-, P-, and B-frames. The arrows indicate the prediction dependencies between frames.

- Predictively coded, or P-frames: the frame is coded based on a previously coded frame
- Bidirectionally predicted frames, or B-frames: the frame is coded using previous as well as future coded frames

Figure 1.5 illustrates the different coded frames and prediction dependencies for an example MPEG group of pictures (GOP). The selection of prediction dependencies between frames can have a significant effect on video streaming performance, e.g., in terms of compression efficiency and error resilience.

Current video compression standards achieve compression by applying the same basic principles.[24,25] The temporal redundancy is exploited by applying MC prediction and the spatial redundancy is exploited by applying the DCT; the color space redundancy is exploited by a color space conversion. The resulting DCT coefficients are quantized, and the nonzero quantized DCT coefficients are runlength and Huffman coded to produce the compressed bitstream.

1.2.2   ISO and ITU Compression Standards

Compression standards provide a number of benefits, fore-most of which is ensuring interoperability or communication between encoders and decoders made by different people or different companies. In this way, standards lower the risk for consumer and manufacturer, which can lead to quicker acceptance and widespread use. In addition, these standards are designed for a large variety of applications, and the resulting economies of scale lead to reduced cost and further widespread use. Currently, several standard bodies oversee the development of international standards for media compression: the International Telecommunications Union-Telecommunications (ITU-T, formerly the International Telegraph and Telephone Consultative Committee, CCITT), several standards for cellular telephony, and the International Organization for Standardization (ISO). This section continues by briefly reviewing some of the important speech, audio, image, and video compression standards. It concludes by focusing on what the standards actually specify in terms of the encoders, compressed bitstreams, and decoders.

Speech Coding Standards. A large number of speech coding standards have been established over the years. Tables 1.4, 1.5, and Table 1.6 summarize speech coding standards for telephone and wideband speech, and capture the bit rate range of 2.4 to 64 kbps described in the prior discussion on speech coding. Further information is available in Goldberg and Riek[20] and Cox.[26]

Audio Coding Standards. A number of audio coding algorithms have been standardized and are listed in Table 1.7. The first audio coding standard was developed in MPEG-1. MPEG-1 consisted of three operating modes, referred to as layer-1, layer-2, and layer-3, which provided different performance/complexity trade-offs. MPEG-1 layer-3 is the highest quality mode and was designed to operate at about 128 kbps for stereo audio. MPEG-1 layer-3 is also commonly known as "MP3." As part of the MPEG-2 effort, a new audio coding standard referred to as MPEG-2 advanced audio coding (AAC) was developed. AAC was designed to provide improved audio

Table 1.4   ITU Waveform Speech Coders

Standards body	ITU	ITU	ITU	ITU
Recommendation	G.711	G.726	G.727	G.722
Type of coder	Companded PCM	ADPCM	ADPCM	SBC/ADPCM
Dates	1972	1990	1990	1988
Bit rate	64 kb/sec	16–40 kb/sec	16–40 kb/sec	48–64 kb/sec
Quality	Toll	Toll	≤ Toll	Commentary
Complexity				
MIPS	<< 1	1	1	10
RAM	1 byte	< 50 bytes	< 50 bytes	1 K
Delay				
Frame size	0.125 ms	0.125 ms	0.125 ms	1.5 ms
Specification type				
Fixed point	Bit exact	Bit exact	Bit exact	Bit exact

Source: Cox, Current Methods in Speech Coding, in Jayant (ed.), *Signal Compression*, World Scientific, 1997.

Table 1.5   ITU Linear Prediction Analysis-by-Synthesis Speech Coders

Standards body	ITU	ITU	ITU
Recommendation	G.728	G.729	G.723.1
Type of coder	LD-CELP	CS-ACELP	MPC-MLQ and ACELP
Dates	1992 and 1994	1995	1995
Bit rate	16 kb/sec	8 kb/sec	6.3 and 5.3 kb/sec
Quality	Toll	Toll	≤ Toll
Complexity			
MIPS	30	≤ 20	≤ 15
RAM	2 K	< 2.5 K	2.2 K
Delay			
Frame size	0.625 ms	10 ms	30 ms
Specification type			
Floating point	Algorithm exact	None	None
Fixed point	Bit exact	Bit exact C	Bit exact C

Source: Cox, Current Methods in Speech Coding, in Jayant (ed.), *Signal Compression*, World Scientific, 1997.

Table 1.6
First- and Second-Generation Digital Cellular Telephony Speech Coders

			Standards body			
	CEPT	ETSI	TIA	TIA	RCR	RCR
Standard name	GSM	GSM 1/2-rate	IS-54	18-96	PDC	PDC 1/2-rate
Type of coder	RPE-LTP	VSELP	VSELP	CELP	VSELP	PSI-CELP
Date	1987	1994	1989	1993	1990	1993
Bit rate	13 kb/sec	5.6 kb/sec	7.95 kb/sec	0.8–8.5 kb/sec	6.7 kb/sec	3.45 kb/sec
Quality	< Toll	= RPE-LTP	= RPE-LTP	< RPE-LTP	< RPE-LTP	= Full rate
Complexity						
MIPS	4.5	30	20	20	20	48 (est.)
RAM	1 K	4 K	2 K	2 K	2 K	4 K
Delay						
Frame size	20 ms	20 ms	20 ms	20 ms	20 ms	40 ms
Look ahead	0	5 ms	5 ms	5 ms	5 ms	10 ms
Specification type						
Fixed point	Bit extract	Bit extract C	Bit stream	Bit stream	Bit stream	Bit stream

Source: Cox, Current Methods in Speech Coding, in Jayant (ed.), *Signal Compression*, World Scientific, 1997.

Table 1.7   Current and Emerging Audio Compression Standards

Audio coding standards	Primary intended applications	Bit rate
MPEG-1 layer-1/2 and layer-3 (MP3)	Compression of wideband audio (32, 44.1, and 48 kHz sampling rates)	MP3: 128 kbps for stereo
MPEG-2 AAC	Improved compression as compared to MP3; improved joint stereo coding capabilities	96 kbps for stereo
MPEG-4 natural audio (AAC, CELP, TwinVQ, etc.)	Coding of natural speech and audio signals at a wide range of bit rates and audio bandwidths, and new functionalities such as scalability and error resilience	Speech at 2–24 kbps; audio at 8–64 kbps
MPEG-4 synthetic audio (TTS, SA)	Text-to-speech; downloadable signal-processing algorithms for synthesizing audio at the receiver and applying postprocessing effects to natural and synthetic audio objects to create the audio scene; also, description of the musical scores to play	Variable

compression performance compared to MP3, as well as improved coding of stereo and multichannel audio signals.

The MPEG-4 audio effort also focused on audio coding; however, unlike the efforts in MPEG-1 and MPEG-2 that primarily focused on improved compression, the emphasis in MPEG-4 was on new functionalities. MPEG-4 audio includes tools designed to cover a very wide range of functionalities, ranging from low-bit-rate intelligible speech to high-quality multichannel audio, for natural and synthetic audio signals. Natural audio is audio captured from the real world (e.g., speech or music) and synthetic audio is computer-generated audio signals.

Specifically, MPEG-4 audio is divided into algorithms for natural audio and algorithms for synthetic audio. To code natural audio, MPEG-4 includes a number of codecs, including parametric coder for 2 to 4 kbps speech at 8-kHz sampling rate or 4 to 16 kbps audio at 8- or 16-kHz sampling rate; CELP coder for speech at 6 to 24 kb/sec at 8 or 16 kHz for narrowband and wideband speech, respectively; and MPEG-4 AAC and twin VQ for general audio signals over a broad range of bit rates. Additional functionalities incorporated in the standard include error resilience and scalability in bit rate, audio bandwidth, and decoder complexity. Note that MPEG-4 AAC is a superset of MPEG-2 AAC, which includes a number of additional functionalities such as perceptual noise substitution, long-term predictor, and different forms of scalability. Therefore, an MPEG-4 AAC decoder can decode MPEG-2 AAC bitstreams, and an MPEG-2 AAC decoder can decode an MPEG-4 AAC bitstream if it is created without the MPEG-4 extensions. Further information MPEG-4 natural audio coding is available in Johnston et al.[22] and Brandenburg et al.[27]

MPEG-4 synthetic audio includes a test-to-speech (TTS) decoder for converting text to spoken speech and the structured audio (SA) synthesis tools.[12] MPEG-4 structured audio is a signal-processing language for describing algorithms that represent musical instruments, as well as the scores or inputs that drive these musical instruments. These algorithms are transmitted to the receiver, which uses them to synthesize the audio. In addition, MPEG-4 SA allows one to describe algorithms that take multiple audio tracks and mix them together, with possible effects processing such as reverberation and spatialization in 3-D, to compose a desired audio scene.

Image Coding Standards. The most widely used image compression algorithm is the JPEG algorithm, named after the Joint Photographic Experts Group, the ISO working group that developed it. The JPEG algorithm 23 is based on the 8×8–pixel 2-D DCT, scalar quantization, runlength, and Huffman coding, as described earlier in this section. This algorithm also standardized a number of other tools for scalable coding, etc.; however, most of the tools beyond the baseline algorithm are not widely used. The JPEG algorithm typically compresses a color (RGB) image with 24 bits/pixel by a factor of 24 (very good quality) to 48 (good quality) corresponding to bit rates in the range of 0.5 to 1.0 b/pixel.

Lossless compression of images is very important in a number of applications, including compression of medical images and compression of non-natural images such as computer graphics or text. Although the original JPEG standard included a lossless coding algorithm, that algorithm achieved limited performance and was not widely used. To provide an improved solution for lossless image compression, the JPEG-LS standard was created. JPEG-LS was designed to provide lossless compression of continuous-tone images with minimal complexity. In addition, JPEG-LS included a "near-lossless" mode for applications that could tolerate a known maximum error for any given pixel value. JPEG-LS is based on predictive coding of pixel values followed by context adaptive coding of the prediction error for each pixel.[28] An important note is that lossless compression of natural images typically achieves a compression ratio of less than 2:1; non-natural images such as computer graphics or text can achieve much higher compression ratios.

The newest image compression standard, JPEG-2000, was designed to provide improved compression performance and new functionalities compared to the original JPEG standard. Specifically, JPEG-2000 was designed to provide high compression performance; various forms of scalability, including bit rate scalability, spatial resolution scalability, and SNR (amplitude resolution or quality) scalability; and support high bit rates up to lossless compression, large bit depths per pixel, many color components (for multispectral imaging), random access, and error resilience. For example, JPEG-2000 supports high-quality compression of very large images and quick random access for remote browsing of portions of these images.

An important element in the design of JPEG-2000 was the separation of the coding and the bitstream ordering operations in the compression system. This enables much greater flexibility in the functionalities provided. For example, an

Table 1.8   Current and Emerging Image Compression Standards

Image coding standard	Primary intended application	Bit rate
JPEG	Coding of continuous-tone images	0.5–1.0 bit/pixel
JPEG-LS	Lossless compression of continuous-tone images	Highly variable
JPEG-2000	Coding of continuous-tone images, various forms of scalability, browsing over the Internet, multichannel and high-bit-depth images	0.1–0.5 bit/pixel

image can be compressed once, and then a different individual bitstream can be organized and delivered for each requesting client; the bitstream is matched to the application goals and constraints of the client, e.g., spatial resolution or bit rate or region of interest in the image. A detailed overview of JPEG-2000 is available in Taubman and Marcellin.[29] The image compression standards are listed in Table 1.8.

Video Coding Standards. The first video compression standard to gain widespread acceptance was the ITU H.261,[30] which was designed for videoconferencing over the integrated services digital network (ISDN) and adopted as a standard in 1990. It was designed to operate at p = 1, 2,...30 multiples of the baseline ISDN data rate, or p × 64 kb/sec. In 1993, the ITU-T initiated a standardization effort with the primary goal of videotelephony over the public switched telephone network (PSTN) (conventional analog telephone lines), where the total available data rate is only about 33.6 kb/sec. The video compression portion of the standard is H.263 and its first phase was adopted in 1996.[31] An enhanced H.263, H.263 Version 2 (V2), was finalized in 1997, and a completely new algorithm, originally referred to as H.26L, has recently been finalized as H.264/MPEG-4 part 10 advanced video coding (AVC).

The Moving Pictures Expert Group (MPEG) was established by the ISO in 1988 to develop a standard for compressing moving pictures (video) and associated audio on digital storage media (CD-ROM). The resulting standard, commonly known as MPEG-1, was finalized in 1991 and achieves approximately VHS quality video and audio at about 1.5 Mb/sec.[32,35] A second phase of their work, commonly known as MPEG-2, was an extension of MPEG-1 developed for application toward digital television and for higher bit rates.[33,36] A third standard, to be called MPEG-3, was originally envisioned for higher bit rate applications such as HDTV; however, it was recognized that those applications could also be addressed within the context of MPEG-2 so those goals were wrapped into MPEG-2 (consequently, there is no MPEG-3 standard).

Currently, the video portion of digital television (DTV) and high-definition television (HDTV) standards for large portions of North America, Europe, and Asia is based on MPEG-2. A third phase of work, known as MPEG-4, was designed to provide improved compression efficiency and error resilience features, as well as increased functionality, including object-based processing, integration of natural and synthetic (computer generated) content, and content-based interactivity.[34]

The newest video compression standard, known as H.264/MPEG-4 Part 10 AVC, was created by the Joint Video Team (JVT) comprising ITU and ISO MPEG. It achieves a significant improvement in compression over all prior video coding standards and has been adopted by ITU and ISO. It is called H.264 and MPEG-4 part 10 AVC, respectively, in the two standardization organizations.[35] The video compression standards are listed in Table 1.9. Currently, the video compression standards primarily used for video communication and video streaming are H.263 V2, MPEG-4, and H.264/MPEG-4 part 10 AVC.

1.2.3 What Do the Standards Specify?

An important question concerns the scope of the compression standards, or what they actually specify. A compression system is composed of an encoder and a decoder with a common interpretation for compressed bitstreams. The encoder takes the original media signal and compresses it to a bitstream. The decoder takes a compressed bitstream and decodes it to

Table 1.9   Current and Emerging Video Compression Standards

Video coding standard	Primary intended applications	Bit rate
H.261	Videotelephony and teleconferencing over ISDN	$p \times 64$ kb/sec
MPEG-1	Video on digital storage media (CD-ROM)	1.5 Mb/sec
MPEG-2	Digital television	2–20 Mb/sec
H.263	Video telephony over PSTN	33.6 kb/sec and up
MPEG-4	Object-based coding, synthetic content, interactivity, video streaming	Variable
H.264/MPEG-4 Part 10 (AVC)	Improved video compression	10s of kb/sec to 10s of Mb/sec

produce the reconstructed signal. One possibility is that a standard would specify the encoder as well as the decoder. However, this approach turns out to be overly restrictive. Instead, the standards have a limited scope to ensure interoperability while enabling as much differentiation as possible.

The standards do not specify the encoder or the decoder; instead, they specify the bitstream syntax and the decoding process. The bitstream syntax is the format for representing the compressed data. The decoding process is the set of rules for interpreting the bitstream. Note that specifying the decoding process is different from specifying a specific decoder implementation. For example, the standard may specify that the decoder use an inverse discrete cosine transform (IDCT), but not how to implement the IDCT. The IDCT may be implemented in a direct form, or using a fast algorithm similar to the FFT, or using MMX instructions. The specific implementation is not standardized, thus allowing different designers and manufacturers to provide standard-compatible enhancements and differentiate their work.

The encoder process is deliberately not standardized. For example, more sophisticated encoders can be designed that provide improved performance over baseline low-complexity encoders. More importantly, improvements can be incorporated even after a standard is finalized, e.g., improved algorithms for motion estimation or bit allocation or new perceptual models may be incorporated in a standard-compatible manner. The only constraint is that the encoder produce a syntactically correct bitstream that can be properly decoded by a standard-compatible decoder.

Limiting the scope of standardization to the bitstream syntax and decoding process enables improved encoding and decoding strategies to be employed in a standard-compatible manner, thereby ensuring interoperability while enabling manufacturers to differentiate themselves. As a result, it is important to remember that not all encoders "are created equal," even if they correspond to the same standard.

1.2.4 Evaluating Media Compression Algorithms

Media compression algorithms are evaluated based on their perceived quality, bit rate, delay, and complexity. An ideal compression algorithm would achieve very high perceived quality at a very low bit rate with low delay and low complexity. Perceived quality is a highly subjective metric and requires a significant amount of testing with a large number of human test subjects individually evaluating the compressed media. Objective measures are significantly easier to use, but in some cases may not accurately describe the perceived quality. In the following, different subjective and objective methods for evaluating media compression algorithm performance are briefly highlighted.

Subjective Tests. Typically, the most useful evaluation of a media compression algorithm's performance is via subjective tests because, in the end, a human user consumes the media. In subjective tests, a large number of expert and/or nonexpert viewers/listeners view or listen to and evaluate the reconstructed quality of the media, and the final evaluation corresponds to some form of averaging of the individual evaluations. The evaluation is often performed using mean opinion scores (MOSs). In the absolute category rating (ACR) for evaluating speech coding, the quality of the speech is evaluated on an absolute scale of 5 to 1 (5 = excellent; 4 = good; 3 = fair; 2 = poor; and 1 = bad quality).

In another approach, the degradation category rating (DCR), the quality of the reconstructed speech is compared to that of the original speech. In this case, the speech samples are presented to the listener as (original speech, processed speech) pairs allowing the listeners to compare the two directly. Each listener rates the relative degradation on a 5 to 1 scale (5 = inaudible; 4 = audible but not annoying; 3 = slightly annoying; 2 = annoying; and 1 = very annoying, respectively). These scores are referred to as degradation mean opinion scores (DMOSs). Subjective tests for speech are detailed in ITU-T Recommendation P.800.[36]

A related evaluation method is the perceptual evaluation of speech quality (PESQ), adopted as ITU-T Recommendation P.862,[37] which is an objective method for predicting the subjective MOS scores (predicting perceived quality) for narrowband speech in a P.800 listening set-up as described previously. Unlike the tests described earlier, such as P.800, in which the primary goal is evaluating compression performance, PESQ also attempts to account for network effects such as variable delays, transcoding, channel errors, and concealment, etc. The PESQ score is in the range of 4.5 to –0.5, but generally provides a MOS-like score between 4.5 and 1.0. PESQ provides a number of important capabilities for roughly evaluating the perceptual quality of speech over a network, e.g., VoIP settings; however, although PESQ is an important and useful tool, it is not a replacement for conventional subjective tests of perceived speech quality.

Objective Tests. It is very valuable to have an objective criterion for measuring the difference between the original and the reconstructed signal, where the measure ideally correlates well with the perceived difference between the two signals. However, this is extremely difficult because modeling human perception is quite challenging, especially modeling human visual perception. The most popular metrics for measuring quality are the mean squared error (MSE), the signal-to-noise ratio (SNR), and the peak signal-to-noise ratio (PSNR). MSE and SNR are the conventional metrics used throughout signal processing, estimation, and communication.

MSE is the energy in the error signal or the energy in the distortion.

SNR is the conventional metric that compares signal energy (or variance) to error energy. Although SNR is convenient for evaluating zero-mean signals such as speech or audio, it is not as convenient for evaluating nonzero-mean signals such as images or video. Specifically, image and video signals have non-negative pixel amplitudes (typically all positive) and as a result these signals have nonzero means. On the other hand, the range of the pixel values is nearly always given by 8 b/color component or [0,255]. As a result, to evaluate image or video signals, it is customary to replace the signal energy in SNR by the energy of the peak pixel-amplitude value to give peak SNR or PSNR. For example, consider a monochrome digital video signal in which the pixel values are given by $f(n,m,k)$, where n and m express the horizontal and vertical position of the pixel, and k the frame number, and where each frame is of size $N \times M$ and there are K frames in the video sequence. Let $g(n,m,k)$ express the reconstructed pixel values for the video signal. The MSE is defined as:

$$MSE = \frac{1}{N \cdot M \cdot K} \sum_{n=1}^{N} \sum_{m=1}^{M} \sum_{k=1}^{K} \left(f(n,m,k) - g(n,m,k) \right)^2$$

The PSNR, expressed in decibels (dB), is defined as:

$$PSNR(dB) = 10 \log_{10} \left(\frac{f_{peak}^2}{MSE} \right)$$

where $f_{peak} = 255$ is the peak pixel value, e.g., for 8-b pixels with amplitude range [0,255].

For a color image or video, the MSE and associated PSNR are computed separately for each color component. It is well known that MSE and PSNR do not correlate very well with perceived distortion in an image or video. However, they are the primary objective measures used for evaluating the distortion in image or video signals because they are simple and mathematically tractable and because better alternatives do not currently exist.

1.2.5 Additional Issues in Media Compression

A number of important problems arise in the context of media applications that relate to media compression. In the following, we briefly comment on two of them.

1.2.5.1 Adapting the Compressed Media Bit Rate

In many instances it is necessary to reduce the bit rate of a compressed media stream. A prime example occurs when media are precompressed and stored, such as for video on demand (VoD), and the original compressed bit rate is larger than the available channel bandwidth. Three approaches to adapt the compressed media to match the available channel bandwidth are through the use of transcoding, multiple-file switching, and scalable coding.

Transcoding. A conceptually straightforward approach to reduce the media bit rate is to decode and then re-encode the media at the desired bit rate. This approach has two drawbacks: (1) decoding/re-encoding results in a significant drop in quality compared to coding the original media directly at the desired bit rate; and (2) this approach is very complex, e.g., video encoding is very complex because it requires motion estimation, which is very complex. These problems can be dramatically reduced by compressed-domain transcoding, in which sophisticated algorithms that operate in the compressed domain (e.g., motion vector and DCT domain for video) are used to perform the transcoding.

These algorithms carefully use information in the original compressed bitstream (e.g., motion vectors) and thereby dramatically reduce the required computation and storage requirements; they also can lead to final transcoded quality much closer to that achievable by directly coding the original media at the desired rate. Important transcoding operations include bit rate reduction, spatial downsampling, frame rate reduction, interlaced to progressive conversion, and changing compression formats (e.g., MPEG-2 to H.263 or MPEG-4).[38]

Multiple-File Switching. In multiple-file switching, the original content is coded at a number of different strategically chosen bit rates, and these multiple compressed versions of the same content are stored at the sender. The early versions of multiple-file switching coded the original media at a few strategic rates targeted for common connection speeds (e.g., one for dial-up modem and one for DSL/cable); the client selected the appropriate media rate (and therefore media file) at the beginning of the session and this rate was used throughout the session. This approach had the drawback that the media rate could only be adapted once, at the beginning of the session, and if the network conditions and available bandwidth changed, this caused problems (even with a large client buffer to overcome short-term variations).

Recent advances in multiple-file switching, such as S-frames or SP-frames in H.264 and intelligent streaming from Microsoft and SureStream from Real Networks, enable dynamic switching between different media rates at selected times in the media session. This midsession switching between different media rates enables much better adaptation in time to the available bandwidth. In addition, in certain contexts it may be desirable to adapt to the available bandwidth and to the packet loss rate; therefore, multiple encodings of the same content optimized for different (available bandwidth, packet loss rate) operating conditions are also possible, as illustrated in Table 1.10. However, typically, multiple-file switching is only used to switch between multiple coded versions of the same content at different bit rates.

Multiple-file switching requires lower complexity than transcoding and also provides (slightly) better compression performance at each of the selected bit rates. On the other hand, multiple copies of the same media need to be stored, one for each selected bit rate. In addition, performance depends on how closely the available bit rate and the selected media rate for transmission match. The practical need to limit storage requirements leads to a coarse granularity in the number of bit rates coded and stored for each medium, thereby limiting its ability to adapt to varying transmission rates.

Table 1.10 Example of Multiple Encoded Files Created for Single
Video Content

| | Packet loss rate | | |
Bit rate	1%	5%	10%
64 kbps	1	4	7
128 kbps	2	5	8
256 kbps	3	6	9

Notes: Nine different encoded versions of the content (numbered 1 through 9 in the table) are created optimized for three different available bit rates and three different channel packet loss rates. The sender adaptively determines which coded version of the video is best to send, based on the channel's estimated available bit rate and packet loss rate.

Scalable Compression. Another approach that enables adaptation of the media rate to the available bandwidth is based on the use of scalable or layered compression. In scalable coding, the media are coded into a set of ordered bitstreams in which the base bitstream can be used to reconstruct the media at a base quality, and the enhancement bitstreams can be used to reconstruct the media at improved qualities. Therefore, different ordered subsets of these bitstreams can be selected to represent the media at different bit rates.[49]

All recent video compression standards, such as MPEG-2, MPEG-4, and H.263, have scalable coding functionalities; however, scalable video coding has not been widely used because current scalable video coding algorithms incur a significant compression penalty compared to nonscalable video coding approaches. On the other hand, scalable image coding in the form of the recent JPEG-2000 standard provides the scalable functionality while incurring a very small penalty in compression compared to nonscalable image coding approaches. Therefore, JPEG-2000 scalable image coding will probably become quite popular in the future. AAC includes scalable audio coding functionalities, but it is unclear how prevalent their use will be in the future.

Methods for identifying the available bandwidth between a sender and receiver and adapting the media bit rate for the available bandwidth are discussed in Section 1.3.5.2. Furthermore, methods for adapting the media for the available bit rate while preserving end-to-end security are discussed in Section 1.3.6.

1.2.5.2   Transmission Error Concealment for Compressed Media

Compressed media are highly vulnerable to errors or lost information. Because transmission errors or packet loss may afflict the delivery of media over broadband networks, media delivery systems are explicitly designed to combat these problems. We will examine in the next section how media delivery systems are designed to overcome networking problems; here, we briefly highlight the important problem of error concealment for compressed media.

The basic goal of error concealment is to estimate the lost information (e.g., missing audio samples or video pixels) in order to conceal the fact that an error has occurred. Many approaches can be used to do this. In the case of video, the key observation is that video exhibits a significant amount of correlation along the spatial and temporal dimensions. This correlation was used to achieve video compression, and unexploited correlation can also be used to estimate the lost information. Therefore, the basic approach in error concealment is to exploit the correlation by performing some form of spatial and/or temporal interpolation (or extrapolation) to estimate the missing information from the correctly received data. As a very simple example, if a portion of a video frame is missing, the missing pixels may be replaced by the pixels in the same spatial location but in the prior frame. The problem of error concealment can also be formulated as a signal recovery or inverse problem, leading to the design of sophisticated algorithms (typically iterative algorithms) that provide improved error concealment in many cases.

An important property of error concealment is that it is performed at the decoder. As a result, error concealment is outside the scope of media compression standards. Specifically, as improved error concealment algorithms are developed, they can be incorporated as standard-compatible enhancements to conventional decoders.

1.3   APPLICATIONS OF BROADBAND NETWORKS

Broadband networks facilitate a large variety of applications. Some of these applications, such as browsing the Internet or checking email, are also possible with networks that support relatively low bit rates. On the other hand, applications that involve media are generally not possible, or quite unpleasant to use, over networks that only support low bit rates. A large number of applications over broadband involve various forms of media (including speech, audio, image, video, and computer graphics) and these applications may be used for work, entertainment, or education. This section examines some of the applications of broadband networks, with an emphasis on applications that are not adequately supported by low bit rate networks; media applications appear prominently in this context.

This section continues by providing a taxonomy of the different operating conditions for media applications over broadband networks — these different operating conditions are important because they have significant effects on the design of broadband applications and systems. A key operating condition is the required latency for the application, which varies considerably for file downloads, streaming, and interactive or conversational services. This latency spectrum is discussed in Section 1.3.2.

The next three subsections examine important applications areas. Interactive communications, with emphasis on VoIP, is discussed in Section 1.3.3 and P2P networks, including their use for audio file sharing, are discussed in Section 1.3.4. Media delivery, with an emphasis on video streaming, is examined in depth in Section 1.3.5. The problem of supporting end-to-end security while adapting media in the middle of a network is discussed in Section 1.3.6; Section 1.3.7 examines the important network protocols for media delivery of IP networks. The design and use of a content delivery network (CDN) infrastructure to improve application performance is presented in Section 1.3.8. Finally, some thoughts on the landscape of future broadband applications are given in Section 1.3.9. (Terminology note: in this section on applications over broadband networks, the term *bandwidth* is synonymous with bit rate, as opposed to frequency bandwidth at the physical layer.)

1.3.1 Classifying Operating Conditions for Media Applications over Broadband Networks

A diverse range of different media applications as well as networks exists that have very different operating conditions or properties. For example, media communication applications may be for point-to-point communication or for multicast or broadcast communication, and the media may be pre-encoded (stored) or may be encoded in real time (e.g., interactive videophone or video conferencing). The networks may be static or dynamic, packet-switched or circuit-switched, may support a constant or variable bit rate transmission, and may support some form of quality of service (QoS) or may only provide best effort support. The attributes and operating conditions for media applications and networks can be roughly classified into the following, possibly overlapping, groups:

- Point-to-point, multicast, and broadcast communication
- Real-time encoding vs. pre-encoded (stored) media
- Interactive vs. noninteractive applications
- Packet-switched or circuit-switched network
- Static vs. dynamic channels
- Constant-bit-rate (CBR) or variable-bit-rate (VBR) coding
- Constant-bit-rate (CBR) or variable-bit-rate (VBR) channels
- Best-effort (BE) or QoS support

Specific properties of the media application and network strongly influence the design of the system; therefore, we continue by briefly discussing the preceding properties and their effects on media application and system design.

1.3.1.1   Point-to-Point, Multicast, and
          Broadcast Communications

Probably the most popular form of media communication is
one-to-many (basically, one-to-all) communication or broadcast communication; the most well-known examples are television and radio. Broadcast is a very efficient form of
communication for popular content because it can often efficiently deliver popular content to all receivers at the same time.

An important aspect of broadcast communications is that
the system be designed to provide every intended recipient
with the required signal. This is an important issue because
different recipients may experience different channel characteristics (e.g., different available bandwidths) and, as a result,
the system is often designed for the worst-case channel. An
example of this is digital television broadcast in the U.S.,
where the source coding and channel coding were designed
to provide adequate reception to receivers at the fringe of the
required reception area, thereby sacrificing some quality to
receivers in areas with higher quality reception (e.g., in the
center of the city). An important characteristic of broadcast
communication is that, due to the large number of receivers
involved, feedback from receiver to sender is generally infea-
sible, thus limiting the system's ability to adapt.

Another common form of communication is point-to-point
or one-to-one communication; examples include the conventional telephone as well as videophone and unicast video
streaming over the Internet. In point-to-point communications, an important property is whether a back channel is
present between the receiver and sender. If a back channel
exists, the receiver can provide the sender with feedback that
the sender can then use to adapt its processing. On the other
hand, without a back channel, the sender has limited knowledge about the channel.

Another form of communication with properties that lie
between point-to-point and broadcast is multicast, which is a
one-to-many communication, but not one-to-all as in broadcast. An example of multicast is IP-multicast over the Internet. However, as discussed later, IP multicast is currently not

widely available in the Internet, and other approaches are being developed to provide multicast capability, e.g., application-layer multicast via overlay networks. To communicate to multiple receivers, multicast is more efficient than multiple unicast connections (i.e., one dedicated unicast connection to each client); overall, multicast provides many of the same advantages and disadvantages as broadcast.

1.3.1.2   Real-Time Encoding vs. Pre-Encoded (Stored) Media

Media may be captured and encoded for real-time communication or may be pre-encoded and stored for later viewing. Interactive applications are one example of applications that require real-time encoding, e.g., videophone, video conferencing, or interactive games. However, real-time encoding may also be required in applications that are not interactive, e.g., the live broadcast of a sporting event.

In many applications, media content is pre-encoded and stored for later listening or viewing. The media may be stored locally or remotely. Examples of local storage include CD, DVD, and video CD, and examples of remote storage include video on demand (VoD), and audio and video streaming over the Internet. Pre-encoded media have the advantage of not requiring a real-time encoding constraint. This enables more efficient encoding such as the multipass encoding typically performed for DVD content. On the other hand, it provides limited flexibility because, for example, the pre-encoded media cannot be significantly adapted to channels that support different bit rates or to clients that support display capabilities different from that used in the original encoding. The requirement to adapt the media to support diverse clients connected by heterogeneous and time-varying networks leads to use of multiple-file switching, transcoding, and scalable coding, as described in Section 1.3.5.1 and Section 1.3.5.2.

1.3.1.3   Interactive vs. Noninteractive Applications

Interactive applications such as VoIP, videophone, or interactive games have a real-time constraint. Specifically, the information has a time-bounded usefulness, and if the information arrives, but is late, it is useless. This is equivalent to a maximum acceptable end-to-end latency on the transmitted information, where by end-to-end we mean: capture, encode, transmission (including any queuing delays), receive, decode, display. The maximum acceptable latency depends on the application, but often is on the order of 150 msec. Noninteractive applications have looser latency constraints — for example, many seconds or potentially even minutes. Examples of noninteractive applications include multicast of popular events or multicast of a lecture; these applications require timely delivery but have a much looser latency constraint. Note that interactive applications require real-time encoding, as may noninteractive applications; however, the end-to-end latency for noninteractive applications is much looser, and this can dramatically simplify the system design.

1.3.1.4   Packet-Switched or Circuit-Switched Network

A key attribute of broadband networks that affects the design of media delivery systems is whether they are packet switched or circuit switched. Packet-switched networks, such as Ethernet LANs and the Internet, are shared networks in which the individual packets of data may exhibit variable delay, may arrive out of order, or may be completely lost. Alternatively, circuit-switched networks, such as the public switched telephone network (PSTN) or ISDN, reserve resources; the data have a fixed delay and arrive in order. However, the data may still be corrupted by bit or burst errors.

1.3.1.5   Static vs. Dynamic Channels

A second key attribute of broadband networks that affects the design of media delivery systems is whether their characteristics (e.g., bandwidth, delay, and loss) are static or dynamic (time varying). An example of a static channel is ISDN (which provides a fixed bit rate and delay, and a very low loss rate). Examples of dynamic channels include wireless channels or shared networks such as the Internet. Media communication over a dynamic channel is much more difficult than over a static channel, and many of the challenges of media streaming, as are discussed later in this chapter, relate to the dynamic attributes of the channel.

1.3.1.6 Constant-Bit-Rate (CBR) or Variable-Bit-Rate (VBR) Coding

Typically, the complexity of a media signal and, in particular, of a video varies with time. Therefore, compressing the video signal to achieve a constant visual quality requires a variable bit rate. Alternatively, coding for a constant bit rate would produce time-varying quality. For example, DVD video provides high-quality throughout an entire movie by employing VBR encoding in which the instantaneous video bit rate is adapted as a function of time (e.g., 2 to 10 Mb/sec) based on the video content. Clearly, it is very important to match the video bit rate to what the channel or network can support.

A practically important case is transmitting a VBR video over a constant-bit-rate channel. To couple the video to the channel, a buffer is typically employed to receive the variable rate compressed stream and output a constant rate stream that the channel expects. Another practically important case is when there is a low-bit-rate bottleneck link over the last mile or the last meter. To couple the video to the channel (and maximize the reconstructed quality at the receiver), a buffer is typically used and a buffer control mechanism provides feedback based on the buffer fullness to regulate the coarseness/fineness of the quantization and thereby the video bit rate — therefore ensuring that the encoder produces the appropriate bit rate for the current channel conditions.

1.3.1.7 Constant-Bit-Rate (CBR) or Variable-Bit-Rate (VBR) Channels

Some channels support a CBR delivery and some may only support VBR. For example, wireless channels such as cellular or 802.11 wireless LAN provide VBR delivery because the available rate depends on the wireless channel conditions, e.g., interference and other users. Furthermore, channels shared among multiple users generally provide VBR to each individual user because the available rate depends on the other traffic (number of users and time-varying characteristics of each). Media communication or streaming over a VBR channel is challenging, as is discussed shortly, because the available rate must be estimated and the media rate matched to the available rate.

Broadband wired networks are generally shared packet networks that provide a constant bit-rate channel that is shared among all the users. This shared channel provides challenges, as mentioned earlier, but it also provides some advantages for streaming of VBR media streams. Specifically, multiple independent VBR media streams are highly unlikely to require their peak bit rates simultaneously. Therefore, the shared channel aggregates the multiple streams; for a large number of independent streams, the number of streams that can be sent through the channel depends on roughly the average rate of each stream as opposed to its peak rates. The increase in the number of VBR streams that can be sent over a shared channel in this manner is referred to as the statistical multiplexing gain.

1.3.1.8   QoS Support

An important area of network research over the past two decades has been QoS support. QoS is a vague and all-encompassing term used to convey that the network provides some type of performance guarantees (e.g., guarantees on throughput, maximum loss rates, or delay) or preferential delivery service. Network QoS support can greatly facilitate media communication because it can support a number of capabilities, including provisioning for media data, prioritizing delay-sensitive media data over conventional data traffic, prioritizing among the different forms of media data (e.g., speech vs. video), and prioritizing across the different packets within a single media application flow (e.g., some video packets are often more important than others).

Unfortunately, QoS is currently not widely supported in packet-switched networks such as the Internet. The current Internet does not provide any QoS support, and it is often referred to as best effort (BE) because its basic function is to provide simple network connectivity by best-effort (without any guarantees) packet delivery. Different forms of network QoS under consideration for the Internet include differentiated services (DiffServ) and integrated services (IntServ). Network QoS is specified for wireless LANs via 802.11e, which provides classes of service with managed levels of QoS for data, voice, and video over wireless IP.

1.3.2   The Latency Spectrum: Download vs. Streaming vs. Low-Latency Communication

From a system design perspective, one of the most important attributes of an application is its latency requirement. Media applications can be roughly classified into three groups based on the required latency for the application: (1) media delivery by file download; (2) streaming, or the simultaneous delivery and playback of media; and (3) low-latency communication. Media delivery by file download has the loosest latency constraint in the sense that the entire media can be downloaded before playback begins. In streaming, the goal is to enable simultaneous delivery and playback; specifically, the playback begins before the media are downloaded. In low-latency communication, the goal is roughly to minimize the latency as much as possible, and latencies above 100 to 200 msec (depending on the specific application) may render the application unacceptable. These three classifications of media delivery are examined in detail next.

Media Delivery via File Download. The most straightforward approach for media delivery over a network is by file copy or file download. We refer to this as media download to keep in mind that it is a media and not a generic file. Specifically, media download is similar to a file download; however, especially in the case of video, it is a LARGE file. Media download allows the use of established delivery mechanisms, for example, TCP as the transport layer or FTP or HTTP at the higher layers, and this use of established delivery mechanisms can greatly simplify application design.

For images or pre-encoded audio content, media download is probably the preferred approach. A more sophisticated version of media download can be achieved by adapting the download in a media-aware manner. For example, when delivering an image to a cell phone or over the Internet to a Web browser, it is beneficial to order the data so that a low-resolution image can be downloaded and displayed first, while the high resolution information is still being downloaded. (Conceptually, this approach may be thought of as splitting the media into multiple portions, which are ordered in terms of importance and sequentially downloaded.)

Although media download is often the preferred approach for pre-encoded audio and images, in the case of video it has a number of disadvantages. Because videos generally correspond to very large files, the download approach usually requires long download times and large storage spaces. These are important practical constraints. In addition, the entire video must be downloaded before viewing can begin. This requires patience on the viewer's part and also reduces flexibility in certain circumstances, e.g., if the viewer is unsure of whether he or she wants to view the video, he or she must still download the entire video before viewing it and making a decision. Video download may also be impossible in certain situations; for example, portable devices may not have sufficient storage capability to download and store the entire video before playback.

Media Delivery via Streaming. Media delivery by media streaming attempts to overcome the problems associated with file download and also provides a significant number of additional capabilities. The goal of media streaming is to enable simultaneous delivery and playback, in contrast to media download, in which the media are downloaded first in their entirety and then playback occurs. The basic idea of media streaming is to split the media into parts, transmit these parts in succession, and enable the receiver to decode and play back the media as these parts are received, without waiting for the entire media to be delivered. Media streaming can conceptually be thought to consist of the following steps:

1. Partition the compressed media into packets.
2. Start delivery of these packets.

3. Begin decoding and playback at the receiver while the media is still being delivered.

Media streaming enables simultaneous delivery and playback of the media. This is in contrast to media download, when the entire media must be delivered before playback can begin. In media streaming, a short delay (usually on the order of 1 to 15 sec) usually occurs between the start of delivery and the beginning of playback at the client. This delay, referred to as the preroll delay, provides a number of benefits, which are discussed in Section 1.3.5.3.

Media streaming provides a number of benefits, including low delay before viewing starts and low storage requirements, because only a small portion of the media is stored at the client at any point in time. The length of the delay is given by the time duration of the preroll buffer, and the required storage is approximately given by the amount of data in this buffer.

Low-latency media communication. A number of applications require low latency (low end-to-end delay), including conversational applications such as voice over IP, videophone, videoconferencing, and interactive games. These applications also require a streaming solution, but the major difference is that the requirement of low latency imposes significant constraints on the system design. For example, the playback buffer (delay) must be very small, e.g., tens of milliseconds as opposed to multiple seconds. The compression algorithms must also be chosen for low-delay encoding and decoding, e.g., only I- and P-frames for video coding and no B-frames because of the additional delay. Channel coding is also constrained because the selection and effectiveness of FEC and interleaving are severely limited by the delay constraint. In addition, feedback schemes, such as ARQ to recover from packet loss, become much less effective because the round-trip time is generally longer than the required end-to-end delay. These constraints of the system design illustrate how low-latency media communication is much more difficult than media delivery by download, as well as more difficult than streaming, which has a looser latency constraint.

1.3.3   Interactive Two-Way Communications: VoIP

The conventional telephone provides one of the most important forms of communication. The telephone operates on a circuit-switched network and, in recent years, significant effort had been given to providing voice-over packet-switched networks such as the Internet. This is often referred to as VoIP and is also known as Internet telephony.

There are a number of motivations for moving from the conventional telephone network to VoIP. One of the main reasons is that IP switching equipment is becoming less expensive than telephone equipment. Another is that there are many benefits for a business to have only one network, instead of two. Therefore, instead of having a circuit-switched telephone network for telephony and a packet-switched IP network for data, strong motivation exists to move to a single network — an IP network — for delivering data and telephony.

A third motivation results from the dramatic change in the amount of data traffic compared to telephony traffic. To put this in perspective, until about 1999 the telephone network delivered more bits per second than the Internet (specifically, the total speech traffic on the telephone trunks was greater than the total traffic on the Internet backbones). However, around 1999 the Internet delivered more bits per second and, a couple of years later, it delivered about an order of magnitude more traffic than the telephone network. Furthermore, Internet traffic is continuing to grow at a rapid rate while telephone traffic is growing at a few percent per year. As a result, the entire telephone network's traffic is only a small fraction of the traffic carried by the Internet and, in principle, could be easily carried by the Internet. This observation motivated different IP service providers to provide IP telephony services to compete with the telephone companies.

In VoIP, the speech signal is captured, encoded using one of the encoders discussed previously, packetized, and sent over the IP network to the destination where it is received, decoded, and played. Designing VoIP systems has two general approaches. The first is based on ITU's H.323 standard, which provides an architectural overview as well as a family of protocols for providing VoIP and other forms of multimedia communication over best-effort packet networks. The second is based on the IETF's session initiation protocol (SIP) for setting up generic multimedia communication sessions over IP networks. SIP, as well as other important protocols for IP networks, is reviewed in Section 1.3.7.

VoIP requires a relatively small amount of bandwidth; however, the end-to-end latency must be on the order of 100 msec or less to provide good conversational services. Videotelephony over IP, which requires voice as well as video, has similar end-to-end latency constraints, but requires significantly more bandwidth then VoIP. The challenges for voice and video delivery over IP are examined in depth in Section 1.3.5.

1.3.4   Peer-to-Peer Networking: Sharing of Audio Files

Peer-to-peer (P2P) systems are architectures and applications in which distributed resources (e.g., computers) are used to perform a task in a decentralized manner. An early motivation for P2P systems was the realization that many computers are idle for large portions of time, thereby suggesting the idea of using these idle resources to perform some task, such as solving a large computational problem. P2P systems have recently received considerable public attention for file sharing, such as illegal sharing of audio content. Popular P2P systems for file sharing include Napster, Gnutella, and Kazaa; these systems primarily differ in the methods used to locate which peer has the desired file.

P2P systems generally have the following attributes:

- Peers discover one another (they are self-organizing).
- Peers generally have equal standing (peer equality) with symmetric communication between any two peers.
- Peers are generally autonomous and the entire system operates in a decentralized manner.
- Each peer may or may not be available at any point in time (no guarantees on availability or services provided).

Peers generally have equal standing in the sense that each peer can consume and provide content, unlike the client–server paradigm in which the server provides the content and the client consumes it. For example, each peer may host and share files with any other peer. P2P file sharing is quite compelling for certain types of content distribution because the content is directly transferred between peers and there are no central servers. In this manner, P2P networks enable one to exploit the resources of all of the nodes, as opposed to only the resources of the central nodes (e.g., servers) in more conventional architectures. However, P2P networks are not good for all applications. For example, when reliability is critical or cost for resources is not an issue, conventional centralized services are generally better. An insightful discussion on the suitability of a P2P architecture for different applications is given in Roussopoulos et al.[39]

Although the term "P2P networking" has gained considerable attention in recent years because of the popularity of file sharing and audio swapping P2P applications, it is important to realize that P2P networks have been around for many years. For example, the Internet is routed using P2P techniques in which the peers are routers that directly exchange information to adapt their routing tables. In addition, the Internet news (Usenet) is a P2P application that has been running since the early days of the Internet. An overview of P2P systems, including design and implementation issues, is given in Milojicic et al.[40]

Benefits of P2P networks include improved scalability and reliability by avoiding dependency on centralized resources, because clients can directly communicate with each other. Client resources (e.g., memory, computation, or bandwidth) can also be aggregated to help accomplish various tasks. An example is to use the computational resources of a number of distributed computers to solve large computational problems (e.g., the Search for Extraterrestrial Intelligence [SETI] often computes very large Fourier transforms by splitting them across many computers). Another example is to use the distributed storage across multiple nodes to store important content; the distributed storage (plus error-correction coding across the nodes) provides reliability when individual nodes may crash.

In addition, eliminating the need for centralized resources can potentially lead to reduced costs; the idea is that because the distributed nodes are required for other purposes, reusing them in a P2P architecture to accomplish additional tasks provides the desired additional services without requiring significant additional costs. P2P difficulties include that the individual peers may be unreliable or untrustworthy and may be intermittently available (they can connect and disconnect at will). This latter property requires that a P2P application keep track of the peers available at any point in time and support the joining and (unannounced) leaving of nodes.

P2P systems for media delivery generally have two steps: (1) discovery and (2) delivery. In the discovery step, a peer identifies another peer that, for example, has the desired content. Once the second peer is identified, the two peers directly interact and, therefore, the delivery step is decentralized. However, the discovery step is often centralized: the P2P network may use a centralized directory.

P2P networks with centralized directories are only partially decentralized; the delivery is decentralized, but the location of content is centralized, which can lead to reliability problems (single point of failure) as well as performance or scaling problems (because the performance is limited by the centralized discovery). A more decentralized solution can be achieved by distributing the content-location directory across the peers. This was the basic idea behind the popular P2P file-sharing network, Kazaa, in 2001; the network of peers is split into a two-level hierarchy in which a certain subset of the peers are selected as group leaders and hold the directory information for all of the peers associated with them. This two-class system of peers is more decentralized than the centralized system discussed earlier; however, because of the different responsibilities of the different nodes, it is not fully decentralized.

The Gnutella P2P file sharing application that became popular in 2002 uses a fully decentralized approach for the content location. Specifically, the P2P network of nodes has a flat, unstructured topology, in contrast to the star topology of Napster or the two-level hierarchy of Kazaa. Gnutella does not use directories for locating content, but rather uses query flooding to identify nearby nodes that have the desired file. This system is simpler than the centralized directory approach or the distributed directory approach, and all of the nodes are equal to each other. However, the use of query flooding to locate content can lead to large amounts of query traffic across the P2P network, which can limit the scalability of the system. Developing improved, fully distributed methods for identifying content location is an important topic of current research and distributed hash tables (DHTs) are an important tool.

1.3.5 Media Delivery: Video Streaming

Media delivery over networks has been an important problem for many years. Popular applications include video broadcast over digital cable, satellite and terrestrial (over-the-air) digital television (DTV), and video conferencing and videophone over circuit-switched networks. The growth and popularity of the Internet has motivated media delivery over best-effort packet networks, including low-cost voice communication via VoIP and MP3 audio downloads (as discussed in the prior sections), as well as video and audio streaming. Media over best-effort packet networks is complicated by a number of factors, including unknown and time-varying bandwidth, delay, and losses, as well as many additional issues such as how to share the network resources fairly among many flows and how to perform one-to-many communication efficiently for popular content. This section examines the challenges in media delivery over broadband networks — specifically, for streaming, or the simultaneous delivery and playback, of media over these networks.

Section 1.3.1 presented a brief overview of the diverse range of media streaming and communication applications. Understanding the different classes of media applications is important because they provide different sets of constraints and degrees of freedom in system design. Also, Section 1.3.2 discussed the similarities and differences among the three general approaches for delivering media: media download, media streaming, and low-latency media communication. We continue by presenting the three fundamental challenges in media streaming over networks: unknown and time-varying bandwidth, delay jitter, and loss.

These three fundamental problems and practical approaches for overcoming them were examined in detail in Section 1.3.5.2, Section 1.3.5.3, and Section 1.3.5.4, respectively. Further overview articles can be found in references 41 through 45. The important problem of adapting encrypted media while preserving security (i.e., without requiring decryption) is discussed in Section 1.3.6. The most important network protocols for streaming media over IP networks are described in Section 1.3.7. Most of the current section has a focus on operations that can be performed at the end hosts (sender and receiver) for optimizing the media delivery performance. The very important problem of designing the network infrastructure (e.g., the proxies and other nodes between the end hosts) to improve scalability, reliability, and other important performance aspects, is examined in Section 1.3.8.

1.3.5.1 Basic Challenges in Media Streaming

A number of basic challenges afflict media streaming. In the following, we focus on the case of video streaming over best-effort broadband networks, such as the Internet, because it is an important, concrete example that helps to illustrate these problems.

A significant amount of insight can be obtained by expressing the problem of media streaming as a sequence of constraints. Consider the time interval between displayed frames to be denoted by Δ, e.g., Δ is 33 msec for 30 frames/sec media and 100 msec for 10 frames/sec media. Each frame must be delivered and decoded by its playback time; therefore, the sequence of frames has an associated sequence of deliver/decode/display deadlines:

- Frame N must be delivered and decoded by time T_N
- Frame $N + 1$ must be delivered and decoded by time $T_N + \Delta$
- Frame $N + 2$ must be delivered and decoded by time $T_N + 2\Delta$
- Etc.

Any data lost in transmission cannot be used at the receiver. Furthermore, any data that arrive late are also useless. Specifically, any data that arrive after their decoding and display deadline are too late to be displayed. (Note that certain data may still be useful even if they arrive after its display time — for example, if subsequent data depend on these "late" data.) Therefore, an important goal of media streaming is to perform the streaming in a manner so that this sequence of constraints is met. Low-latency (conversational) media communication is similar to media streaming, except that the delay constraint is much tighter. For example, media streaming may afford a deliver/decode/display delay constraint of multiple seconds, but low-latency communication has an encode/deliver/decode/display constraint that is preferably less than 100 msec.

Video streaming over the Internet is difficult because the Internet only offers best-effort service. That is, it provides no guarantees on bandwidth, delay or delay jitter, or loss rate. Specifically, these characteristics are unknown and dynamic. Therefore, a key goal of video streaming is to design a system to deliver high-quality video reliably over the Internet when dealing with unknown and dynamic:

- Bandwidth
- Delay jitter
- Loss rate

In the following paragraphs we briefly examine these three challenges.

The bandwidth available between two points in the Internet is generally unknown and time varying. If the sender transmits faster than the available bandwidth, congestion occurs, packets are lost, and video quality drops severely. If the sender transmits more slowly than the available bandwidth, the receiver produces suboptimal video quality. To overcome the bandwidth problem requires (1) estimating the available bandwidth; and (2) matching the transmitted video bit rate to the available bandwidth. Additional considerations that make the bandwidth problem very challenging include accurately estimating the available bandwidth, matching the pre-encoded video to the estimated channel bandwidth, transmitting at a rate that is fair to other concurrent flows in the Internet, and solving this problem in a multicast situation in which a single sender streams data to multiple receivers where each may have a different available bandwidth. These problems and approaches to overcome them are discussed in Section 1.3.5.2.

Each packet takes a certain amount of time to travel from sender to receiver; this time is referred to as the end-to-end delay. From a sender to a receiver, this delay typically has a minimum delay determined by propagation time and minimum routing/switching times. However, because of congestion in the network (e.g., backed-up queues), the end-to-end delay can often be much larger, typically fluctuating from packet to packet as a function of cross traffic on the network or interference for a wireless link. This variation in end-to-end delay is referred to as the delay jitter and is a problem because the receiver must receive/decode/display frames at a constant rate; because of delay jitter (fluctuations in the end-to-end delay), packets transmitted at a constant rate generally do not arrive at a constant rate. In addition, any packets that arrive after their scheduled delivery deadline (late packets) because of delay jitter can produce problems in the reconstructed video, e.g., jerks in the video. This problem is typically addressed by including a playout buffer at the receiver, as is discussed in Section 1.3.5.3. Although the playout buffer can compensate for the delay jitter, it also introduces additional delay.

The third fundamental problem is losses. A number of different types of losses may occur, depending on the particular network under consideration. For example, wired packet networks such as the Internet are afflicted by packet loss in which an entire packet is erased (lost). On the other hand, wireless channels are typically afflicted by bit errors or burst errors: the packet is received, but its contents are corrupted. These corrupted packets are sometimes referred to as "dirty packets."

Many current systems simply detect and discard the dirty packets, i.e., corrupted packets are treated as lost packets. The rationale is that it is currently easier to pass only clean packets to the application for processing. As a result, the application can expect that, for any packet that it receives, it can trust the contents to be uncorrupted, and therefore all packets are received correctly or not received at all (erased). A practically important consequence of this approach is that the end hosts do not need to know what is inside the network because the links can be wired or wireless (with dropped or corrupted packets, respectively) and, in either case, the end hosts will receive or not receive each packet. This also leads to a clean demarcation between the physical and network layers within an end host; only clean packets are passed between the two, thus allowing the two layers to be designed independently of each other.

Losses can have a very destructive effect on reconstructed video quality. To combat the effect of losses, a video streaming system is designed with error control. Approaches for error control can be roughly grouped into four classes: (1) forward error correction (FEC); (2) retransmissions; (3) error concealment; and (4) error-resilient video coding. These approaches are discussed in Section 1.3.5.4.

The three fundamental problems of unknown and dynamic bandwidth, delay jitter, and loss are highly coupled and some of the coupling is quite clear. For example, transmitting at a media rate higher than the available bandwidth has the consequence that the network will discard packets, resulting in the application seeing packet loss. However, some of these couplings may not be apparent at first. For example, the effect of delay jitter on packet loss is illustrated in Figure 1.6. The probability density function (PDF) of the end-to-end delay for each packet is shown on the left, where the minimum

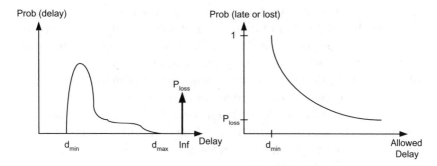

Figure 1.6 Relationship between end-to-end packet delay and packet loss. The probability density function for the packet delay is shown on the left, and the probability of a packet arriving late or being lost as a function of the delay allowed by the application is shown on the right.

delay and maximum delay are identified, and the lost packets are shown as an impulse of area P_{loss} at infinite delay.

This PDF illustrates that most packets typically have a delay slightly larger than the minimum delay, but that some packets can have a much larger delay (identified by the long tail in the PDF). The plot on the right illustrates the probability of a packet arriving late or being lost, Prob (late or lost), as a function of the delay allowed by the application. Note that, in most streaming applications, if a packet arrives late (after its scheduled delivery deadline), it is equivalent to a situation in which it was lost. The plot illustrates that, if the application's allowed end-to-end delay is less than the minimum delay d_{min}, all packets will be lost. As the allowed delay is increased, Prob (late or lost) monotonically decreases and continues to decrease until there are no more late packets, and only lost packets.

As these plots suggest, it is desirable to increase the application's allowed delay so that packets that arrive but have large delay are still received and processed in time and are not counted as late or lost packets. Many streaming applications utilize a large playout delay, as is discussed in Section 1.3.5.3, to achieve this purpose. However, low-latency applications, such as videophone or videoconferencing, cannot have a delay greater than 100 to 200 msec without adversely affecting the interactivity of the application. Therefore, designing applications that provide low-latency and high-quality leads to some of the most challenging system designs for media over networks.

The three fundamental problems of unknown and dynamic *bandwidth, delay jitter, and loss,* are considered in more depth in the following three sections. Each section focuses on one of these problems and discusses various approaches for overcoming it.

1.3.5.2   Rate Control for Streaming Media

This section begins by discussing the need for streaming media systems to control transmission rate adaptively according to prevalent network condition. We then discuss some ways in which appropriate transmission rates can be estimated dynamically at the time of streaming and briefly discuss how media coding has evolved to support such dynamic changes in transmission rates.

1.3.5.2.1   Why Do We Need Rate Control?

The available bandwidth in most current networks is dynamic and the instantaneous available bandwidth is *a priori* unknown. For example, in the Internet it is not possible to reserve bandwidth, and as the offered load on a link exceeds the link's transmission capacity, performance degrades. On a wireless link, such as 3G cellular or 802.11, radio propagation effects and interference also significantly affect the available bandwidth. Therefore, the available bandwidth is a function of the media session's bandwidth, cross traffic, and environmental effects. Transmitting faster than the available bandwidth can lead to congestion, higher delays and delay jitter, packet loss, and severe drop in video quality. As discussed before, these problems lead to significant challenges for streaming media systems.

On the other hand, a too conservative approach that transmits more slowly than the available bandwidth can lead to suboptimal video quality. Therefore, the goal is to match the media bit rate to the available bandwidth. This process is referred to as *rate control* (or congestion control in the network community) and includes the steps of (1) estimating the available bandwidth; and (2) matching the media rate to this available bandwidth.

1.3.5.2.2   Rate Control: Estimating the Available Bandwidth

The most widely used rate control mechanism today is that used in the transmission control protocol (TCP), which is used to deliver Web pages, file downloads, email, and most traffic on the Internet. TCP's rate control is based on a simple rule of "additive increase multiplicative decrease" (AIMD).[46] Specifically, end-to-end observations are used to infer packet losses or congestion: when no congestion is inferred, the transmission is increased at a constant rate (additive increase) and when congestion is inferred, the transmission rate is halved (multiplicative decrease). In addition, TCP provides guaranteed delivery by retransmitting lost packets.

TCP has been empirically proven to provide stability and scalability and thus appears to be a natural choice for streaming. However, this protocol provides a number of disadvantages for streaming media. First, the AIMD rule leads to widely varying transmission rate because of the multiplicative decrease (cutting in half) of the transmission rate. This leads to a saw-tooth pattern for the transmission rate as the transmission slowly increases (additive increase) and then abruptly is cut in half (multiplicative decrease). These saw-tooth oscillations are detrimental for streaming. A second problem is that the guaranteed delivery that TCP provides is achieved by persistent retransmission until the packet is received, which can lead to potentially unbounded delay. As previously discussed, unlike the case of data delivery in which it is important that every packet be delivered, for media, if a packet does not arrive by its delivery deadline, it is useless and we would prefer simply to discard that packet and instead use the available transmission opportunities to deliver other media packets by their delivery deadlines.

To overcome the disadvantages of TCP, current streaming systems for the Internet build upon user datagram protocol (UDP), which is a best effort delivery service (it does not guarantee delivery) that handles rate control and error control at the application layer. This approach allows significantly more flexibility for rate and error controls. For example, unlike the case of TCP, in which persistent retransmissions are performed until a packet is received, selective retransmissions can be performed in which, for each lost media packet, a decision can be made on whether to retransmit based on the importance of the packet and its delivery deadline.

In addition, other error control techniques can also be used, such as FEC or unequal FEC, or a hybrid combination of FEC and retransmission. In the case of rate control, although TCP's AIMD approach leads to a specific profile for rate adaptation, the use of application-layer rate control provides much greater flexibility in rate control to better match the media stream's instantaneous characteristics. For example, by having similar macroscopic (large scale) transmission patterns as TCP, and by adjusting these macroscopic patterns in the same way as TCP does, one can act like a TCP flow in certain ways and potentially inherit some of TCP's properties such as scalability and stability. Specifically, the average throughput of TCP (λ) can be inferred from end-to-end measurements of packet loss rate (ρ), round trip time (RTT), and the network maximum transmittable unit (MTU) or packet size.[47,48]

$$\lambda = 1.22 \times MTU\big/TT \times \sqrt{\rho}$$

Therefore, rate control algorithms can be designed that follow TCP rate control on the macroscopic scale, but without the undesirable microscale saw-tooth fluctuations of AIMD.[49,50] In this manner, these rate control algorithms can match the TCP traffic pattern on a coarser (macroscopic) scale (e.g., same average throughput over a time window) without exactly matching the TCP traffic pattern of AIMD-driven oscillations, which is detrimental for streaming. Another often argued benefit of these rate control algorithms is that they may fairly coexist with other TCP-based applications using the same network, in which fairness is in the sense of fairly sharing the available bandwidth. Because of these benefits, these rate control algorithms are often referred to as TCP-friendly rate control algorithms. Although TCP-friendly rate control provides a number of advantages as highlighted earlier, it also has disadvantages. For example, strictly following the preceding average throughput of TCP based on end-to-end measurements has disadvantages, e.g., it may be undesirable for a streaming session to vary the average rate as a function of the round-trip-time or the packet loss rate.

The preceding discussion has focused on sender-driven rate control because the sender explicitly estimates the available bandwidth and adapts to it, based on feedback from the receiver such as packet loss rate. Another important area is receiver-driven rate control, in which the receiver explicitly adapts the rate. A key application area for this is in multicast, where different receivers may be able to receive and decode at different rates. In this case, the sender codes video with a scalable or layered coder and sends different layers over different multicast groups. Each receiver individually estimates its bandwidth and joins an appropriate number of multicast groups to receive an appropriate number of layers up to its available bandwidth. This approach is referred to as receiver-driven layered multicast.[51] In contrast to sender-driven rate control, in this case each receiver individually performs the rate control by selecting the number of layers to receive and therefore selecting among a number of possible rates.

1.3.5.2.3 Rate Control: Matching the Media Rate to the Available Bandwidth

Given an estimate of the available bandwidth, the media rate must be adapted to match it. When the media are encoded in real time, e.g., videophone or videoconferencing, then adapting the media rate to the available bandwidth is straightforward. However, this problem is much more difficult for pre-encoded content. If the average media rate is less than the average bandwidth available and if buffering is possible, short-term fluctuations in the available bandwidth may be overcome through buffering. However, if the average media rate is larger than the average available bandwidth, or if sufficient buffer time is not available, the media rate must be changed to fit in the available bandwidth.

Three methods to solve the problem of adapting the media bit rate are transcoding, scalable coding, and multiple-file switching, which were examined in Section 1.2.1 and Section 1.2.5.1. In transcoding, one takes the original compressed stream and computes another compressed stream that has the desired property, e.g., that meets the bit rate constraint. In scalable coding, the media are coded into a set of ordered bitstreams, where the base bitstream can be used to reconstruct the media at a base quality, and the enhancement bitstreams can be used to reconstruct the media at improved qualities. Therefore, different ordered subsets of these bitstreams can be selected to represent the media at different bit rates. In multiple-file switching, the original content is coded at a number of different strategically chosen bit rates; these multiple compressed versions of the same content are stored at the sender, and the sender switches between the different compressed bitstreams to stream to the receiver based on the available bandwidth.

Multiple file switching and scalable coding are similar in the sense that, for a given available bit rate, both approaches simply select the appropriate bitstream(s) to transmit. The difference is that multiple-file switching stores an entirely new compressed copy of the media for every selected bit rate, but scalable coding stores an ordered set of bitstreams in which different ordered subsets correspond to the media at different rates. Therefore, multiple-file switching requires more storage than scalable coding, but it is not afflicted by the significant compression penalty incurred by current scalable video coders.

The performance for multiple file switching and scalable coding depends on how closely the available bit rate and the selected media rate for transmission match. The practical need to limit storage requirements leads to a coarse granularity in the number of bit rates coded and stored for each piece of content, thereby limiting the ability of multiple-file switching to adapt to varying transmission rates. Scalable coding of video has similar constraints. Transcoding, on the other hand, provides the greatest flexibility in meeting the available bit rate constraint and also requires the minimum storage, but it is by far the most computationally complex approach.

1.3.5.3 Playout Buffer for Overcoming Delay Jitter

It is common for streaming media clients to have a 1- to 15-sec buffering before playback starts. As we have seen at the beginning of this section, streaming can be viewed as a sequence of constraints for individual media samples. The use of buffering essentially relaxes all the constraints by an identical amount. Buffering is critical to the performance of streaming systems over best-effort networks such as the Internet, and it provides a number of important benefits that are discussed next.

Jitter Reduction. Variations in network conditions cause the amount of time that it takes for packets to travel between identical end hosts to vary with time. Such variations can be due to a number of possible causes, including queuing delays and link-level retransmissions. Jitter can cause jerkiness in playback due to the failure of same samples to meet their presentation deadlines and therefore being skipped or delayed. The use of buffering effectively extends the presentation deadlines for all media samples and, in most cases, practically eliminates playback jerkiness due to delay jitter. The benefits of a playback buffer are illustrated in Figure 1.7, where packets are transmitted and played at a constant rate and the playback buffer reduces the number of packets that arrive after their playback deadline.

Error Recovery through Retransmissions. Extended presentation deadlines for the media samples allow the use of retransmission to recover lost packets. Because compressed media streams are sensitive to errors, the ability to recover losses greatly improves streaming media quality.

Error Resilience through Interleaving and Forward Error Correction. Losses in some media streams, especially audio, can often be better concealed if the losses are isolated instead

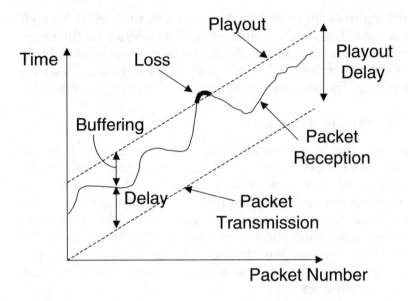

Figure 1.7   Effect of playout buffer on reducing the number of late packets. The packets are transmitted at a constant rate, and played out at a constant rate, at which the playout time of each packet is given by the packet's transmission time plus the playout delay. As long as a packet's delay is less than the playout time, it will be buffered (stored) and then played out at its scheduled playout time. If a packet's delay is longer than the playout time, it will arrive late and will be equivalent to a lost packet. (From Reference 68. Reprinted with permission.)

of concentrated (bursty). The extended presentation deadlines with the use of buffering allow interleaving to transform possible burst loss in the channel into isolated losses, thereby enhancing the concealment of the subsequent losses. As is discussed in the next section, the extended deadlines also allow other forms of error control schemes, such as the use of error control codes, which are particularly effective when used with interleaving.

Smoothing throughput Fluctuation. Because a time-varying channel gives rise to time varying throughput, the buffer can provide needed data to sustain streaming when throughput is low. The benefits of buffering do come at a price. Besides additional storage requirements at the streaming client, buffering also introduces additional delay before playback can begin or resume (after a pause due to buffer depletion). Adaptive media playout (AMP) is a new technique that enables a valuable trade-off between delay and reliability.[44,52]

1.3.5.4 Error Control for Overcoming Transmission Losses

The third fundamental problem that afflicts media streaming is losses, which can have a very destructive effect on the reconstructed quality. If the system is not designed to handle losses, even a single bit error can have a catastrophic effect. A number of different types of losses may occur, depending on the particular network under consideration. For example, wired packet networks such as the Internet are afflicted by packet loss, in which congestion may cause an entire packet to be discarded (lost). In this case, the receiver will receive a packet in its entirety or completely lose a packet.

On the other hand, wireless channels are typically afflicted by bit errors or burst errors at the physical layer. These may be passed up from the physical layer to the application as bit or burst errors or, alternatively, entire packets may be discarded when any errors are detected in these packets. Therefore, depending on the interlayer communication, a video decoder may always expect to receive "clean" packets (without any errors) or it may receive "dirty" packets (with errors). The loss rate can vary widely depending on the particular network, as well as for a given network, depending on the amount of cross traffic. For example, for video streaming over the Internet, one may see a packet loss rate of less than 1% or sometimes greater than 5 to 10%.

A media streaming system is designed with error control to combat the effect of losses. Approaches for error control fall into four rough classes: (1) retransmissions; (2) forward error correction (FEC); (3) error concealment; and (4) error-resilient media coding. The first two classes of approaches can be thought of as channel coding approaches for error control, and the last two are source coding approaches for error control.

The first three approaches for error control (retransmission, FEC, and error concealment) were discussed in Section 1.1.1.3. This section continues by discussing why compressed video is vulnerable to losses and then examines the fourth approach to error control: the design of error-resilient video coding systems. In addition, after identifying video's vulnerability to losses, we examine joint source/channel coding — that is, the joint design of source coding and channel coding (retransmission and/or FEC) to maximize the system's reliability while efficiently using the available resources. Note that a media streaming system is often designed using a number of these four different approaches for error control. Additional information and specifics for H.263 and MPEG-4 are available in Wang and Zhu,[53] Färber et al.,[54] and Talluri.[55]

Error-Resilient Video Coding. Compressed media are highly vulnerable to errors. The goal of error-resilient source coding is to design the compression algorithm and the compressed bitstream so that it is resilient to specific types of errors. We continue by providing a brief overview of error-resilient video compression, briefly identifying the basic problems introduced by errors and then discussing the approaches developed to overcome these problems.

Most video compression systems possess a similar architecture based on motion-compensated (MC) prediction between frames and block-DCT (or other spatial transform) of the prediction error, followed by entropy coding (e.g., run-length and Huffman coding) of the parameters. The two basic error-induced problems that afflict a system based on this architecture are: (1) loss of bitstream synchronization; and (2) incorrect state and error propagation. The former refers to the case when an error can cause the decoder to become confused and lose synchronization with the bitstream, i.e., the decoder may lose track of which bits correspond to which parameters. The latter refers to what happens when a loss afflicts a system that uses predictive coding.

Overcoming Loss of Bitstream Synchronization. When an error causes the decoder to loss track of which bits correspond to which parameters, loss of bitstream synchronization occurs. For example, consider what happens when a bit error afflicts a Huffman codeword or other variable length codeword (VLC). Not only would the codeword be incorrectly decoded by the decoder, but also, because of the variable length nature of the codeword, it is highly probable that the codeword would be incorrectly decoded to a codeword of a different length; thus, all the subsequent bits in the bitstream (until the next resync) would be misinterpreted. Even a *single* bit error can lead to significant subsequent loss of information.

The key to overcoming the problem of loss of bitstream synchronization is to provide mechanisms that enable the decoder quickly to isolate the problem and resynchronize to the bitstream after an error has occurred. Four mechanisms that enable bitstream resynchronization are (1) the use of strategically placed resync markers; (2) reversible variable length codewords (RVLCs); (3) data partitioning with resynchronization at each partition; and (4) application-level framing. We briefly discuss application-level framing next because it is one of the most successful approaches for overcoming the bitstream synchronization problem.

Application-Aware Packetization or Application Level Framing (ALF). When communication takes place over a packet network, the losses have an important structure that can be exploited. Specifically, a packet is accurately received in its entirety or it is completely lost. This means that the boundaries for lost information are exactly determined by the packet boundaries. This motivates the idea that, to combat packet loss, one should design (frame) the packet payload to minimize the effect of loss. This idea was crystallized in the application level framing (ALF principle presented in Clark and Tennenhouse,[56] who basically said that the "application knows best" how to handle packet loss, out-of-order delivery, and delay, and therefore the application should design the packet payloads and related processing). For example, if the video encoder knows the packet size for the network, it can design the packet payloads so that each packet is independently decodable, i.e., bitstream resynchronization is supported at the packet level so that each correctly received packet can be straightforwardly parsed and decoded. MPEG-4 H.263V2 and H.264/MPEG-4 AVC support the creation of

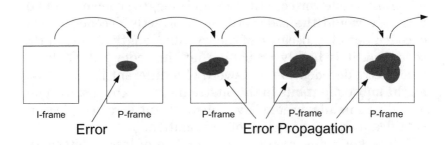

Error   Error Propagation

Figure 1.8 Example of error propagation that can result from a single error.

different forms of independently decodable video packets. As a result, careful use of the application level framing principle can often overcome the bitstream synchronization problem. Therefore, the major obstacle for reliable video streaming over lossy packet networks such as the Internet is the error propagation problem, which is discussed next.

Overcoming Incorrect State and Error Propagation in Video. If a loss has occurred, and even if the bitstream has been resynchronized, another crucial problem is that the state of the representation at the decoder may be different from the state at the encoder. In particular, when using MC prediction an error causes the reconstructed frame (state) at the decoder to be incorrect. The decoder's state is then different from the encoder's, leading to incorrect (mismatched) predictions and, frequently, significant error propagation that can afflict many subsequent frames, as illustrated in Figure 1.8. We refer to this problem as having incorrect (or mismatched) state at the decoder because the state of the representation at the decoder (the previous coded frame) is not the same as the state at the encoder. This problem also arises in other contexts (e.g., random access for DVDs or channel acquisition for digital TV) in which a decoder attempts to decode beginning at an arbitrary position in the bitstream.

A number of approaches have been developed over the years to overcome this problem with the common goal of trying to limit the effect of error propagation. The simplest approach to overcome this problem is by only using I-frames. Clearly by not using any temporal prediction, this approach avoids the error propagation problem; however, it also provides very poor compression and therefore is generally not an appropriate streaming solution. Another approach is to use periodic I-frames, e.g., the MPEG GOP as discussed in Section 1.2.1. For example, with a 15-frame GOP, there is an I-frame every 15 frames and this periodic reinitialization of the prediction loop limits error propagation to a maximum of one GOP (15 frames in this example). This approach is used in DVDs to provide random access and digital TV to provide rapid channel acquisition. However, the use of periodic I-frames limits the compression, so this approach is often inappropriate for very low bit rate video, e.g., video over wireless channels or over the Internet.

More sophisticated methods of intracoding often apply partial intracoding of each frame, where individual macroblocks (MBs) are intracoded as opposed to entire frames. The simplest approach of this form is periodic intracoding of all MBs: $1/N$ of the MBs in each frame are intracoded in a predefined order, and after N frames all the MBs have been intracoded. A more effective method is pre-emptive intracoding, in which one optimizes the intra–inter mode decision for each macroblock based on the macroblock's content, channel loss model, and the estimated vulnerability to losses.

The use of intracoding to reduce the error propagation problem has a number of advantages and disadvantages. The advantages include: (1) intracoding does successfully limit error propagation by reinitializing the prediction loop; (2) the sophistication is at the encoder and the decoder is quite simple; and (3) the intra–inter mode decisions are outside the scope of the standards and more sophisticated algorithms may be incorporated in a standard-compatible manner. However, intracoding also has disadvantages, including: (1) it requires a significantly higher bit rate than intercoding, leading to a sizable compression penalty; and (2) optimal intrausage depends on accurate knowledge of channel characteristics. Although intracoding limits error propagation, the required high bit rate limits its use in many applications.

The special case of point-to-point transmission with a back channel and with real-time encoding facilitates additional approaches for overcoming the error propagation problem.[57] For example, when a loss occurs, the decoder can notify the encoder of the loss and tell the encoder to reinitialize the prediction loop by coding the next frame as an I-frame. This approach uses I-frames to overcome error propagation (similar to the previous approaches described above); however, the key is that I-frames are *only used* when necessary. Furthermore, this approach can be extended to provide improved compression efficiency by using P-frames as opposed to I-frames to overcome the error propagation. The basic idea is that the encoder and the decoder store multiple previously coded frames. When a loss occurs, the decoder notifies the encoder as to which frames were correctly/erroneously received and therefore which frame should be used as the reference for the next prediction. These capabilities are provided by the reference picture selection (RPS) in H.263 V2 and NewPred in MPEG-4 V2. To summarize, for point-to-point communications with real-time encoding and a reliable back channel with a sufficiently short round-trip-time (RTT), feedback-based approaches provide a very powerful approach for overcoming channel losses. However, the effectiveness of this approach decreases as the RTT increases (measured in terms of frame intervals), and the visual degradation can be quite significant for large RTTs.[58]

Summary of Error-Resilient Coding Approaches. This section discussed the two major classes of problems that afflict compressed video communication in error-prone environments: (1) bitstream synchronization; and (2) incorrect state and error propagation. The bitstream synchronization problem can often be overcome through appropriate algorithm and system design based on the application level framing principle. However, the error propagation problem remains a major obstacle for reliable video communication over lossy packet networks such as the Internet. Although this problem can be overcome in certain special cases (e.g., point-to-point communication with a back channel and with sufficiently short and reliable RTT), many important applications do not have a back channel, or the back channel may have a long RTT, thereby severely limiting effectiveness. Therefore, it is important to be able to overcome the error propagation problem in the feedback-free case, when a back channel between the decoder and encoder does not exist, e.g., broadcast, multicast, or point-to-point with unreliable or long-RTT back channel. This is an active area of research, and a number of novel approaches, such as multiple description coding, have been developed in an attempt to solve this problem.

1.3.5.4.1 Joint Source/Channel Coding

Data and video communication are fundamentally different. In data communication all data bits are equally important and *must* be reliably delivered, though timeliness of delivery may be less important. In contrast, for video communication, some bits are more important than other bits, and often it is *not* necessary for all bits to be reliably delivered. However, timeliness of delivery is often critical for video communication. Examples of coded video data with different importance include the different frame types in MPEG video (i.e., I-frames are most important, P-frames have medium importance, and B-frames have the least importance) and the different layers in a scalable coding (i.e., base layer is critically important and each of the enhancement layers is of successively lower importance).

A basic goal is then to exploit the differing importance of video data, and one of the motivations of joint source/channel coding is to design the source coding and the channel coding jointly to exploit this difference in importance. This has been an important area of research for many years, but the limited space here prohibits a detailed discussion; therefore, we only present two illustrative examples of how error control can be adapted based on the importance of the video data. For example, for data communication, all bits are of equal importance and FEC is designed to provide equal error protection for every bit. However, for video data of unequal importance, it is desirable to have unequal error protection (UEP) as shown in Table 1.11 and also discussed in Section 1.1.1.3.

Table 1.11 Adapting Error Control Based on
Differing Importance of Video Data: Unequal Error
Protection and Unequal (Prioritized) Retransmission
Based on Coded Frame Type

	I-frame	P-frame	B-frame
FEC	Maximum	Medium	Minimum (or none)
Retransmit	Maximum	Medium	Can discard

Similarly, instead of a common retransmit strategy for all data bits, it is desirable to have unequal (or prioritized) retransmit strategies for video data. A number of important variations on retransmission-based schemes exist. For example, video streaming of time-sensitive data may use delay-constrained retransmission, in which packets are only retransmitted if they can arrive by their time deadline, or priority-based retransmission, in which more important packets are retransmitted before less important packets. These ideas lead to interesting scheduling problems, such as which packet should be transmitted next (e.g., see Girod et al.[44] and Chou and Miao[59]).

1.3.6   End-to-End Security and Midnetwork Media Transcoding

A media delivery system may need to deliver a media stream to a number of clients with diverse device capabilities (e.g., screen display size, computation or memory resources) and connection qualities (e.g., available bandwidth). This may require midnetwork nodes, or proxies, to perform stream adaptation, or transcoding, to adapt streams for downstream client capabilities and time-varying network conditions. These intermediate nodes can collect updated statistics about local and downstream network conditions and client capabilities and then act as control points by transcoding compressed streams to best match them to these downstream network conditions and client capabilities.

Another important property is security to protect content from eavesdroppers. This makes it necessary to transport

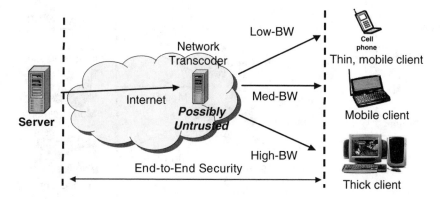

Figure 1.9 Challenge of transcoding while preserving end-to-end security: The media should be encrypted at the sender and decrypted only at the receiver; it should remain encrypted at all points in between. In conventional systems, network transcoding poses a potential security threat because it requires decryption.

streams in encrypted form. The problem of adapting to downstream conditions while also providing security is particularly acute for mobile, wireless clients because the available bandwidth can be highly dynamic and the wireless communication makes the transmission highly susceptible to eavesdroppers. This general situation is illustrated in Figure 1.9.

In this context, conventional midnetwork transcoding poses a serious security threat because it requires decrypting the stream, transcoding the decrypted stream, and re-encrypting the result. Because every transcoder must decrypt the stream (and therefore requires the key), each network transcoding node presents a possible breach to the security of the entire system — not an acceptable solution in situations that require end-to-end security. Therefore, it is desirable to have the ability to transcode or adapt streams at a midnetwork node or proxy while preserving end-to-end security. We refer to this capability as *secure transcoding*, in order to stress that the transcoding is performed without unprotecting the content (e.g., without decryption) and thereby preserving the end-to-end security.

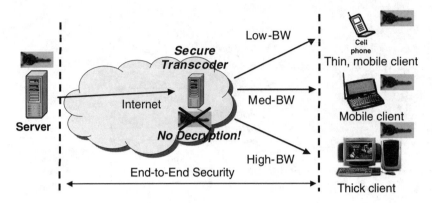

Figure 1.10 Example of end-to-end security and midnetwork secure transcoding. The media are encrypted at the sender and decrypted only at the receiver; they remain encrypted at all points in between. The midnetwork node or proxy performs secure transcoding to adapt the protected content for each receiving client. It performs this secure transcoding without unprotecting the content (e.g., without decryption) and thereby preserves the end-to-end security.

The desired capability of secure transcoding can be achieved by a framework for codesigning the compression, encryption, and packetization referred to as secure scalable streaming (SSS). Further details on this approach are available,[60,61] as well as its application to JPEG-2000 compressed images[62] and incorporation in the JPEG-2000 security standard (JPSEC).[43] Figure 1.10 shows an example use-scenario that is analogous to that presented in Figure 1.9. Specifically, in this case a sender transmits encrypted content to a midnetwork node or proxy, which performs a secure transcoding operation to adapt the received protected data for each of three clients: a low-bandwidth client, a medium-bandwidth client, and a high-bandwidth client. The midnetwork node performs secure transcoding, that is, transcoding without decryption, and therefore preserves the end-to-end security — i.e., the content is protected from the server to each receiving client. In effect, the transcoding is performed in the encrypted domain. Note that the encryption keys are only available to the sender and the receiving clients, and not to the midnetwork node performing the secure transcoding operation.

For simplicity, the preceding discussion has focused on the security service of confidentiality (provided by encryption); however, additional security services such as authentication and digital signatures are also supported by the SSS framework.

1.3.7 Networking Protocols for Streaming over IP Networks

This section briefly describes the network protocols for media streaming over wired and wireless IP networks such as the Internet. In addition, it highlights some of the current popular standards and specifications for video streaming over wired and wireless networks.

1.3.7.1 Protocols for Video Streaming over the Internet

This section briefly highlights the network protocols for video streaming over the Internet. First, we review the important Internet protocols of IP, TCP, and UDP. This is followed by the media delivery and control protocols.

1.3.7.1.1 Internet Protocols: TCP, UDP, IP

The Internet was developed to connect a heterogeneous mix of networks that employ different packet switching technologies. The Internet protocol (IP) provides baseline, best-effort network delivery for all hosts in the network: providing addressing, best-effort routing, and a global format that can be interpreted by everyone. On top of IP are the end-to-end transport protocols, in which transmission control protocol (TCP) and user datagram protocol (UDP) are the most important. TCP provides reliable byte-stream services and guarantees delivery via retransmissions and acknowledgments. However, UDP is simply a user interface to IP and is therefore unreliable and connectionless. Additional services provided by UDP include checksum and port-numbering for demultiplexing traffic sent to the same destination. Some of the differences between TCP and UDP that affect streaming applications are:

- TCP operates on a byte stream but UDP is packet oriented.
- TCP guarantees delivery via retransmissions but, because of the retransmissions, its delay is unbounded. UDP does not guarantee delivery but, for those packets delivered, the delay is more predictable (i.e., one-way delay) and smaller.
- TCP provides flow control and congestion control; UDP provides neither. This offers more flexibility for the application to determine the appropriate flow control and congestion control procedures.
- TCP requires a back channel for the acknowledgments. UDP does not require a back channel.

Web and data traffic are delivered with TCP/IP because guaranteed delivery is far more important than delay or delay jitter. For media streaming, the uncontrollable delay of TCP is unacceptable and compressed media data are usually transmitted via UDP/IP, with the exception of control information, which is usually transmitted via TCP/IP.

1.3.7.1.2 Media Delivery and Control Protocols

The IETF has specified a number of protocols for media delivery, control, and description over the Internet. Figure 1.11 illustrates the protocols used for streaming between a server and a client. The typical stacking of MPEG and IETF protocols for media control and delivery is illustrated in Figure 1.12.

Media Delivery. The real-time transport protocol (RTP) and real-time control protocol (RTCP) are IETF protocols designed to *support streaming media*. RTP is designed for data transfer and RTCP for control messages. Note that these protocols do *not* enable real-time services — only the underlying network can do this — however, they do provide functionalities that

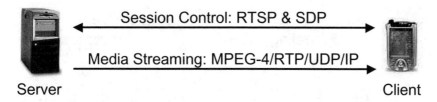

Figure 1.11 Example of media streaming between a server and a client.

support real-time services. RTP does not guarantee QoS or reliable delivery, but provides support for applications with time constraints by providing a standardized framework for common functionalities such as time stamps, sequence numbering, and payload specification. RTP enables detection of lost packets and provides feedback on quality of data delivery. It provides QoS feedback in terms of number of lost packets, interarrival jitter, delay, etc.

RTCP specifies periodic feedback packets in which the feedback uses no more than 5% of the total session bandwidth and at least one feedback message occurs every 5 sec. The sender can use the feedback to adjust its operation, e.g., adapt its bit rate. The conventional approach for media streaming is to use RTP/UDP for the media data and RTCP/TCP or RTCP/UDP for the control. Often, RTCP is supplemented by another feedback mechanism explicitly designed to provide the desired feedback information for the specific media streaming application. Other useful functionalities facilitated by RTCP include interstream synchronization and round-trip time measurement.

Media Control. Media control is provided by either of two session control protocols: real-time streaming protocol (RTSP) or session initiation protocol (SIP). RTSP is commonly used in video streaming to establish a session. It also supports basic VCR functionalities such as play, pause, seek, and record. SIP is commonly used in VoIP, and is similar to RTSP, but in addition it can support user mobility and a number of additional functionalities.

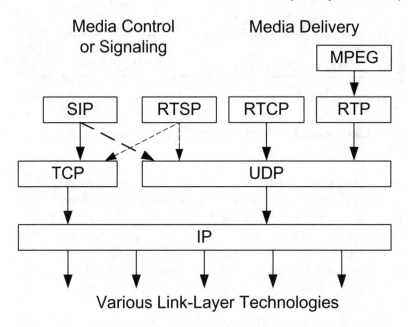

Figure 1.12 Stacking of MPEG and IETF protocols for media control and media delivery. The media control protocols, RTSP and SIP, may be sent over UDP or TCP, and the media delivery protocols are generally sent over UDP.

Media Description and Announcement. The session description protocol (SDP) provides information describing a session — for example, whether it is video or audio, the specific codec, bit rate, duration, etc. SDP is a common exchange format used by RTSP for content description purposes, e.g., in 3G wireless systems. It has also been used with the session announcement protocol (SAP) to announce the availability of multicast programs.

1.3.7.2 Media Streaming Standards and Specifications

Standards-based media streaming systems — for example, as specified by the third-generation partnership project (3GPP) for media over 3G cellular[64] and the Internet streaming media alliance (ISMA) for streaming over the Internet[65] — employ the following protocols:

- Media encoding
  - MPEG-4 video and audio (AMR for 3GPP), H.263
- Media transport
  - RTP for data, usually over UDP/IP
  - RTCP for control messages, usually over UDP/IP
- Media session control
  - RTSP
- Media description and announcement
  - SDP

The streaming standards do not specify the storage format for the compressed media, but the MP4 file format has been widely used. One advantage of MP4 file format is the ability to include "hint tracks" that simplify various aspects of streaming by providing hints such as packetization boundaries, RTP headers, and recommended transmission times.

1.3.8 Content Delivery Networks: The Core, the Edge, and the Last Mile

The discussion in this and the prior section has been on media compression and media delivery over broadband networks, with particular focus on the operations performed at the end hosts (sender and receiver). We now turn our attention to the important problem of network infrastructure design. Specifically, we examine the question of how to design the infrastructure to better support media delivery over broadband networks. As part of this discussion, we provide a brief overview of the emerging streaming media content delivery networks (SM-CDNs).

When the Internet first became popular (mid-to-late 90s), when a client requested a particular piece of content, the request went to the origin server that hosted the content, which then directly served the request to the client. This straightforward approach works for a small number of clients requesting the same or different content. However, it has important scaling problems, in the sense that it cannot scale to support a large number of clients requesting the same content (from same origin server). For example, even large-scale delivery of web pages resulted in network congestion, annoying delay, and server overload — and a web page corresponds to a relatively small amount of data compared to that in a video.

A number of bottlenecks arose for content delivery. The chain from the origin server to the client is composed of the origin server, the first mile, the core or backbone network, the last mile, and, finally, the client. The origin server was typically a server farm designed to support some number of simultaneous sessions. However, the origin servers could be overloaded for popular content, leading to low-quality delivery (e.g., large delays, missed delivery deadlines of the video, etc.) or even server crash. The first mile connecting the origin server to the core network could also be overloaded, leading to late or lost packets.

On the other hand, perhaps surprisingly, the core or backbone network has been highly overprovisioned, and as of this writing is not a bottleneck. For example, tier-one Internet service providers (ISPs) can support high-bandwidth delivery from the east coast of the U.S. to the west coast with less than a 35-msec coast-to-coast delay on the backbone and much less than 1% average packet loss rate. However, the last mile is once again a bottleneck. Therefore, the primary bottlenecks for content delivery relate to the origin server, first mile, and last mile. Content delivery networks were developed to overcome the problems of the origin server and the first mile.

1.3.8.1 Content Delivery Networks (CDNs)

Content delivery networks were developed to overcome performance problems, such as network congestion and server overload, that arise when many users access popular content. CDNs improve end-user performance by replicating (caching) popular content at multiple strategically located servers placed close to the end users. Because the content is cached on servers at the "edge" of the network, these servers are often called edge servers. Specifically, the edge of the network is conceptually at the boundary between the core or backbone network and the last mile — as close to the end users as possible.

Caching popular content on multiple edge servers close to the end users provides a number of benefits. First, it helps prevent server overload because the content is replicated on multiple edge servers (as opposed to only on a single origin server) and therefore multiple servers are available to deliver the same content to the requesting end users. For example, by caching a particular piece of content on N edge servers, one can increase the number of users that can simultaneously be served that content by about a factor of N. Furthermore, because the content is delivered from the closest edge server and not from the generally much more distant origin server, the content is sent over a shorter network path, thus reducing the request response time, probability of packet loss, and total network resource usage. Although CDNs were originally intended for static web content, the same benefits are exploited by state-of-the-art streaming video systems.

1.3.8.1.1 Streaming Media Content Delivery Network: Basic Components and Functionalities

A CDN corresponds to an overlay network, where nodes are placed on top of (overlaid on) an existing network in order to provide new or improved services while leveraging the existing network infrastructure. These nodes are referred to as overlay nodes. For example, the baseline IP network of the Internet provides connectivity between end hosts; however, it often does not provide good delivery performance (in terms of bandwidth, delay, or loss rate) between the end hosts. By placing the overlay network on top of the existing network infrastructure, one can exploit the connectivity provided by the existing network while using the overlay nodes to provide improved performance or new services. The overlay nodes have compute and storage resources that can be used for streaming, caching, relaying, and various forms of media processing.

A streaming media CDN corresponds to an overlay network designed for streaming media delivery. It is composed of a number of overlay nodes placed at strategic locations in the network — generally, at the edge of the network close to the end users. Each overlay node consists of one or more edge

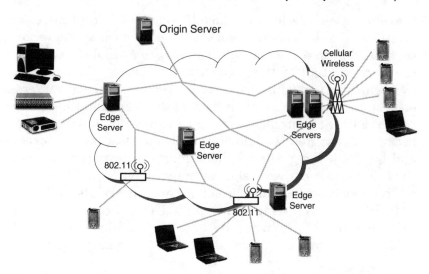

Figure 1.13 Example of a streaming media content delivery network (SM-CDN). A variety of diverse clients (e.g., PCs, laptops, PDAs) is connected to the network by wired and wireless links (cellular and 802.11 wireless links in this figure). The content is taken from the origin server at the top of the figure and replicated at the five edge servers that form the SM-CDN. These edge servers are located close to the requesting clients, thereby providing significantly better delivery performance than if the content were streamed from the distant origin server to each of the clients. Furthermore, the five edge servers provide much better scalability because, compared to what the sole origin server could deliver, they allow five times as many clients to be streamed the same content. Note that a SM-CDN may consist of 10s, 100s, or even 1000s of edge servers.

servers, and a manager that may be part of an edge server or may be a separate entity. Each edge server provides a number of capabilities including, for example, caching, streaming, resource monitoring and management, and media processing. Figure 1.13 provides an illustrative example of a streaming media CDN.

Some of the key functionalities provided by a streaming media CDN include:

- Caching (and content distribution to the caches)
- Server selection or redirection
- Streaming
- Application-level multicast
- Transcoding
- Midsession handoff between streaming servers or transcoding servers
- Other media services

Caching or replicating popular content at multiple servers improves the scalability of the system — that is, the number of possible simultaneous media deliveries for that content. An important related problem is the predistribution of the popular content to the caches before the content is requested. By selecting the best node to serve the content for each requesting client, the CDN can achieve significant streaming performance improvements for streaming to the client, e.g., higher available bandwidth, lower delay and delay jitter, and fewer lost packets. The process of selecting the best node or best server is often referred to as server selection. This process is also often referred to as redirection because the original request for a particular content from the origin server is redirected to another nearby server that can best deliver the content. In the preceding discussion, the term "best" is defined based on a number of metrics such as available server and network resources. Intuitively, the server selection generally leads to selecting a server (node) that has the requested content and is at the edge of the network and close to the requesting client.

One of the most challenging problems in large-scale media delivery is to deliver highly popular live content to very large numbers of users simultaneously. A prime example of this type of live content is sporting events. IP multicast is a natural approach for solving this problem; however, IP multicast is typically not supported in most networks. Fortunately, the SM-CDN can perform application-level multicast, in which the overlay nodes form the multicast tree and forward the packets among the overlay nodes and finally to the end users. Providing multicast capability for popular content through the use of application-level multicast in the overlay is one of the most important functionalities of the SM-CDN.

Another important functionality of an SM-CDN is providing transcoding within the network. Specifically, an overlay node may receive a stream (from an origin server or a multicast session, or maybe reading from its cache) and can then transcode the stream for downstream client capabilities (e.g., computation, storage, or display capabilities) or downstream available bandwidth.

As described previously, a carefully designed and operated streaming media content delivery network (the network infrastructure between the end hosts) can lead to significantly improved scalability, reliability, and other important performance aspects. Further information on SM-CDN design and additional references are available in Wee et al.[66]

1.3.9 Future Applications

Although it is imprudent to be prophetic about future applications of broadband, it may be useful to talk about expectations and trends, some of which are summarized here.

- Future applications are likely to be increasingly bandwidth demanding and increasingly demanding of the uplink in the access part of the network, as in telecollaboration from the home and interactive games. This does not imply that downlink-sensitive or narrowband applications will become less important. Downlink-centric video distribution and telehealth services are examples of the latter classes, and these applications are likely to be extremely important parts of the mix in the future.

- Future applications are expected to deal with content that will be not only increasingly rich, but also increasingly diverse, personalized, customized, and context sensitive (and therefore *smart*). Among other things, they will more fully utilize the rich framework for content distribution, management, indexing, search, retrieval, electronic commerce, and consumption that is supported by, for example, the MPEG standards.

- Future applications will address the intelligent and efficient sharing of information as much as the delivery of it.
- Contemporary applications are highly asymmetric ones in which a small number of people create content and a large number of people consume it; however, future applications will be more symmetric because a large fraction of the people will be able to create and share content with everyone else.
- Contemporary applications involving media generally use natural media — that is, media captured from the real world using cameras or microphones. Future applications will also use high-quality synthetic media, created in real-time as the application runs, and thus providing a significant increase in the richness and flexibility of the applications. Furthermore, the synthetic media may be synthesized at the receiving client, with only signaling and control information actually sent over the network from sender to receiver.
- Digital rights management will become increasing important in the future, and a rich and diverse range of rights management policies, business strategies, and solutions will most likely be developed and used.
- Future applications will include security as a fundamental design goal, in contrast with conventional application designs in which security is generally an afterthought.
- Future applications will increasingly address multi-modal services. However, to the extent to which these are still audiovisual, they will be various linear and nonlinear combinations of the core speech, audio, image, and video (and synthesized versions of these) media applications that this introduction has already attempted to quantify.

Of course, future applications will include those that have yet to be invented, and perhaps a single killer application for broadband will be developed.

1.4 SUMMARY

This introductory chapter provided an overview of last mile broadband, including two elastic definitions for what is meant by "broadband." It is impossible to predict killer broadband application(s) in the future, but a number of broadband applications that will likely be important in the future were identified. Because compression was identified as a fundamental enabler of broadband applications, media compression was briefly reviewed, with an emphasis on media compression standards important for future broadband media applications.

This chapter also provided an overview of media delivery over broadband networks as an illustrative example of a set of important broadband applications, broadband network challenges, and methods for overcoming these challenges. The chapter concluded by discussing cutting-edge approaches for designing the network infrastructure to provide improved scalability and performance for content delivery. In particular, it showed that the design of content delivery networks overcomes many of the bottlenecks in content delivery, including the origin server, first and second miles, and the core network, leaving the remaining bottleneck as the last mile. The following chapters in this book examine broadband last mile technologies that can overcome the last mile bottleneck — by bringing broadband to the last mile.

These chapters of the book discuss broadband access along the following broad themes:

- *The backbone, the edge, and the last mile (Chapter 2).* For completeness, this chapter provides a review of the backbone network, although the book's focus is on the access piece of the network. The chapter also describes the increasing importance of the edge of the network — the demarcation between the backbone from the access mile, a notion already introduced in Chapter 1. It concludes by rearticulating the importance of the access network, which is becoming increasingly broadband in wired and wireless domains.

- *The broadband channels and networks in the last mile (Chapter 3 through Chapter 7).* These chapters provide specialized treatments of the copper, cable, wireless, satellite, and optical last miles. In each case, discussions generally begin with the exposition of the physical channel, bandwidth characteristics, and illustrative modems and digital speeds available. Also discussed as appropriate are network layer technologies and issues related to quality of service.
- *Network management and broadband services (Chapter 8 and Chapter 9).* These chapters discuss the protocols for managing parameters associated with the data link layer or above. The treatment is largely unspecific to the physical medium, but special considerations are mentioned when appropriate. The chapter on services provides closure to the collection by touching upon a variety of service and user elements, including quality of service, on-demand models, operations support, and the residential information gateway.

GLOSSARY

AAC	adaptive audio coding
ACELP	adaptive code excited linear prediction
ACK	acknowledgment
ACR	absolute category rating
ADPCM	adaptive differential pulse code modulation
AIMD	additive increase multiplicative decrease rate control
ALF	application level framing
AMD	adaptive media display
BBM	broadband margin
CC	channel coding
CDN	content distribution network
CELP	code excited linear prediction
DCF	discrete coordination function
DCT	discrete cosine transform
DMOS	degradation mean opinion score

DTV	digital television
DVD	digital video disk
EEP	equal error protection
FEC	forward error correction
FTP	file transfer protocol
Gbps	gigabits per second
GOP	group of pictures
HCF	hybrid coordination function
HDTV	high-definition television
HTTP	hypertext transfer protocol
IDCT	inverse discrete cosine transform
IETF	Internet engineering task forum
IP	Internet protocol
ISDN	integrated services digital network
ITU-T	International Telecommunication Union-Transmission (Group)
JPEG	Joint Picture Experts Group
JVT	Joint Video Team (ITU + MPEG)
kbps	kilobits per second
LAN	local area network
LD-CELP	low delay CELP
LDPC	low density parity check
LPC	linear predictive coding
MAC	medium access control
Mbps	megabits per second
MC	modulation coding
MC-P	motion compensated prediction
ME	motion estimation
MIMO	multiple input multiple output
MOS	mean opinion score
MPEG	Motion Picture Experts Group
MSE	mean squared error
MTU	minimum transmittable unit
NC	network coding
NG	next generation
NRC	National Research Council
PCF	point coordination function
PESQ	perceptual evaluation of speech quality

PHY	physical (layer)
PSNR	peak signal-to-noise ratio
PSTN	public switched telephone network
PSI-CELP	pitch synchronous innovations CELP coder
P2P	peer-to-peer
QAM	quadrature amplitude modulation
QoS	quality of service
RCPC	rate compatible punctured convolutional (code)
ROM	read-only memory
RPE	residual pulse excitation
RTCP	real-time transmission control protocol
RTP	real-time transport protocol
RTSP	real-time streaming protocol
RTT	round trip time
SC	source coding
SIP	session initialization protocol
SM	streaming media
TCP	transmission control protocol
TTS	text to speech
UDP	user datagram protocol
UEP	unequal error protection
VLC	variable length coding
VoD	video on demand
VoIP	voice over IP
VSELP	vector sum excited linear prediction

REFERENCES

1. U.S. National Research Council, Broadband: bringing home the bits, N. Jayant, Ed., 2002.

2. E. Lee and D. Messerschmitt, *Digital Communications*, 2nd ed., Kluwer Academic Publishers, Norwell, MA, 1994.

3. J. Proakis, *Digital Communications*, 2nd ed., McGraw-Hill, New York, 1989.

4. T. Rappaport, *Wireless Communications: Principles and Practice*, 2nd ed., Prentice Hall PTR, Upper Saddle River, NJ, 2002.

5. D.J. Goodman, *Wireless Personal Communications Systems*, Addison-Wesley, Reading, MA, 1997.

6. P. Garg, R. Doshi, R. Greene, M. Baker, M. Malek, and X. Cheng, Using IEEE 802.11e MAC for QoS over wireless, IPCCC, 2003.

7. J. Hagenauer, Rate compatible punctured convolutional (RCPC) codes and their applications, *IEEE Trans. Commun.*, April, 1988, pp. 389–400.

8. J. Apostolopoulos and J. Lim, Video compression for digital advanced television systems, in *Motion Analysis and Image Sequence Processing*, M. Sezan and R. Lagendijk, Eds., Chapter 15, Kluwer Academic Publishers, Norwell, MA, 1993.

9. www.microsoft.com/windows/windowsmedia

10. www.realnetworks.com

11. N. Jayant, J. Johnston, and R. Safranek, Signal compression based on models of human perception, *Proc. IEEE*, October, 1993, pp. 1385–1422.

12. N.S. Jayant and P. Noll, *Digital Coding of Waveforms*, Prentice Hall, Upper Saddle River, NJ, 1984.

13. A.N. Netravali and B.G.Haskell, *Digital Pictures*, Plenum Press, New York, 1988, 1995.

14. N. Jayant, Ed., *Signal Compression*, World Scientific Press, Singapore, 1997.

15. J. Mitchell, W. Pennebaker, C. Fogg, and D. LeGall, *MPEG Video Compression Standard*, Chapman & Hall, New York, 1997.

16. B. Haskell, A. Puri, and A. Netravali, *Digital Video: An Introduction to MPEG-2*, Chapman & Hall, New York, 1997.

17. A. Puri and T. Chen, Eds., *Multimedia Systems, Standards, and Networks,* Marcel Dekker, New York, 2000.

18. J. Watkinson, *The MPEG Handbook*, Focal Press, Oxford, 2001.

19. D. Hoang and J. Vitter, *Efficient Algorithms for MPEG Video Compression*, John Wiley & Sons, New York, 2002.

20. R. Goldberg and L. Riek, *A Practical Handbook of Speech Coders*, CRC Press, Boca Raton, FL, 2000.

21. V. Madisetti and D. Williams, Eds., *The Digital Signal Processing Handbook*, CRC Press, Boca Raton, FL, and IEEE Press, 1997.

22. J. Johnston, S. Quackenbush, J. Herre, and B. Grill, Review of MPEG-4 general audio coding, in *Multimedia Systems, Standards, and Networks*, A. Puri and T. Chen, Eds., Chapter 5, Marcel Dekker, New York 2000.

23. G.K. Wallace, The JPEG still picture compression standard, *Commun. ACM*, April, 1991.

24. V. Bhaskaran and K. Konstantinides, *Image and Video Compression Standards: Algorithms and Architectures*, Kluwer Academic Publishers, Boston, MA, 1997.

25. J. Apostolopoulos and S. Wee, Video compression standards, *Wiley Encyclopedia of Electrical and Electronics Engineering*, John Wiley & Sons, Inc., New York, 1999.

26. R.V. Cox, Speech coding, in *Signal Compression*, N. Jayant, Ed., World Scientific Press, Singapore, 1998.

27. K. Brandenburg, O. Kunz, and A. Sugiyama, MPEG-4 natural audio coding, *Signal Processing: Image Communication*, 2000, pp. 423–444.

28. M. Weinberger, G. Seroussi, and G. Sapiro, LOCO-1: a low complexity, context-based, lossless image compression algorithm, *Proc. IEEE Data Compression Conf.*, March, 1996.

29. D. Taubman and M. Marcellin, *JPEG2000: Image Compression Fundamentals, Standards, and Practice*, Kluwer Academic Publishers, Boston, MA, 2002.

30. Video codec for audiovisual services at p × 64 kbits/s, ITU-T Recommendation H.261, International Telecommunication Union, 1993.

31. Video coding for low bit rate communication, ITU-T Rec. H.263, International Telecommunication Union, version 1, 1996; version 2, 1997.

32. ISO/IEC 11172. Coding of moving pictures and associated audio for digital storage media at up to about 1.5 Mb/s. International Organization for Standardization (ISO), 1993.

33. ISO/IEC 13818. Generic coding of moving pictures and associated audio information. International Organization for Standardization (ISO), 1996.

34. ISO/IEC 14496. Coding of audio-visual objects. International Organization for Standardization (ISO), 1999.

35. ITU-T recommendation H.264, Video coding for low bit-rate communication, March, 2003.

36. ITU-T recommendation P.800, Methods for subjective determination of transmission quality, August, 1996.

37. ITU-T recommendation P.862, Perceptual evaluation of speech quality (PESQ), an objective method for end-to-end speech quality assessment of narrow-band telephone networks and speech coders, February, 2001.

38. S. Wee, J. Apostolopoulos, and N. Feamster, Field-to-frame transcoding with temporal and spatial downsampling, *IEEE Int. Conf. Image Process.*, October, 1999.

39. M. Roussopoulos, M. Baker, D. Rosenthal, T. Giuli, P. Maniatis, and J. Mogul, 2 P2P or Not 2 P2P, *3rd Int. Workshop Peer-to-Peer Syst. (IPTPS)*, February, 2004.

40. D. Milojicic, V. Kalogeraki, R. Lukose, K. Nagaraja, J. Pruyne, B. Richard, S. Rollins, and Z. Xu, Peer-to-peer computing, HP labs technical report 2002-57R1, Palo Alto, CA, 2002.

41. M.-T. Sun and A. Reibman, Eds., *Compressed Video over Networks*, Marcel Dekker, New York, 2001.

42. G. Conklin, G. Greenbaum, K. Lillevold, A. Lippman, and Y. Reznik, Video coding for streaming media delivery on the Internet, *IEEE Trans. Circuits Syst. Video Technol.*, March, 2001.

43. D. Wu, Y. Hou, W. Zhu, Y.-Q. Zhang, and J. Peha, Streaming video over the Internet: approaches and directions, *IEEE Trans. Circuits Syst. Video Technol.*, March, 2001.

44. B. Girod, J. Chakareski, M. Kalman, Y. Liang, E. Setton, and R. Zhang, Advances in network-adaptive video streaming, *2002 Tyrrhenian Int. Workshop Digital Commun.*, September, 2002.

45. Y. Wang, J. Ostermann, and Y.-Q. Zhang, *Video Processing and Communications,* Prentice Hall, Upper Saddle River, NJ, 2002.

46. V. Jacobson, Congestion avoidance and control, *ACM SIGCOMM,* August, 1988.

47. M. Mathis et al., The macroscopic behavior of the TCP congestion avoidance algorithm, *ACM Computer Commun. Rev.*, July, 1997.

48. J. Padhye et al., Modeling TCP Reno performance: a simple model and its empirical validation, *IEEE/ACM Trans. Networking,* April, 2000.

49. W. Tan and A. Zakhor, Real-time Internet video using error-resilient scalable compression and TCP-friendly transport protocol, *IEEE Trans. Multimedia,* June, 1999.

50. S. Floyd et al., Equation-based congestion control for unicast applications, *ACM SIGCOMM,* August, 2000.

51. S. McCanne, V. Jacobsen, and M. Vetterli, Receiver-driven layered multicast, *ACM SIGCOMM,* August, 1996.

52. M. Kalman, E. Steinbach, and B. Girod, Adaptive media playout for low delay video streaming over error-prone channels, *IEEE Trans. Circuits Syst. Video Technol.,* June, 2004, pp. 841–851.

53. Y. Wang and Q. Zhu, Error control and concealment for video communications: a review, *Proc. IEEE,* May, 1998.

54. N. Färber, B. Girod, and J. Villasenor, Extension of ITU-T recommendation H.324 for error-resilient video transmission, *IEEE Commun. Mag.,* June, 1998.

55. R. Talluri, Error-resilient video coding in the ISO MPEG-4 standard, *IEEE Commun. Mag.,* June, 1998.

56. D. Clark and D. Tennenhouse, Architectural considerations for a new generation of protocols, *ACM SIGCOMM,* September, 1990.

57. B. Girod and N. Färber, Feedback-based error control for mobile video transmission, *Proc. IEEE,* October, 1999.

58. S. Fukunaga, T. Nakai, and H. Inoue, Error-resilient video coding by dynamic replacing of reference pictures, *GLOBECOM,* November, 1996.

59. P. Chou and Z. Miao, Rate-distortion optimized streaming of packetized media, *IEEE Trans. Multimedia,* submitted Feb. 2001.

60. S. Wee and J. Apostolopoulos, Secure scalable video streaming for wireless networks, *IEEE Int. Conf. Acoustics, Speech, Signal Process. (ICASSP),* March, 2001.

61. S. Wee and J. Apostolopoulos, Secure scalable streaming enabling transcoding without decryption, *IEEE Int. Conf. Image Process. (ICIP)*, September, 2001.

62. S. Wee and J. Apostolopoulos, Secure scalable streaming and secure transcoding with JPEG-2000, *IEEE Int. Conf. Image Process. (ICIP)*, September, 2003.

63. ISO/IEC JPEG-2000 Security (JPSEC) Committee Draft, T. Ebrahimi and C. Rollin, Eds., April, 2004.

64. The 3rd Generation Partnership Project (3GPP), www.3gpp.org.

65. Internet Streaming Media Alliance, www.isma.tv.

66. S. Wee, J. Apostolopoulos, W. Tan, and S. Roy, Research and design challenges for mobile streaming media content delivery networks (MSM-CDNs), *IEEE ICME*, July, 2003.

67. J. Ha and S. McLaughlin, Punctured LPDCs for AWGN channels, *IEEE Int. Conf. Commun.*, 5, 3110–3114, Anchorage, AK, May, 2003.

68. J.G. Apostolopoulos, W. Tan, S. Wee, Video streaming: Concepts, algorithms, and systems, in *Handbook of Video Databases*, B. Furht and O. Marques, Eds., CRC Press, 2004.

2

The Last Mile, the Edge, and the Backbone

BENNY BING AND G.K. CHANG

2.1 INTRODUCTION

Current last mile (also known as first mile) technologies represent a significant bottleneck in bandwidth and service quality between a high-speed residential/enterprise network and a largely overbuilt core backbone network. Backbone networks are provisioned for operation under worst-case scenarios of link failures, and thus backbone links are lightly loaded most of the time. In addition, high-capacity routers and ultra-high-capacity fiber links have created a true broadband architecture. However, large backbones are not the whole equation; distribution of that connectivity to individual enterprises and homes is just as critical for meeting the huge demand for more bandwidth (Figure 2.1). Unfortunately, the cost of deploying true broadband access networks with current technologies remains prohibitive. This in turn makes it difficult to support end-to-end quality of service (QoS) for a wide variety of applications, particularly nonelastic applications such as voice, video, and multimedia that cannot tolerate variable or excessive delay or data loss.

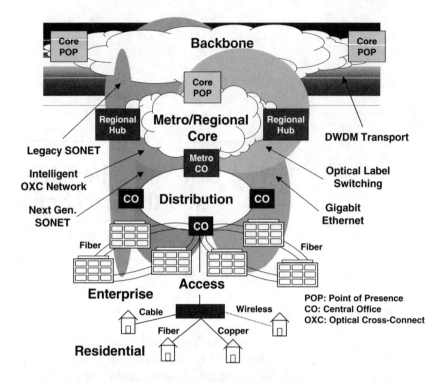

Figure 2.1 Distributing optical backbone connectivity to enterprises and homes.

In this chapter, innovative approaches that can overcome the cost and monopoly barriers of cable and DSL to help realize ubiquitous broadband access are concisely presented. Although many broadband access options are available, including powerline (world's largest infrastructure), free-space optics, and broadcast satellite access (attractive for rural as well as urban areas and even isolated areas not serviceable by terrestrial methods), we have omitted them in this chapter. Our goal is to provide insights into the research issues that have an impact on the approaches proposed. In doing so, we hope the reader will be able to understand the key technical problems related to ubiquitous broadband access and be able to stimulate innovative solutions for these problems.

The chapter is organized as follows. We begin by describing the limitations of current access technologies and explaining why existing access infrastructures are poorly equipped for transport of future data traffic. We then highlight the importance of dynamic QoS provisioning for broadband access, including traffic prioritization schemes, resource reservation protocols, subscriber-based service level agreements, and policy-based approaches for QoS specification, classification, and maintenance. This is followed by a case study on the practical application of traffic prioritization and bandwidth scheduling in QoS resource allocation. We base our study on the DOCSIS cable standard, the first Ethernet-based last mile standard to provide a comprehensive suite of QoS features.

Section 2.5 covers the activities of Ethernet for the first mile, which is gaining popularity and wide industry support due to the ubiquity of Ethernet deployment, including gigabit Ethernet LANs, metro Ethernet, and wireless Ethernet such as Wi-Fi. Section 2.6 and Section 2.7 focus on various aspects of broadband wireless and optical access, respectively — two complementary technologies that the authors believe may eventually play important roles in ensuring widespread broadband access connectivity. In Section 2.8, emerging optical metro and backbone technologies are discussed and the importance of optical label switching as a unifying transport technology for IP and Ethernet traffic is emphasized. In the final section, we examine the need for an efficient and fast file download mechanism to service future broadband applications.

2.2 EXISTING LAST MILE SOLUTIONS

When it comes to last mile access, network operators have a difficult choice among competing technologies: digital subscriber line (DSL), cable, optical, and fixed wireless. Key considerations for the choice include deployment cost and time, service range, and performance. The most widely deployed solutions today are DSL and cable modem networks, which had a combined total of roughly 25 million users by the end of 2003. Although they offer better performance over 56-kb/s dial-up telephone lines, they are not true broadband solutions for several reasons. For instance, they may not be able to provide enough bandwidth for emerging services such as content-rich services, media storage, peer-to-peer services, multiplayer games with audio/video chat to teammates,[27] streaming content, on-line collaboration, high-definition video on demand, and interactive TV services.

In addition, fast Web-page download still poses a significant challenge, particularly with rich, engaging, and value-added information involving high-resolution DVD video streaming, multimedia animation, or photo-quality images. Finally, only a handful of users can access multimedia files at the same time, in stark contrast to direct broadcast TV services. To encourage broad use, a true broadband solution must be scalable to thousands of users and must have the ability to create an ultrafast Web-page download effect — superior to turning the pages of a book or flipping program channels on a TV — regardless of the content.

A major weakness of DSL and cable modem technologies is that they are built on top of existing access infrastructures not optimized for data traffic.[1] In cable modem networks, RF channels remaining after accommodating legacy analog TV services are dedicated for data. DSL networks do not allow sufficient data rates at required distances due to signal distortion and crosstalk. Most network operators have come to realize that a new, data-centric solution is necessary, most likely optimized over the Internet protocol (IP) platform. The new solution should be inexpensive, simple, scalable, and capable of delivering integrated voice, video, and data services to the end-user over a single network.

2.3 QUALITY OF SERVICE PROVISIONING FOR BROADBAND ACCESS

Just as a computer will run faster when its resources are optimized, well-specified QoS parameters help network planning through measurement and management of traffic flows (or traffic streams), thus enabling more efficient bandwidth allocation and utilization. For these reasons, QoS assurance has become important, and the contracts that specify it (called service level agreements or SLAs) are becoming very common. The purpose of specifying an SLA is to ensure that subscribers receive network services of adequate quality, including performance guarantees such as minimum bandwidth, maximum packet loss, maximum delay, etc. The usual agreement specifies the end-to-end performance to which the subscriber is entitled over a specified period of time. Defining QoS parameters is only a first step. Mechanisms must also be available to label traffic flows with respect to their priorities (to distinguish the traffic characteristics) and for the network to recognize and act on those labels (in order to service connections with different QoS requirements).[12]

2.3.1 QoS for IP Networks

Unlike backbone networks, an access network does not deal with highly aggregated traffic. Instead, *individual* guarantees are required by each subscriber, as specified by individual SLA contracts. As a result, access networks require resource allocation and SLA enforcement algorithms different from those designed for backbone networks due to large numbers of subscribers requiring individual guarantees. The problem is particularly acute for optical access systems because the extended reach allowed by the optical fiber increases the concentration ratio between the number of subscribers and the central office dramatically.

To deliver QoS requires guarantees in the maximum delay and data rate required by the user process. Guaranteeing a rate requires a priority queue and a way to ensure that it is not oversubscribed.[17] Although IP networks have no mechanism to avoid oversubscription, the Internet Engineering Task Force, or IETF (www.ietf.org), has nevertheless proposed several methods for improving IP QoS, including integrated service (IntServ) and differentiated service (DiffServ).

2.3.2 IntServ

IntServ allows the network to support multimedia traffic using IETF's SLAs. These SLAs provide explicit reservation and scheduling of traffic on a per-flow basis, effectively assigning a specific flow of data to a traffic class, which defines a certain level of service. Another form of IntServ supports traffic scheduling on a per-packet (rather than per-flow) basis. This form allows greater control over the differentiation of traffic types than that provided by simple prioritization while avoiding the complexity of flow scheduling.

To determine *a priori* if sufficient bandwidth is available in the network to handle a traffic flow with QoS requirements, a reservation protocol (most predominantly, resource reservation protocol, or RSVP) is used to communicate information about a flow to the admission controller and then to the packet scheduler after connection is established. RSVP operates on the premise that options for QoS are needed but not as a replacement for best effort service because many applications cannot predict their bandwidth requirements. Once the data flow is assigned a class, a path message is forwarded to the destination to determine whether the network has sufficient resources to support that specific class of service. If *all* devices along the path are found capable of providing the required resources, the receiver generates a reservation message and returns it to the source. The procedure is repeated continually to verify that the necessary resources remain available. If the required resources are not available, the receiver sends an RSVP error message to the source.

IntServ has several problems. There is no guarantee that the necessary resources will be available when desired. It also does not scale well because it reserves network resources on a per-flow basis. If multiple flows from an aggregation point require the same resources, the flows will be treated independently and the reservation message must be sent separately for each flow. Nevertheless, IntServ is popular in access network deployment. For instance, it has been adopted by the cable network community in an initiative called PacketCable, which defines a multimedia system architecture that overlays a high-speed cable modem access network. The aim is to enable a broad range of IP-based multimedia services, such as voice over IP (VoIP), online messaging, and video conferencing.

PacketCable partitions resource management into distinct segments: the access and backbone networks.[23] Access network QoS is based on IntServ, where QoS is reserved and scheduled for individual traffic flows; the DiffServ model (described in the next section) is employed to support QoS for the backbone network. For the access network, when the cable modem termination system (CMTS) at the headend receives a RSVP request, it will perform admission control to determine whether resources should be committed to the QoS request.

2.3.3 DiffServ

In DiffServ, packets are grouped into class types, which are specified in the IP header. The class type specifies how packets are treated in each router. For example, packets marked "expedited routing" are handled in a separate queue and routed as quickly as possible. On the other hand, packets marked "assured forwarding" are buffered at the router with a specific drop preference. In DiffServ, any traffic management or bandwidth control mechanism that treats different flows differently (e.g., weighted fair queuing, RSVP, or any lightweight mechanism that does not depend entirely on per-flow resource reservation) can be employed. A short tag is appended to each packet depending on its service class. Traffic flows having the same resource requirements may then be aggregated on the basis of their tags when they arrive at the edge routers. The routers at the core then use the tag information to forward the flows to their destinations. This is done without examining the individual packet headers in detail, so the core network can run much faster. The present trend in providing QoS for backbone IP networks is to use DiffServ complemented by some of the resource reservation capabilities of RSVP. For example, in PacketCable, the CMTS can act as an edge router, performing per-flow RSVP admission control, DiffServ code point marking, and traffic policing.

2.3.4 Policy Support from Higher Layers

DiffServ and IntServ need support from layers above the transport layer in order to specify or maintain QoS. The upper layers tag the traffic appropriately or create a path for the data flow and specify path parameters. Thus, the method of identifying and classifying users, applications, and network resources is important. A policy-based approach typically employs the lightweight directory access protocol (LDAP) or IETF's common open policy service (COPS) standard. A client can add advanced services easily using COPS, which specifies a service in unequivocal terms and allocates the resources required to deliver that service. The tool is more adaptable to a customer's requirements, allowing them to vary with time of the day, application, or even user session. The requirements and rules for resource allocation (known as policies) are decided in advance. Once a user has put such a policy in place, network parameters can be configured to meet customer-initiated QoS requirements. COPS is used in PacketCable for policy enforcement between the gate controller and the CMTS.

2.4 QoS CASE STUDY BASED ON THE DOCSIS CABLE STANDARD

The hybrid-fiber coax (HFC) network is a vast improvement of the older, one-way CATV distribution plant and is capable of supporting a wide variety of services for the last mile. CATV was originally designed to extend broadcast analog TV signals to homes obstructed from line-of-sight reception,[18] which resulted in viewers not being able to pick up a good signal. CATV overcame terrain issues by erecting mountain-top sites and tall towers in strategic locations. The HFC network has been extensively deployed in over 70% of all U.S. homes. An HFC network uses a bidirectional shared-coaxial cable medium in a tree-branch architecture combined with fiber to increase the bandwidth capabilities of the access system. The combination of the shared medium and tree-branch topology allows a high degree of aggregation of user's traffic; this is not possible in dedicated telephone connections using traditional twisted-pair local loops between each home and the central office.

The data over cable service interface specification (DOCSIS) standard is a *de facto* HFC standard developed by Cable-Labs. DOCSIS enables high-speed Ethernet-based services, including packet telephony, video conferencing, and multimedia Internet services. The upstream and downstream data are transferred between the cable modem termination system (CMTS) at the headend and the cable modems (CMs) at the subscriber premises. The CMTS can manipulate its internal queues to give certain traffic classes preferential treatment. For instance, data traffic going to specific network addresses or to specific service-based addressing (associated with DOCSIS service IDs) can be given priority or have internal buffers or bandwidth reserved for them. Although such prioritization is not as functional as latency and error rate QoS guarantees, it can be usable for service-level agreements when combined with restricted entry (e.g., attempted connections at high priority are rejected if they increase the traffic load to the point at which the entire network will slow down).

In this section, the measured QoS performance metrics of a commercial DOCSIS cable network are presented, including throughput, average delay, and packet loss rate under different degrees of network loading and prioritization profiles. For multimedia applications, such metrics play important roles in providing various levels of service quality. Based on the measurement results obtained, a QoS model for the CMTS scheduler is developed using the weighted round-robin algorithm. Simulation results from OPNET demonstrate a close match between this model and the measured results.

2.4.1 Experimental Setup

The experimental setup is shown in Figure 2.2. Four Arris Touchstone CM300A CMs are connected to an Arris Cornerstone 1500 CMTS using a tap and a power splitter. A computer running the dynamic host configuration protocol (DHCP) and trivial file transfer protocol (TFTP) server programs is connected to the CMTS through an Ethernet switch to provide IP addresses and configuration profiles for the CMs in the initialization phase. The Spirent SmartBits 6000B is a network performance analysis tool that can test, simulate, analyze, troubleshoot, and certify the performance of network equipment. A six-port 10/100 Mb/s Ethernet card of the

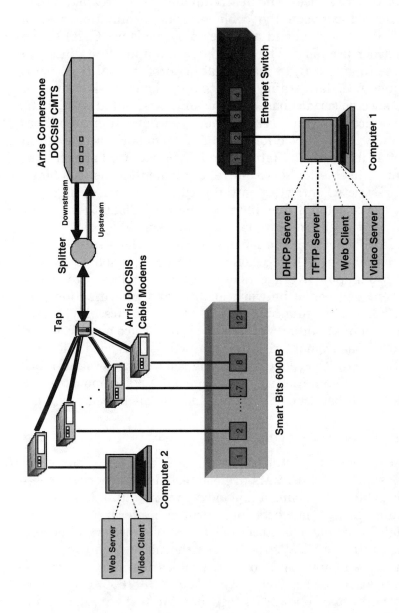

Figure 2.2 Experimental setup.

SmartBits 6000B is used in this experimental setup. Four ports from the card are used to generate IP traffic for the CMs. Another port is used to collect and analyze the received data packets, as well as to generate measurement reports. QPSK and 64-QAM modulations are used for the upstream and downstream links, respectively.

2.4.2 Measured Results

By keeping only one CM active, contention among CMs is removed and the maximum upstream and downstream throughput of the CM can therefore be measured. The Smart-Bits 6000B generates a constant bit rate (CBR) traffic flow with varying transmission rates and packet lengths. As shown in Figure 2.3, upstream and downstream throughputs increase with increasing packet length. The throughput refers to the bit rate associated with the transmission of the data packet, which is an Ethernet frame. A maximum throughput of about 2 Mb/s can be obtained for the upstream link (from CM to CMTS) and the maximum downstream throughput (from CMTS to CM) is about 25.7 Mb/s. In both cases, the maximum throughput occurs when packets with the longest length (i.e., 1500 bytes) are transmitted. For all packet lengths, the saturation throughput occurs when the upstream or downstream link reaches capacity and packet discard occurs.

We now measure the prioritized upstream performance by keeping eight CMs active with varying priorities and sending rates. The SmartBits 6000B generates CBR traffic flows with varying rates but the packet length is fixed at 1500 bytes. The results are shown in Figure 2.4. The sending rate refers to the data rate of each individual CM; the delay is computed from the time a packet arrives at the CM to the time at which the packet is received by the CMTS. The loss rate is normalized to the ratio of the number of discarded packets against the total number of packets presented to the CM.

It can be observed that when the sending rate is low, the throughput and loss rate of each prioritized packet stream

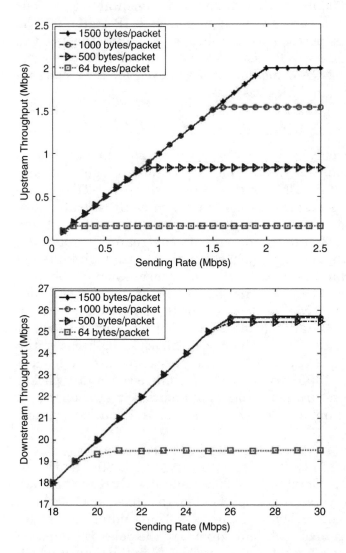

Figure 2.3 Upstream and downstream throughputs of one CM with different packet lengths.

are virtually the same. The distinction between different prioritized streams arises when the sending rate increases and three saturation points appear. Beyond the saturation points,

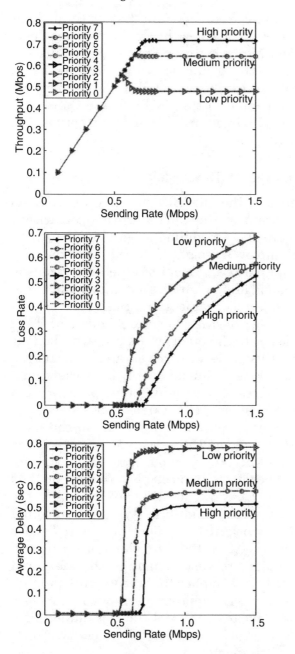

Figure 2.4 Upstream performance of eight CMs with different priorities and fixed packet length.

the higher priority traffic streams obtain higher throughput and experience smaller delay. When the throughput and delay start to saturate, the CM reaches capacity and packet discard occurs (as evident in the graph showing the loss rate). Although eight priorities can be allocated to CMs, only three QoS levels are implemented in the CMTS. These QoS levels can be grouped under high (priority 7), medium (priority 6, 5, 4), and low (priority 3, 2, 1, 0).

2.4.3   QoS Model for the CMTS Scheduler

The QoS performance of the DOCSIS standard depends on the scheduling algorithm in the CMTS. In this section, we attempt to investigate the scheduling scheme implemented by the CMTS. We employ a weighted round-robin queuing algorithm,[3] which computes and monitors the delay requirement of every packet and allows fairness for on-demand bandwidth assignments to bursty traffic. Here, the CMTS acts as a server serving multiple prioritized queues (Figure 2.5). Each CM sends a data packet to the queue of its priority. The four input parameters for this model are arrival rate, λ (packets/s), departure rate, μ (packets/s), queue size L in packets, and scheduling weights representing the different priorities. An OPNET simulation for this model is carried out.

We employ simulation parameters similar to the measurements. The CMTS bandwidth is set to 4.5364 Mb/s. The arrival rate (λ) varies with the sending rate; the departure rate (μ) is set to 378 packet/s with packet length fixed at 1500 bytes; the queue size (L) is 134 packets per CM; and the scheduling weights are 0.157 (high priority), 0.141 (medium priority), and 0.105 (low priority). As shown in Figure 2.6, for eight CMs with priorities 0 to 7, the throughput, loss rate and delay simulation results are very similar to the measured results from Figure 2.4. This implies that the weighted round-robin queuing algorithm with appropriate input parameters provides a convenient means to model the scheduling algorithm of the CMTS accurately without the necessity of performing physical measurements.

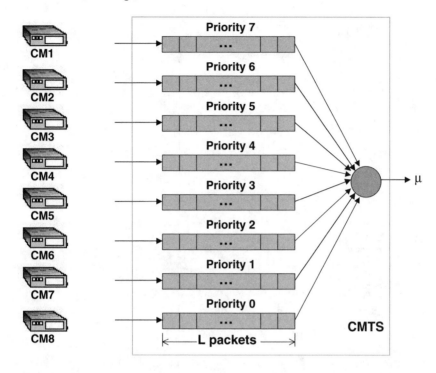

Figure 2.5 Weighted round-robin queuing model for the CMTS scheduler.

2.5 ETHERNET FOR THE FIRST MILE

Ethernet for the first mile (EFM) is an effort to extend Ethernet's reach over the first mile access link between end-users and carriers and make Ethernet a low-cost broadband alternative to technologies such as DSL and cable. The motivation for doing this is sound because over 500 million Ethernet ports are currently deployed globally and it is advantageous to preserve the native Ethernet frame format rather than terminate it and remap its payload into another layer two protocol (e.g., point-to-point protocol, PPP). The EFM specifications are developed by the IEEE 802.3ah Task Force (http://www.ieee802.org/3/efm) formed in November 2000. The draft standard (version 3.0 was issued in January 2004)

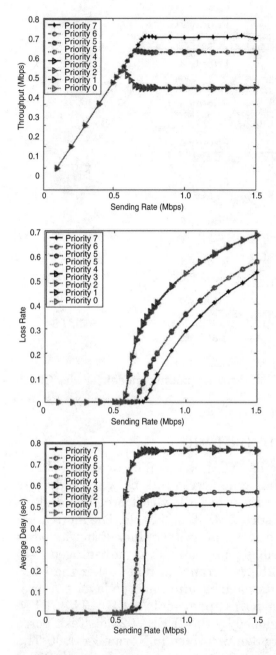

Figure 2.6 OPNET simulation of the weighted round-robin CMTS scheduler.

includes *physical layer* specifications for copper, fiber point-to-point, and fiber point-to-multipoint topologies. It is supported by the EFM Alliance or EFMA (www.efmalliance.org), a vendor consortium formed in December 2001 to:

- Promote industry awareness and acceptance of EFM standards
- Contribute technical resources to facilitate standard development
- Provide resources for multivendor interoperability

2.5.1 EFM Topologies

As shown in Figure 2.7, the IEEE 802.3ah EFM Task Force defines three main access topologies and physical layers:

- EFM copper (EFMC): existing copper wire (Cat 3) at more than 10 Mb/s for a range of up to at least 750 m
- EFM fiber (EFMF): single-mode fiber at 0.1 to 1 Gb/s for a range of at least 10 km
- EFM PON (EFMP): optical fiber at 1 Gb/s for a range of up to 20 km
- EFM hybrid (EFMH): topologies that combine the preceding three topologies

The EFM draft standard also defines operations, administration, and maintenance (OAM) aspects of the technology, which local carriers and network operators will use to monitor, manage, and troubleshoot access networks. The same management protocols and architecture work across all EFM topologies.

2.5.2 EFM Copper, EFM Fiber, EFM PON

EFMC exploits existing twisted-pair, voice-grade copper wiring infrastructure, which dominates the first mile (e.g., within residential neighborhoods as well as within buildings). The spectrum is compatible with PSTN, ISDN, and ADSL, enabling all technologies to coexist in the same wirings. EFMF specifies a 1-Gb/s, full-duplex, single-mode fiber transport for the access network of up to 20 km and also supports single

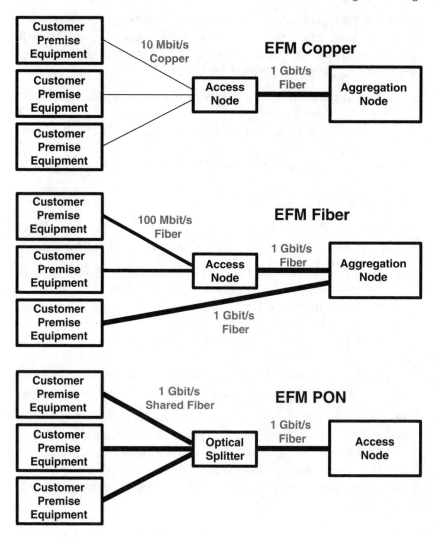

Figure 2.7   EFM topologies.

and dual point-to-point fiber for 100 Mb/s access. It is a candidate for fiber to the home (FTTH) applications and aims to reduce the costs of single-mode fiber access to a point at which it will replace multimode fiber.

As the name implies, EFMP uses a passive optical network (PON), which is a single, shared optical fiber with passive optical splitters that split a single fiber into a multipoint network feeding each subscriber. Subscribers are connected via dedicated fibers but share the optical distribution network (ODN) trunk fiber back to a central office (CO). They are called "passive" because, other than at the CO and subscriber endpoints, there are no active electronics within the access network. Another advantage is the reduction in the amount of deployed fiber compared to point-to-point fiber topologies. EFMP (also known as Ethernet PON or EPON) supports speeds of 1 Gb/s for a distance of up to 20 km.

2.5.3 QoS Provisioning and Bandwidth Allocation in EFM

Ethernet was originally developed as a data-centric protocol; however, it has evolved to support a full range of services, including real-time services with stringent bandwidth, delay, and loss requirements. The features of a number of existing and future IEEE standards have been employed (e.g., prioritization, virtual LAN or VLAN tagging, traffic shaping, bandwidth management, and resource reservation). In addition to these QoS mechanisms, EFM provides the required bandwidth for voice, video, and high-speed data access. For instance, EFM supports speeds of up to 1 Gb/s, which is significantly higher than DSL and cable. By removing protocol conversions and low-bandwidth links using a universal technology, EFM is future-proofed for any application. By specifying a common and simple OAM method, EFM reduces network management complexity when integrating mixed services.

2.6 BROADBAND WIRELESS ACCESS

Wireless solutions for the last mile have been labeled as "disruptive" technologies because many phone companies are losing their landline business to wireless (just as they are losing business to VoIP).[28] However, wireless solutions cannot compete with fiber solutions in terms of bandwidth provisioning, particularly as the operational range increases. They are also more prone to eavesdropping, security attacks, and possible traffic analysis. On the other hand, the broadcast nature of the wireless medium offers ubiquity, user mobility, and immediate access.

In addition to offering a quick, low-cost alternative to installing cable or leased telephone lines and allowing fixed line operators to extend their broadband networks, wireless solutions also allow a long-distance carrier to bypass the local service provider, thereby cutting down subscriber costs. They are indispensable when wired interconnections are impractical, e.g., rivers, rough terrain, private property, and highways.

When implemented effectively, wireless approaches can greatly simplify network management by reducing network configuration complexity and dependence on gateways, and eliminating overheads associated with moves, adds, and changes.[1]

2.6.1   Cost Reduction and Improved Reliability

The use of wireless technology has the potential to reduce the cost of residential broadband access significantly. A substantial portion of the cost of last mile deployment is in interconnecting the customer's premise to the central office. A typical CATV connection, for example, requires long cascades of some 30+ trunk and line amplifiers between the headend and the customer premise, in addition to numerous passive taps. Reliability is a serious problem because all amplifiers are in series and failure of any one device will result in a downstream signal outage from that point.[19]

A wireless solution not only removes the labor, material, and equipment costs associated with cabling but also offers the flexibility to reconfigure or add more subscribers to the network without much planning effort and the cost of recabling, thereby making future expansion and growth inexpensive and easy. Moreover, the inherent flexibility of wireless communications to reconfigure quickly is an important consideration for increasing the reliability of network connectivity during emergencies and catastrophic events. Another key advantage of wireless technology is the speed of deployment. In addition, a large portion of the deployment costs is incurred only when a customer signs up for service. Note that the performance of wireless solutions is fundamentally limited by signal inter-ference among concurrent transmissions, which is a function of the timing and spatial separation of transmissions. As such, a combination of wireless and wired infrastructure may still be required for wide coverage areas.

Some emerging broadband wireless technologies have realized impressive efficiencies of the order 10 b/s/Hz, and unprecedented levels of individual and aggregate capacities in the order of Gb/s wireless data rates.[21,22] This is in contrast to current Wi-Fi technologies (maximum efficiencies of 0.5 b/s/Hz for 802.11b and 2.7 b/s/Hz for 802.11a/g). Radio spec-trum is also increasingly deregulated; this can potentially lead to an abundance of bandwidth when spectrum is used (and reused) more efficiently and cooperatively.[26] The capacity increase and efficiency are a direct result of being able to switch between different idle channels only for the period of usage.

In the next few sections, we focus on the key character-istics of multihop (or mesh) and long-range wireless technol-ogies, which have a tremendous potential in enabling *pervasive* broadband access. Multihop wireless simplifies pri-vate network deployment for residential users and long-range wireless can be implemented as an overlay over other access technologies, including multihop technologies. Interestingly, long-range wireless solutions can also compete with local wireless hotspots as well as 3G and integrated cell phone/Wi-Fi technologies for revenue.

2.6.2 Advantages of Multihop Wireless Solutions

Currently, the prevalent model of a residential broadband access network is that of a star topology with a hub at a central office location and individual homes connected to the central office with an access line. Multihop wireless topologies are different and offer several unique advantages.

In multihop wireless access, a fixed wireless access point is typically mounted on the rooftop of the subscriber's home. Each access point creates a small wireless coverage area called a "hop" and acts much like a router on the Internet, automatically discovering neighboring access points and

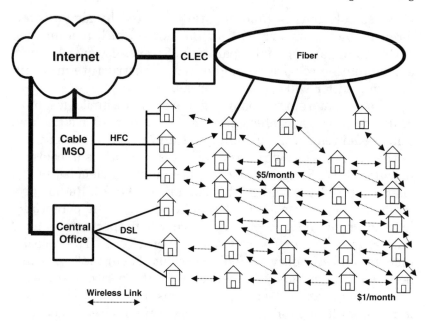

Figure 2.8   A multihop wireless access architecture.

relaying packets across several wireless hops. With multihop wireless, subscribers can create local community networks on demand, allowing residences to communicate directly with neighbors and enabling broadband applications between homes (e.g., neighborhood watchdog applications, medical and emergency response tasks, etc.). Thus, the ownership of the access network becomes distributed across the residences. Such decentralized topologies not only provide a degree of autonomy (just as network domains or autonomous systems do in the Internet today) but can also lower subscriber costs substantially through shared services and resources (Figure 2.8).

Although there is a need for innovative approaches to digital rights management (which will ultimately help to increase the availability of high fidelity multimedia content), the ability to share resources is a key driver in reducing network service costs. Economies of scale provided by enterprise LANs in the 1980s, the public Internet in the 1990s, and, more recently, peer-to-peer file sharing applications such as KaZaA (a dominant online music swapping platform) have forced big recording labels to agree to offer music downloads for as little as $1 a song.[4]

Other key advantages of operating distributed, peer-to-peer multihop wireless systems are lower transmit power (because it is not necessary to transmit information all the way back to the ultimate destination, which also results in a corresponding reduction in interference) and the ability to reuse limited radio spectrum efficiently. These advantages help to increase the capacity of the wireless access network as well as improve range performance. For example, the network coverage of the highly popular 802.11 wireless Ethernet (http://www.ieee802.org/11), commercially known as Wi-Fi (www.wi-fi.org), can be extended in this manner and may ultimately play a part in reducing the costs for residential broadband access networks.

2.6.3 Disadvantages of Multihop Wireless Solutions

In wireless home networks, some unsophisticated subscribers may allow their networks to be "shared" unknowingly; other, more knowledgeable subscribers may wish to protect their networks through a variety of security mechanisms. In multihop wireless access networks, cooperation among subscribers is clearly required because the access point in each home must be "always on" in order for transit traffic to be forwarded. Although subscribers have a strong incentive to cooperate due to a reduction in service costs, unfair treatment may arise because subscribers located far away from the service provider can potentially enjoy lower costs while subscribers closer to the service provider suffer more severe network congestion and interference.

In the former case, the problem arises because it is likely that not all neighbors in the entire service domain will agree on the shared cost, especially when the domain is large. It is more likely that a subset of the total number of subscribers (e.g., neighbors who know each other) participates in cost sharing and some of these subscribers can further auction their service to another group to reduce the shared cost. In the latter case, the problem is a result of congestion and interference generated by intermediate relay nodes in upstream and downstream directions; this becomes acute for nodes residing nearer to the service provider in the relay chain. A well designed multihop routing algorithm needs to alleviate as much traffic as possible from these nodes, while taking into account current interference conditions, thereby reducing the possibility of congestion and improving the overall QoS and fairness of the network.

Additional practical concerns exist. Because a relay node in a wireless multihop network is typically half-duplex in nature, entire packets will need to be stored before being forwarded. Clearly, this will increase the latency compared to a single-hop link and may degrade the performance of time-sensitive traffic such as voice and video. One way of reducing the store-and-forward delay is to have dual-radio systems in which two separate frequency bands are dedicated to the reception and transmission of packets. In this case, the destination address of the packet is read by the relay node, a routing decision is made, and the packet is forwarded even as the rest of the packet is received, creating a pipelining effect similar to wide-area packet switching networks. Because such a dual-band radio system is expensive, and more complex to manage due to the need for frequency planning, other innovative methods should be explored to reduce the latency in the forwarding mechanism.

In the case of security, it is clearly more effective to secure, on an end-to-end basis, a single wireless link in a conventional last mile architecture than multiple links in a wireless multihop network. The problem is magnified by the fact that the wireless transmission for the last mile is performed outdoors, thus making it easier for an intruder to masquerade as one of the wireless relays forwarding packets. This will in turn aggravate the difficult security problems currently facing Wi-Fi implementations: namely, the proliferation of rogue access points (i.e., unauthorized access points plugged into a legitimate wireless network) and soft access points (i.e., notebook computers and mobile devices masquerading as access

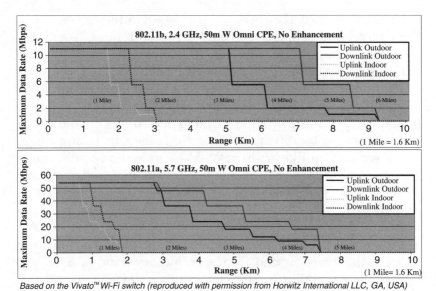

Based on the Vivato™ Wi-Fi switch (reproduced with permission from Horwitz International LLC, GA, USA)

Figure 2.9 Long-range 802.11 (Wi-Fi) for wireless last mile operation.

points) can easily generate a security hole and compromise the entire network connected to these access points.[6,14] Solutions to these problems are challenging; existing rogue mitigation techniques (e.g., triangulation methods) that detect, locate, and shut down the offending access points may not be directly applicable to a multihop wireless network.

Thus, although a wireless multihop last mile solution provides unique benefits, it also poses tough challenges in guaranteeing security and QoS associated with always-on connections. Alternative solutions of increasing the wireless coverage without resorting to a multihop architecture should also be encouraged. For example, some commercial long-range Wi-Fi products can provide an omnidirectional 5-mile radius coverage using sensitive antenna arrays mounted on fixed locations and only tens of milliwatts of transmit power (Figure 2.9). Base stations that service several miles/kilometers can be located on a tower, a rooftop of a tall building, or on another elevated structure. If directional high-gain antennas are used, the operational range can extend up to 30 miles.

Other long-range wireless solutions are based on the IEEE 802.16 standard (http://www.ieee802.org/16), commercially known as Wi-MAX (www.wimaxforum.org), and wireless DOCSIS. Wi-MAX can potentially offer a data rate of up to 150 Mb/s using 25 MHz of radio spectrum. This data rate can be lowered to 100 or 50 Mb/s if a longer operating range is desired and, unlike Wi-Fi, a Wi-MAX link allows full-duplex communication. Wi-MAX focuses on frequency bands between 10 and 66 GHz (line of sight) and between 2 to 11 GHz (non-line of sight). The 2- to 11-GHz standard (802.16a) employs unlicensed and licensed bands. Wi-MAX operates with a TDMA MAC protocol that supports multiple physical layer specifications customized for the frequency band of use.

Currently, efforts are underway to develop an 802.16 client standard. Wireless DOCSIS is a proprietary long-range wireless access method that extends the features of the DOCSIS standard for wireless access. A key advantage of doing this is the minimization of overheads associated with protocol translation because many DOCSIS features (such as modulation methods and MAC protocol) can be retained for wireless access. The main requirement involves the need to translate the frequency to the appropriate RF band. Most wireless DOCSIS products operate with sectored antennas in the licensed 2.5-GHz or the unlicensed 5-GHz U-NII bands. Data rates are similar to DOCSIS data rates and typically range from 27 to 36 Mb/s (6- to 8-MHz bandwidth) on the downstream and 256 kb/s to 10 Mb/s (200-kHz to 3.2-MHz bandwidth) on the upstream.

Wi-Fi access solutions, however, are currently more attractive than other long-range solutions because they can be easily integrated with Wi-Fi home networks, which are growing at a staggering pace. According to In-Stat/MDR, shipments of Wi-Fi home adapters and access points increased 214% in 2003 — equivalent to sales of 22.7 million units. Nevertheless, all long-range wireless access solutions should be explored and enhanced to complement the strengths of multihop wireless technologies.

2.7 BROADBAND OPTICAL ACCESS

DSL or cable-modem access provides benefits of installed infrastructure, virtually eliminating deployment costs. If fixed wireless access is chosen, network providers gain the benefit of quick and flexible deployment. However, these access methods may suffer bottlenecks in bandwidth-on-demand performance and service range. For example, cable networks are susceptible to ingress noise, DSL systems can be plagued with significant crosstalk, and unprotected broadcast wireless links are prone to security breach and interference. Furthermore, current DSL and cable deployments tend to have a much higher transmission rate on the downstream link, which restricts Internet applications to mostly Web browsing and file downloads.

Although wireless access is excellent for bandwidth scalability in terms of the number of users, optical access is excellent for bandwidth provisioning per user. Furthermore, the longer reach offered by optical access potentially leads to more subscribers.

Optical access networks offer symmetrical data transmission on upstream and downstream links, allowing the end-user to provide Internet services such as music/video file sharing and Web hosting. In addition to providing a good alternative, such networks represent an excellent evolutionary path for current access technologies. These costs still require laying fiber, which makes optical access networks more expensive to install. However, because fiber is not bandwidth limited but loss limited (as opposed to copper wires, cable, and wireless), the potential performance gains and long-term prospects make optical access networks well-suited for new neighborhoods or installations.

In addition, there are innovative solutions for deploying fiber in the last mile, even in established neighborhoods. For example, instead of investing in expensive dedicated fiber conduits, existing sanitary sewers, storm drains, waterlines, and natural gas lines that reach the premises of many end-users can be exploited.[5] Fiber can be housed in these utilities by forming creative business partnerships among optical fiber owners, service providers, utility pipe owners, vendors, and city municipalities.

Access networks should be scalable in bandwidth provisionable per user. To be scalable with a number of subscribers, it is highly important to identify architectures that allow low equipment cost per subscriber. As new applications appear and demand higher bandwidth, the network should be gracefully upgraded. It is important to be able to perform an incremental upgrade in which only subscribers requiring higher bandwidth are upgraded, not the entire network. Because the life span of optical fiber plant is longer compared to copper or coaxial cables, it is expected that the optical network will be upgraded multiple times during its lifetime. Therefore, it is important to design access architectures that allow seamless upgrade. In addition, the deployment of fibers between residences can be used to connect end-users directly, forming an autonomous communication network among residential end-users and thereby improving the overall service reliability through provision of redundant data paths similar to the multihop wireless architecture described in the previous section.

2.7.1 Passive Optical Networks

Passive optical network (PON) is a technology viewed by many as an attractive solution to the last mile problem because PONs can provide reliable yet integrated data, voice, and video services to end-users at bandwidths far exceeding current access technologies. Unlike other access networks, PONs are point to multipoint networks capable of transmitting over 20 km of single-mode fiber. As shown in Figure 2.10, a PON minimizes the number of optical transceivers, central office terminations, and fiber deployment compared to point-to-point and curb-switched fiber solutions. By using passive components (such as optical splitters and couplers) and eliminating regenerators and active equipment normally used in fiber networks (e.g., curb switches, optical amplifiers), PONs reduce installation and maintenance costs of fiber as well as connector termination space.

The general PON architecture consists of the optical line terminator (OLT) on the service provider side and optical network unit (ONU) (or sometimes the optical network terminal) on the user side (Figure 2.11). The ONUs are connected to the OLT through one shared fiber and can take different

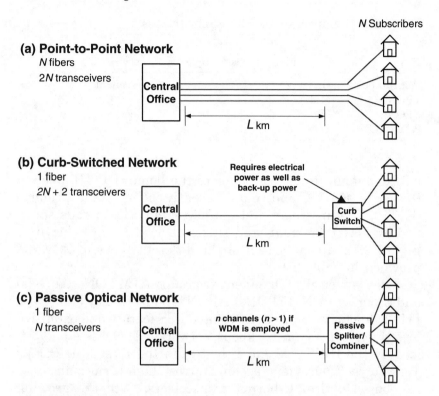

Figure 2.10 Fiber to the home (FTTH) deployment scenarios.

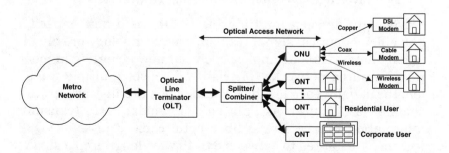

Figure 2.11 A typical passive optical access network.

Table 2.1 Comparison of PON Architectures

Type	Fiber deployment	ONU cost/complexity	Upstream service
Point-to-point	High	High	Excellent
APON	Mid	Mid	Good
EPON	Mid	Mid	Good
WDM PON	Low	Low	Excellent

FTTx configurations, e.g., fiber to the home (FTTH), fiber to the curb (FTTC), and, more recently, fiber to the premise (FTTP). The upstream and downstream optical bands specified by ITU-T for dual- and single-fiber PONs are shown in Figure 2.12; a comparison of the different types of PONs is provided in Table 2.1.

PONs typically fall under two groups: ATM PONs (APONs) and Ethernet PONs (EPONs). APON is supported by FSAN and ITU-T due to its connection-oriented QoS feature and extensive legacy deployment in backbone networks. EPON is standardized by the IEEE 802.3ah Ethernet in the First Mile (EFM) Task Force. EPONs leverage on low-cost, high-performance, silicon-based optical Ethernet transceivers. With the growing trend of GigE and 10 GigE in the metro and local area networks, EPONs ensure that IP/Ethernet packets start and terminate as IP/Ethernet packets without expensive and time-consuming protocol conversion, or tedious connection setup.

2.7.2 Wavelength Division Multiplexing Optical Access

Wavelength division multiplexing (WDM) is a high-capacity and efficient optical signal transmission technology prevalent in long-haul backbone applications, but now emerging in metropolitan area networks (MANs). WDM uses multiple wavelengths of light, each of which corresponds to a distinct optical channel (also known as lightpath or lamda, λ), to transmit information over a single fiber optic cable. Current WDM systems are limited to between 8 to 40 wavelengths on a single fiber. This is an economical alternative to installing more fibers and a means of improving data rates dramatically.[16]

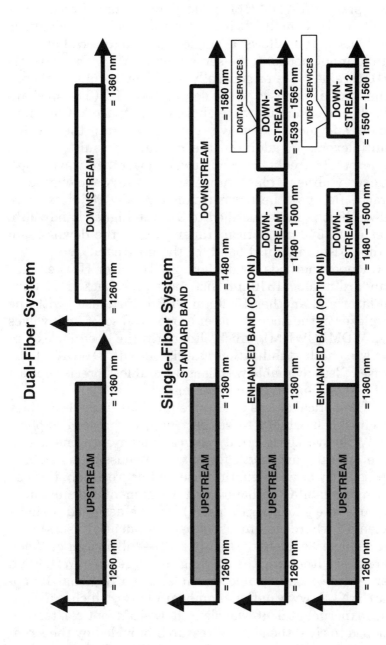

Figure 2.12 Upstream and downstream optical bands for dual and single-fiber PONs.

WDM has been considered as a transition path for HFC cable systems.[8] A WDM HFC cable plant in smaller node areas presents many benefits. It increases the bandwidth available to each user and allows a single wavelength (rather than multiple input nodes) to be served by a CMTS. Dense WDM (DWDM) cable systems generally use 200-GHz (1.6-nm) optical channels as opposed to the denser 100-GHz (0.8-nm channels). The closer the wavelengths are spaced, the more difficult (and more expensive) it is to separate them using demultiplexers. In addition, the optical transmitters must provide stable frequency and lower power per wavelength (because nonlinear glass properties cause wavelengths to interact). Because the downstream typically requires more bandwidth, it is more justified to use the higher bandwidth, higher cost 1550-nm optical transmitter in the downstream and the 1310-nm optical band for the upstream.

Cheaper methods for deploying WDM over HFC employ a standard analog 1310-nm distributed feedback (DFB) forward laser and another 1550-nm DFB return laser, with the multiplexer in the node. Another economical solution involves coarse WDM (CWDM), which eliminates the need for expensive wavelength stabilization associated with DWDM at the expense of less available wavelengths and less precise optical bands (8- to 50-nm channel spacing).

WDM optical access is a future-proof last mile technology with enough flexibility to support new, unforeseen applications. WDM switching can dynamically offer each end-user a unique optical wavelength for data transmission as well as the possibility of wavelength reuse and aggregation, thereby ensuring scalability in bandwidth assignment. For instance, heavy users (e.g., corporate users) may be assigned a single wavelength whereas light users (e.g., residential users) may share a single wavelength (Figure 2.13) — all on a single fiber. We are also witnessing the exciting convergence of WDM and Ethernet; the most notable example is the National LambdaRail or NLR (www.nationallambdarail.org), which is a high-speed, experimental 40-wavelength DWDM optical testbed developed to rival the scale of research provided by the Arpanet (the Internet's precursor) in the 1960s. NLR is the first

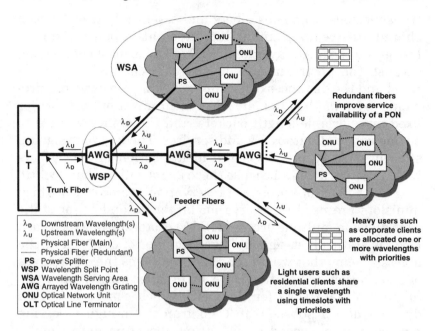

Figure 2.13 A WDM optical access architecture.

wide-area use of 10-Gb/s switched Ethernet and is based on a routed IP network. It is owned by the university community, thanks to the plunge in dark fiber prices over the last 3 years.

2.7.3 WDM Broadband Optical Access Protocol

The benefits of PONs can be combined with WDM, giving rise to WDM PONs that provide increased bandwidth and allow scalability in bandwidth assignment. Key metrics in the physical layer performance of WDM PONs include latency, link budget, transmitter power and passband, receiver sensitivity, number of serviceable wavelengths, and distance reachable.

In this section, we examine how the high fiber capacity in WDM PONs can be shared efficiently among distributed residential users to further help reduce the cost per user and how the medium access control (MAC) protocol can fulfill this purpose in the time and wavelength domains. The MAC protocol helps to resolve access contentions among the upstream transmissions from subscribers and essentially transforms a shared upstream channel into a point-to-point channel. Designing the MAC layer for a WDM PON is most challenging because the upstream and downstream channels are separated and the distances can go up to 20 km. The basic requirements are a method to maintain synchronization, support for fixed and variable-length packets, and the ability to scale to a wide range of data rates and services. The key to designing efficient MAC protocols is the identification of subscribers with data to transmit because these users will use bandwidth.

Most PON protocols use reservation protocols (in which minislots are used to reserve for larger data slots) or static time division multiple access (TDMA) protocols (in which each ONU is statically assigned a fixed number of data slots). The access scheme that we present here combines the advantages of reservation and static TDMA by preallocating a minimum number of data slots for the ONU, which can be increased dynamically in subsequent TDM frames through reserving minislots on a needed basis. By preallocating data slots, a data packet can be transmitted immediately instead of waiting for a duration equivalent to the two-way propagation delay associated with the request and grant mechanism in reservation protocols.

However, the number of data slots that can be preallocated must be kept to a minimum because if more slots are preallocated, it is possible that slots could become wasted when the network load is low; this is the main disadvantage of static TDMA. Other novelties of the scheme are that it not only arbitrates upstream transmission and prevents optical collisions, but also varies bandwidth according to QoS demand and priority; it accounts for varying delays caused by physical fiber length difference and handles the addition/reconfiguration of network nodes efficiently.

We employ simulation parameters consistent with the ITU-T standards and they are described as follows. We consider ATM and Ethernet traffic types at each ONU. In the case of ATM traffic, an arrival corresponds to a single ATM cell; however, for Ethernet, we modeled each variable-size

Ethernet packet as a batch (bulk) arrival of ATM cells. For the single packet arrival case, arrival pattern is Poisson with the link rate chosen to be 155.52 Mb/s shared by 10 ONUs. A data cell comprises 3 header bytes and 53 data bytes, and a TDM frame comprises 53 data slots (equivalent to the duration of 53 data cells of 56 bytes each) with 1 data slot allocated for minislot reservation (i.e., 52 slots allocated for data cells). This gives a TDM frame duration of $(53 \times 56 \times 8 \text{ b})/155.52$ Mb/s = 152.67 μs and a slot time of 152.67 μs/53 = 2.8807 μs. Note that the TDM frame duration is typically designed for the maximum round-trip (two-way propagation) delay, so our TDM frame duration is equivalent to a maximum distance of $(152.67 \times 10^{-6} \text{ s}) \times (2 \times 10^8 \text{ m/s})/2$ or roughly 15 km. The traffic load refers to the ratio of the data generation rate over the link rate. The cell arrival interval in each ONU is exponential distributed with mean $10 \times (53 \times 8)/(\text{link rate} \times \text{load})$. The bandwidth is dynamically allocated using the request and grant mechanism.

For the batch packet arrival case, the TDM frame and network settings are the same as before. The only change is the traffic pattern, in which we have simulated Poisson arrivals at each ONU with batch size uniformly distributed between 1 and 30 data cells (30 data cells is roughly equivalent to one maximum-length Ethernet packet of 1500 bytes). The batch arrival interval is exponential distributed with mean $10 \times (53 \times 8)/(\text{link rate} \times \text{load} \times \text{average batch size})$. For the preallocation case, each ONU is preassigned a minimum of one data slot. A request packet transmitted on a minislot increases this minimum number to a number indicated in the request packet. For the case without preallocation, all data slots are assigned only after a request is made on a minislot. For static TDMA, five cells are allocated for each ONU with the remaining three slots of the TDM frame wasted.

The performance of the case when each ONU has a single packet arrival is shown in Figure 2.14a. Preallocation of data slots clearly reduces the average delay, even when the network is heavily loaded. Under high load, the performance of both protocols converges, which is expected because all data

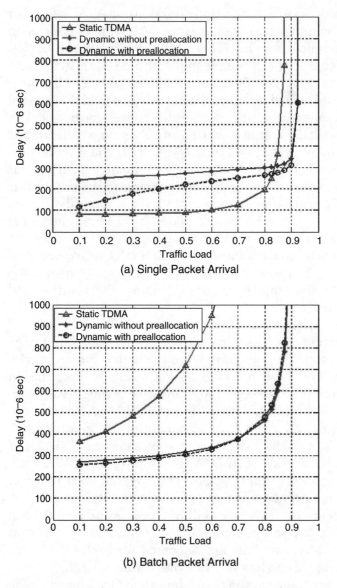

(a) Single Packet Arrival

(b) Batch Packet Arrival

Figure 2.14 Average delay vs. normalized traffic load comparison.

slots in a TDM frame tend to become filled continuously. Static TDMA performs best under low load due to the single arrival assumption. As can be seen from Figure 2.14b, static TDMA performs poorly because the average batch size is 15.5 cells per packet, which implies that an average of three TDM frames are needed to transmit all cells in a packet. Because each packet arrives randomly, in static TDMA, ONUs are not able to request more bandwidth appropriate for the batch size. It is also interesting to note that the performance of the preallocation algorithm is now slightly better than without preallocation under light load. Under heavy load, the preallocation algorithm incurs slightly more overheads.

2.7.4 Quality of Service Provisioning for Broadband Optical Access Networks

Unlike metro and long-haul networks, optical access networks must serve a more diverse and cost-sensitive customer base. End-users may range from individual homes to corporate premises, hotels, and schools. Services must therefore be provisioned accordingly. Data, voice, and video must be offered over the same high-speed connection with guarantees on QoS, and the ability to upgrade bandwidth and purchase content on an as-needed basis.

QoS provisioning must aim to match the vision of agile, high-capacity, and low-cost metro optical networks against the practical operational reality of existing infrastructures deployed by telecom carriers. There is also a strong imperative for metro optical networks to extend broadband connectivity to end-users. This can be accomplished by augmenting capacity, but, more importantly, by introducing new technologies with strong price/service benefits that can support emerging broadband data services in a scalable and cost-effective manner, such as virtual private networks (VPNs), voice over IP (VoIP), and virtual leased lines (VLLs) among others. A number of new solutions have been proposed to enable the deployment of simple, cost-effective, and bandwidth-efficient metro optical networks. Some of these solutions are geared towards enhancing and adapting existing synchronous optical network/synchronous digital hierarchy (SONET/SDH) ring technologies in the backbone network, while others are designed specifically to compete against SONET/SDH.

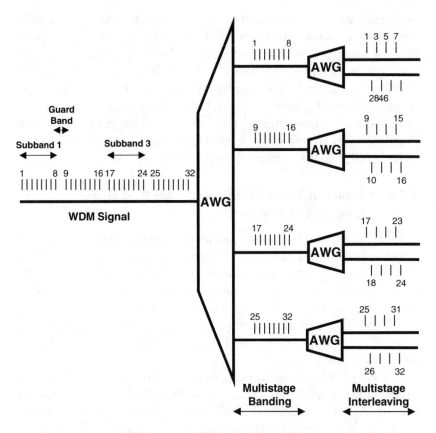

Figure 2.15 Multiplexing in WDM PONs.

2.7.5 Wavelength Multiplexing in WDM Optical Access Networks

The multiplexing/demultiplexing method employed by WDM optical access networks determines the maximum number of wavelengths that can be distributed among subscribers. The number of wavelengths that can be handled by a single arrayed wavelength grating (AWG) multiplexer/demultiplexer is limited. This limitation can be overcome by using multistage banding and interleaving (Figure 2.15). Multistage banding is a divide-and-conquer approach, subdividing the total number of wavelengths into smaller subbands so that they can be managed by individual AWGs. Guard bands between subbands are necessary for optical filters to provide adequate crosstalk suppression while retaining low insertion loss.[7] The use of guard bands is eliminated by employing multistage interleaving. Each interleaving stage doubles the optical passband, allowing broader filters to extract the individual channels. Any periodic filter (e.g., Mach–Zehnder interferometer) can be used as an interleaver by matching its period to the desired channel spacing.

As the number of wavelengths increases in a WDM optical access network, the number of AWGs to handle a larger number of wavelengths cannot be increased in an arbitrary manner. This is because the addition of an AWG stage increases the wavelength count at the expense of introducing some power insertion loss. Thus, innovative approaches are needed to evaluate the optimal balance between number of AWG stages and number of wavelengths that can be serviced for a WDM optical access network.

2.8 NETWORK EDGE AND THE BACKBONE NETWORK

Optical backbone networking technologies have made phenomenal progress recently. The MONET project sponsored by DARPA is an excellent demonstration of a vertical integrated optical networking field trial.[10] The grand vision is to create a next-generation optical Internet, bypassing unnecessary electronic processing and network protocols.[11–13] It will provide a revolutionary networking solution that combines the advantages of packet-switched WDM optical routers (for handling fine-grained data bursts) and circuit-switched WDM cross-connects (for handling coarse-grained end-to-end traffic).

To have seamless integration between the access network and the high-speed backbone network, the edge interface deserves careful investigation. Because of the mismatch in data rates, congestion arises at the edge, leading to queuing delay, which is the main component of delay responsible for variations. To overcome this drawback and provide QoS to delay-sensitive (or nonelastic) traffic to be carried by the access network, the merits of packet concatenation need to be studied in relation to performance tradeoffs in terms of packet overheads, buffer requirements and prioritization, switching efficiency, and link utilization.

The previous section discussed the importance of WDM for broadband optical access. WDM technology plays an equally important role in the optical backbone network. The three key elements in a WDM backbone network include optical line terminators (OLTs), optical add/drop multiplexers (OADMs), and optical cross-connects (OXCs). OLTs are used at the edge of point-to-to-point optical links. OLTs multiplex/demultiplex a composite WDM signal into individual wavelengths, but OADMs and OXCs separate only a fraction of the wavelengths. OXCs are able to serve a larger number of wavelengths than OADMs and are typically deployed in large optical backbones. Reconfigurable OADMs (ROADMs) allow desired wavelengths to be dropped or added dynamically — a very desirable attribute for flexible network deployment.

2.8.1   SONET/SDH for Multimedia and Ethernet Transport

Digital transmission using SONET/SDH technology is critical to building high-speed multimedia access networks. However, many broadband operators install two parallel networks: a SONET/SDH system for voice and data and an analog or proprietary digital system for video.[20] This is done for two reasons: SONET/SDH is not efficient for video transport and SONET/SDH circuits cannot monitor video performance. To address this transport issue, video-optimized SONET/SDH multiplexers are currently being investigated.

Perhaps the most significant development in backbone transport technologies is the evolution of SONET/SDH to make Ethernet "carrier worthy" and more transparent for backbone access. Transparency is advantageous for a carrier because it typically reduces the amount of provisioning required at the end points of the carrier network. This has prompted the establishment of the Metro Ethernet Forum or MEF (http://www.metroethernetforum.org) to address various aspects of Ethernet WAN transport. Two of the key enhancements in the SONET standard involve virtual concatenation (VCAT) and generic frame procedure (GFP).

The main idea of VCAT is to repackage Ethernet (and other protocols) efficiently into SONET transport networks with flexible channel capacities. High-order VCAT is achieved by aggregating sets of STS-1 or STS-3 data paths (e.g., an STS-3-7 V path corresponds to a data rate of 1.08864 Gb/s, which is suited for 1 Gb/s Ethernet). Low-order VCAT groups VT1.5 data paths together (e.g., a VT1.5-7 V path corresponds to a data rate of 10.5 Mb/s, which is suited for 10 Mb/s Ethernet).

GFP provides a frame-encapsulation method that allows complete Ethernet frames to be packaged together with a GFP header. GFP can also be made transparent by aggregating 8B/10B traffic streams into 64B/65B superblocks for transport over SONET/SDH. Aggregated Ethernet flows can be distinguished by inspecting parts of the Ethernet frame (e.g., VLAN/MPLS tags, IP type of service, DiffServ code, etc.). A simple table lookup can then be used to determine the SONET/SDH VCAT group to encapsulate the flow. Stackable tags using the IEEE 802.1q standard for virtual LANs or MPLS labels should be encouraged to avoid data coordination problems when the same tags are used at the ingress and egress end points of the backbone network. To preserve a loss-free environment, sufficient buffering to hold up to three jumbo Ethernet frames (9600 bytes) per end point to accommodate a span of 10 km is necessary.[9]

2.8.2 Evolution of Packet Switching and Routing

Switching and routing technologies have evolved rapidly in the last three decades. Many LAN connections and the public Internet now consist of hybrid switches and routers based on packet technologies such as Ethernet. When routers were first designed in the 1970s, they were used for a simple function: forward individual packets on a link-by-link and best-effort basis. The speed of these devices was then improved significantly for LAN applications in the 1980s, with increased router complexity due to the need to handle routing as well as bridging protocols. The growth of the Internet then led to the development of many routing algorithms, including RIP, OSPF, EGP, BGP, and multicast routing.

As a result, routing tables became huge and required cached routes for improved performance. Routers therefore became bigger, faster, more complex and thus more difficult to administer. In a parallel development, switching fabrics have also become increasingly complex. The end points of a circuit switched network, however, are simple (e.g., the PSTN); they require no local processing, memory, or intelligence and thereby allow a cheap, reliable, and scalable network to be designed.

2.8.3   Emergence of Tag Switching

Routers introduce delay as they perform the tasks of address resolution, route determination, and packet filtering. If the network becomes congested, the data packets may encounter more router hops to reach the destination, which in turn increases the aggregate delay caused by these routers. In addition, the delay variation associated with the transmission of each packet is not deterministic in nature. When routers started carrying real-time voice and video traffic with demanding QoS requirements, these limitations became evident; switches had no such limitations. However, it is hard to replace routers entirely with switches because end-to-end connections are inefficient for short bursts of data, which typify e-mail, Web, and LAN communications.

To combine the benefits of packet and circuit switching, many network vendors in the last decade attempted to merge switching and routing, first using ATM virtual circuits and then employing tag (label) switching, most notably, IETF's MPLS. A virtual circuit forces all packets belonging to the same flow (typically from the same user) to follow the same path in the network, thereby allowing better allocation of resources in the network to meet certain QoS requirements. Unlike a circuit-switched network, a virtual circuit does not guarantee a fixed bandwidth because multiple virtual circuits can be statistically multiplexed within the network.

The idea behind tag switching in MPLS is to move routing to the network edge and rely on faster, simpler switching in the core. Like routers, MPLS nodes still inspect each packet and route it on a connectionless basis. Like switches, these nodes employ dedicated hardware for information transfer along the fastest path in the core. Unlike a pure-router network, MPLS improves manageability and scalability of IP networks by reducing the number of devices that perform network layer route calculation. This in turn reduces the complexity of the edge devices. The IETF is also working on transporting Ethernet frames through layer-2 encapsulation in MPLS frames.

2.8.4   Current and Future Switching and Routing Technologies

A number of vendors currently build routers to handle over 100 million packets per second using one protocol, TCP/IP, and dedicating hardware to get the necessary performance. Other vendors offer high-powered network processors, including tens of gigabits per second chipsets that can switch packets with per-flow queuing, multicast, filtering, and traffic management. The emergence of traffic managers and load balancers allows multiclass requests based on class-based and per-flow queuing to be intelligently distributed across multiple Web servers.

It can be envisioned that routing, switching, and flow management can eventually be handled individually for each user and Web URL.[24] Some parts of a Website may be set for different service classes (e.g., audio and video traffic require very low latency), while other sections get normal service. Because Web sites typically comprise servers in different locations, future routers and switches will need to use class of service and QoS to determine the paths to specific Web pages for specific end-users, requiring intelligent cooperation of higher networking layers, including the transport and application layers.

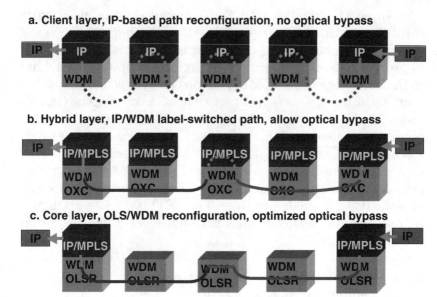

a. Client layer, IP-based path reconfiguration, no optical bypass

b. Hybrid layer, IP/WDM label-switched path, allow optical bypass

c. Core layer, OLS/WDM reconfiguration, optimized optical bypass

Figure 2.16 Optical label switching.

2.8.5 Optical Label Switching for Dynamic Optical Backbone

Optical label switching (OLS) is a unified wavelength routing and packet switching platform that provides robust management and intelligent control for IP and WDM switching (Figure 2.16). The current industry trend in GMPLS/MPλS (MPLS in the optical domain) paves the way towards OLS. The many advantages of OLS include:

- Removing layering overhead and simplifying operations
- Providing a plug-and-play interworking module with existing infrastructure
- Providing fast dynamic circuit switching service creation, provisioning, and protection
- Enabling flexible burst switching service bandwidth granularity

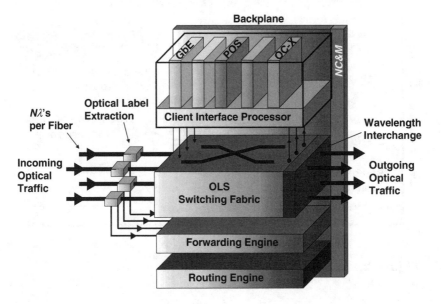

Figure 2.17 Components of the next-generation Internet optical label switching router.

The new concept of optical label switch router (OLSR) (Figure 2.17) can attain low latency and high end-to-end throughput while supporting various types of host applications. OLSR enables transfer of diverse data types across an optical Internet, where information in units of flows, bursts, or packets is distributed efficiently independent of data payload. The approach will provide high-capacity, bandwidth-on-demand, burst-mode services efficiently in optical internetworking systems. It will affect future networking systems well beyond what is available with emerging high-performance IP/GMPLS routers by dynamically sensing, controlling, and routing traffic.

2.8.6 Unified Optical Backbone Architecture

A unified and dynamic network architecture (Figure 2.18) that automatically integrates an optical packet and circuit

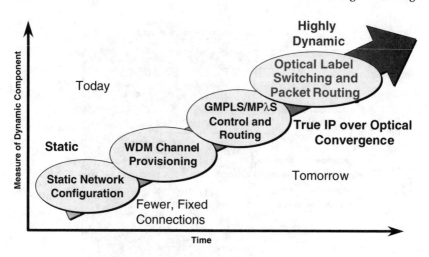

Figure 2.18 Evolution of agile optical networks.

switched network in an adaptive traffic engineering framework independent of data types is a primary goal for ongoing
research activities. It supports smart traffic bypass across fast
switching fabric that allows the traffic to hop on circuit-
switched paths while guaranteeing contention-free flow,
resulting in higher overall throughput. The optical Internet
can be extended to optical LAN through the access card for
burst traffic users. There are also strong contributions and
breakthroughs in these areas: high-speed opto-electronic systems for lightwave communications, optical add/drop multiplexers, optical cross-connects, optical MPLS routers, and
network control algorithms and protocols.

We envision a national backbone optical Internet infrastructure based on the principles of OLSRs. This will replace
the combination of reconfigurable WDM OXCs and high-performance GMPLS routers with integrated OLSRs to perform
fine-grained packet and coarse-grained circuit switch functions
for next-generation Internet. Bandwidth-starved customers will
benefit from innovations to provide low-cost and efficient bandwidth-on-demand wide-area metro services independent of data
format and bit rates. These include long-haul carriers (IXCs);

traditional Bell operating companies (ILECs); and the new entrants of competitive local exchange carriers (CLECs), Internet service providers (ISPs), and integrated communications providers, as well as DSL, cable, wireless, and fiber access companies. In addition, carriers can eliminate lengthy planning overhead for user services, drastically improving the provisioning speed to microseconds, and scale to support large networks while simultaneously maintaining the low latency and high throughput of the network.

2.9 PEER-TO-PEER FILE SHARING AND SOFTWARE DOWNLOAD

Although much debate has centered on the killer application for broadband access, possibly the most important technology that can allow any future killer application to happen is the downloading of content. Downloading content is prevalent in the Internet and many companies are working on ways to improve the performance of peer-to-peer file sharing and software download over the Internet. The number of applications that can leverage on an improved downloading mechanism is virtually limitless. For example, cost-efficient next-generation phone systems may take the shape of decentralized, peer-to-peer file-sharing networks running over the Internet and eliminate phone company middlemen.

The free Skype software program (http://www.skype.com) represents a big step in that direction. It allows users of the same application to make unlimited, high-quality, encrypted voice calls over the Internet and works with all firewalls, network address translators (NATs), and routers. With Skype, no VoIP media gateways are required because the peer-to-peer mechanism effectively allows a voice packet to be downloaded quickly by the recipient from multiple sources. The success of programs like Skype is also boosted by a recent FCC ruling in February 2004 that voice communications flowing *entirely* over the Internet (i.e., VoIP services that do not interconnect the telephone system) are no different from e-mail and other peer-to-peer applications blossoming on the Internet; therefore, they are not subject to traditional government regulations and taxes applied to public-switched telephone networks. Other popular applications that can benefit from an improved download mechanism include interactive gaming and media streaming.

The download algorithm can potentially be applied to a wider context such as updates for new software versions, device drivers, or even configuration files for Internet appliances, networked sensors, and programmable silicon radio devices (i.e., wireless devices with radio chips that emulate the flexibility of the silicon chip in its ability to run different applications on a common hardware[14,25]). The key to providing updates is to calculate the difference (or delta) corresponding to new information between the old and new versions of the file transmitted (i.e., the entire file need not be downloaded). The technique offers the possibility of better performance than the individual compression of each file version (many service providers currently use individual file compression technology in Web browsers to achieve DSL-like performance with only dial-up speeds).

A form of security is inherent in delta compression: only the client with the original version can successfully generate the new version. In addition, delta compression removes the need to set up and activate separate compression algorithms for specific applications. Because delta files are typically small and can be transmitted quickly, one can envision that the updating process can potentially be made much faster and efficient by allowing different delta files to be downloaded on demand over the Internet from an extensive network of file-sharing servers.

2.9.1 Delta Compression Performance

The delta compression concept can be applied on a packet-to-packet basis to improve the performance of the WDM PON MAC protocol discussed in Section 2.7. The improvement is particularly significant at high traffic loads or when users have sudden bursts of data. For example, by using delta compression to compute the delta between packets (given one or

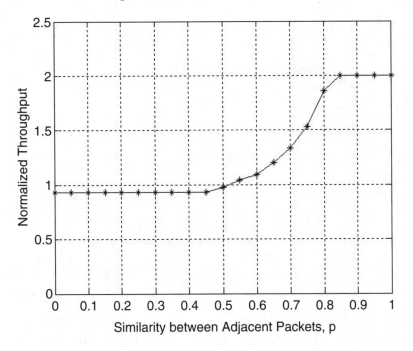

Figure 2.19 MAC-layer throughput gain using delta compression.

more known packets) and then transmitting only the delta, the need to reserve a large number of data slots is avoided (recall that reservation protocols incur a mandatory request and grant delay corresponding to the two-way propagation delay). Moreover, the number of preallocated time slots is kept to a minimum (preallocating a large number of slots will result in a performance similar to static TDMA, which is poor at high loads).

The throughput gain for the batch arrival case with delta compression is shown in Figure 2.19. Here, the delta compression scheme is lossless and is applied for values of p between 0.5 and 1, where p represents the similarity between adjacent packets. A higher value for p indicates a high correlation between adjacent packets, and therefore a higher compression gain is possible. The normalized sending rate is 2 (i.e., overload case). Without delta compression, the normalized

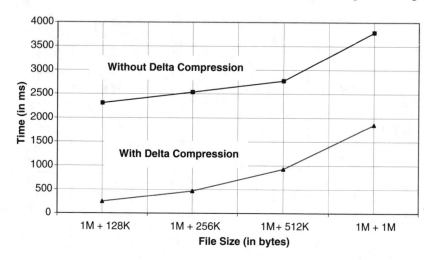

Figure 2.20   Delta compression for a wireless link.

throughput is about 0.928 (same as Figure 2.14). With delta compression, the effective normalized throughput can be increased according to the compression ratio. Note that the delta compression technique is generic enough to be applied to any MAC protocol, including the reservation and static TDMA protocols described in Section 2.7.

In Figure 2.20, the performance gains of using a delta compression scheme are illustrated for a point-to-point 11 Mb/s 802.11 wireless link.[15] The syntax 1M + xK in the figure means that the old file is 1 Mbyte and x kbytes have been inserted to create the new file.

Figure 2.21 shows how delta compression can be employed to achieve faster Web page download. Our experimental setup is a DOCSIS cable network depicted in Figure 2.2 and we employ an optimal method (not necessarily the fastest method) to perform lossless delta compression. Because we wish to evaluate the worst-case performance of the scheme, we employ a forum Web page where real-time messages with random contents and varying lengths are posted one after the other. In this case, the reference message changes and is always the message immediately preceding

	Uncompressed	Delta Compressed	Winzip Compressed	
File Size	44 KB	3 KB	7 KB	**Nonreal-Time**
Throughput	218 Kbps	551 Kbps	N/A	**Real-Time**

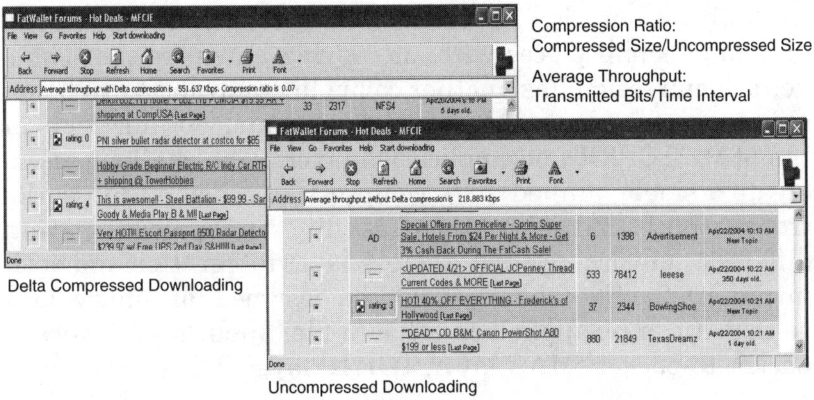

Delta Compressed Downloading

Uncompressed Downloading

Compression Ratio:
Compressed Size/Uncompressed Size

Average Throughput:
Transmitted Bits/Time Interval

Figure 2.21 Fast Web page download using delta compression.

the next message. The messages are converted to binary format before the delta compression algorithm is applied. The real-time throughput improvement using the optimal delta compression is more than twice the case when there is no compression. This takes into account the delta file creation, transmission, and decoding. The run-time speed and real-time throughput can be improved using suboptimal delta compression schemes, although these schemes cannot reach the static compression bound provided by the optimal method (i.e., a compression ratio of about 12 times, as shown in Figure 2.21).

2.9.2 Delta Compression as Unifying Cross-Layer Technology

In addition to wireless and optical access networks, delta compression can help reduce the need to transmit many packets across slow, bandwidth-constrained dial-up links, as well as other access methods, thus potentially providing a unified approach to improve the performance of broadband access over heterogeneous last mile solutions. This idea has been applied pervasively in different forms, but optimized for specific applications, including MPEG coding of video sequences, software patch update, Web-page refresh/caching, storage backups, TCP/IP header suppression, and even genetic sequences.

TCP/IP also provides a file access service (known as network file system or NFS) that allows multiple clients to copy or change small pieces without copying an entire file. Thus, delta compression has another compelling advantage of being able to perform cross-layer integration and optimization in the MAC, TCP/IP, and application layers. The challenge is to create a single, unified delta compression scheme that works transparently and efficiently across different applications and different network layers (i.e., cross-layer integration and optimization). Such a scheme should be designed for binary data so that it can be applied on a file-to-file (application layer) or packet-to-packet (MAC layer, TCP/IP layer) basis.

2.10 CONCLUSIONS

Although radical changes have been made in recent years to increase the capacity of the Internet, the majority of U.S. households still lack the infrastructure to support high-speed, quality service and high-resolution content. This could be due to prevailing paradigms for last mile access that are not sufficient for addressing the scale and unique requirements inherent in the support of pervasive broadband applications. The authors believe that the presence of multiple wireless access providers and optical fiber-based services will result in more choices for residential broadband access solutions and that the increased competition will drive down the prices for broadband access services. This departure from the monolithic, operator-centric business model allows the residential user to employ a multihoming approach, commonly adopted by enterprises, in which more than one Internet access service are subscribed, providing redundancy, diversity, and increased reliability.

The emergence of distributed, autonomous topologies (wireless and optical) as well as effective, peer-to-peer content download mechanisms can help ensure that broadband access technologies are available to all for broad use and will ultimately lead to innovative applications. Wireless and optical access technologies can complement each other and enhance other access methods. For instance, HFC is already employing fiber for improved range and reliability and hybrid Wi-Fi HFC and wireless DOCSIS networks are starting to appear.

The authors also advocate the use of seamless, end-to-end IP QoS on an Ethernet platform, a vision upon which researchers and practitioners should focus when designing a new generation of last mile, edge, and backbone solutions. The key to success will be the integration of resource allocation algorithms and individual SLA enforcement across heterogeneous access networks, as well as core backbone networks and services.

REFERENCES

1. NSF Workshop Report, Residential broadband revisited: research challenges in residential networks, broadband access, and applications, January 20, 2004.

2. C. Xiao, B. Bing, G.K. Chang, "An Efficient MAC Protocol with Preallocation for High-Speed WDM PONs," *IEEE Infocom*, March 13–17, 2005.

3. C. Xiao and B. Bing, Measured QoS performance of the DOCSIS hybrid-fiber coax cable network, *13th IEEE Workshop Local Metropolitan Area Networks*, San Francisco, April 25–28, 2004.

4. Net telephony as file-trading, *Business Week Online*, January 6, 2004.

5. J.K. Jeyapalan, Municipal optical fiber through existing sewers, storm drains, waterlines, and gas pipes may complete the last mile, available from: http://www.agc.org/content/public/PDF/Municipal_Utilities/Municipal_fiber.pdf.

6. B. Bing (Ed.), *Wireless Local Area Networks: The New Wireless Revolution*, John Wiley & Sons, New York, 2002.

7. R. Ramaswami and K. Sivarajan, *Optical Networks: A Practical Perspective*, Morgan Kaufmann, San Francisco, 2002.

8. R. Howald, HFC's transition, *Commun. Syst. Design*, July/August 2003, 7–9.

9. D. Dubois and M. Fauber, Blast through the barriers to Ethernet in the metro, *Commun. Syst. Design,* July/August 2003, 20–27.

10. W.T. Anderson, J. Jackel, G.K. Chang et. al., The MONET Project — a final report, *IEEE J. Lightwave Technol.,* December 2000, 1988–2009.

11. B. Meager, G.K. Chang, G. Ellinas et. al., Design and implementation of ultra-low latency optical label switching for packet-switched WDM networks, *IEEE J. Lightwave Technol.,* December 2000, 1978–1987.

12. J. Yu and G.K. Chang, A novel technique for optical label and payload generation and multiplexing using optical carrier suppression and separation, *IEEE Photonic Technol. Lett.,* January 2004, 320–322.

13. J. Yu and G.K. Chang, Spectral efficient DWDM optical label generation and transport for next generation Internet, invited paper, W2A-(13)-1, *IEEE Conf. Laser Electro-Opt.,* Pacific Rim, December 2003.

14. T.-K. Tan and B. Bing, *The World-Wide Wi-Fi,* John Wiley & Sons, New York, 2003.

15. B. Bing, Wireless software download for reconfigurable mobile devices, Georgia Tech Broadband Institute Technical Report, 2004.

16. W. Ciciora, J. Farmer, D. Large, and M. Adams, *Modern Cable Television Technology,* 2nd ed., Morgan Kaufmann, San Francisco, 2004.

17. L. Roberts, Judgment call: quality IP, *Data Commun.,* 28(6), April 1999, 64.

18. S. Dravida, D. Gupta, S. Nanda, K. Rege, J. Strombosky, and M. Tandon, Broadband access over cable for next-generation services: a distributed switch architecture, *IEEE Commun. Mag.,* 40(8), August 2002, 116–124.

19. R. Green, The emergence of integrated broadband cable networks, *IEEE Commun. Mag.,* 39(6), June 2001, 77–78.

20. G. Donaldson and D. Jones, Cable television broadband network architectures, *IEEE Commun. Mag.,* 39(6), June 2001, 122–126.

21. N. Jayant, Scanning the issue: special issue on gigabit wireless, *Proc. IEEE*, 92(2), February 2004, 1–3.

22. A. Polydoros et. al., WIND-FLEX: developing a novel testbed for exploring flexible radio concepts in an indoor environment, *IEEE Commun. Mag.*, 41(7), July 2003, 116–122.

23. E. Miller, F. Andreasen, and G. Russell, The PacketCable architecture, *IEEE Commun. Magazine*, 39(6), June 2001, 90–96.

24. J. McQuillan, Routers and switches converge, *Data Commun.*, 26(14), October 21, 1997, 120–124.

25. B. Bing and N. Jayant, A cellphone for all standards, *IEEE Spectrum*, May 2002, 39(5), 34–39.

26. G. Staple and K. Werbach, The end of spectrum scarcity, *IEEE Spectrum*, March 2004, 41(3), 49–52.

27. R. Yassini, *Planet Broadband*, Cisco Press, Indianapolis, IN, 2004.

28. R. Lucky, Where is the vision for telecom? *IEEE Spectrum*, May 2004, 41(5), 72.

3

Last Mile Copper Access

KRISTA S. JACOBSEN

3.1 INTRODUCTION

This chapter addresses various aspects of last mile data transport on copper wire networks. The transport is dubbed "last mile" even though the distances over which these systems operate can be well over a mile. The focus is on transmission over public telephone networks, not only because these networks are ubiquitous in most of the world, but also because the conditions on phone lines — many of which are several decades old — present a great challenge to modem designers. High-speed transmission on telephone networks requires design of a sophisticated yet cost-effective system, which in turn requires ingenuity and creativity from modem designers.

Telephone networks were originally deployed to support transmission of voice signals between a telephone company's central office (CO) and customers. The portion of the network between the CO and customers consists of individual loops, which are composed of twisted-pair lines. Loops* are so named because current flows through a looped circuit from the CO on one wire and returns on another wire.

---

* In this chapter, as in the DSL industry, the terms "loop" and "line" are used interchangeably.

"Plain old telephone service," which goes by the unfortunate acronym POTS, occupies the telephone line bandwidth from DC to approximately 4 kHz. In most regions of the world, residential telephony is provided using POTS. Early Internet access relied primarily on voiceband modems, which operate within the same bandwidth as POTS (instead of POTS) and support bit rates up to 56 kilobits per second (kbps). As the Internet has become more popular and content has become more sophisticated, data rate needs have grown. Today, voiceband modem speeds are woefully inadequate, and demand for technologies and services that provide broadband access has increased dramatically.

Integrated services digital network (ISDN) is a newer mechanism to provide voice and data services. ISDN occupies a wider bandwidth than POTS and is a fully digital service, which means the telephone at the subscriber's premises digitizes voice signals prior to transmitting them on the telephone line. ISDN also provides data channels along with the voice channel, which enables simultaneous use of the telephone and data transmission. These data channels typically support higher bit rates than a voiceband modem, for example, 128 kbps. However, one disadvantage of ISDN is cost. New subscribers may need to get a new telephone to use the voice services of ISDN, and the cost of service is typically high (often $80 per month or more). Another problem with ISDN is that the maximum bit rate is 144 kbps, which is inadequate to support many high-speed services.

ISDN was the precursor to the digital subscriber line (DSL) technologies becoming prevalent today. Like ISDN, the most popular DSL services allow subscribers to use, simultaneously, the telephone and the modem. However, in contrast to ISDN, which might require a new telephone that can digitize voice signals, residential DSL occupies the bandwidth above the voiceband. Thus, subscribers can keep their POTS service and their existing telephones. Furthermore, DSL occupies a wider bandwidth than ISDN, which allows transmission at higher bit rates, typically from several hundred kilobits per second to tens of megabits per second (Mbps). Best of all, the amount consumers pay for DSL service is less than the cost of ISDN service, a paradox that results from consumer demand and competition among service providers.

One advantage of DSL relative to other means of providing broadband in the last mile, such as cable modems or wireless access, is that each subscriber has a dedicated transmission medium — the phone line. Therefore, DSL avoids the use of protocols to coordinate bandwidth use, which provides some level of system simplification. Furthermore, because phone lines are already installed and in use for POTS (or ISDN), broadband service can be enabled without operators needing to make significant changes to the loop plant.* Telephone lines are connected to nearly every home and business in the developed world, while coaxial facilities are not present in many regions. Furthermore, because subscribers of cable television services are residential customers, coaxial networks are almost never deployed in business districts. Therefore, cable modem service is not available to many attractive potential broadband customers: i.e., businesses.

However, DSL is not without issues. DSL bit rates depend on the length of loop on which service is provided. Because the attenuation of frequencies used in DSL increases with increasing loop length and increasing frequency, physics dictates that achievable bit rates are lower on longer lines. Practically, the dependence of bit rates on loop length causes provisioning headaches for network operators, who tend to prefer to offer the same service to all customers.

A second issue in DSL is, ironically, interference from transmissions by other subscribers. The interference results because telephone lines are tightly packed together in cables in the last mile, and although individual lines are twisted to reduce interference, crosstalk does still occur. Crosstalk noise reduces the bit rate that a line can support, thus exacerbating operators' provisioning difficulties.

---

* Some telephone lines, particularly lines that are very long, have loading coils (inductors placed in series with the loop at 5000-ft intervals) installed to improve the quality of the bandwidth at voiceband frequencies at the expense of frequencies above the voice band. Loading coils block transmission at DSL frequencies, and therefore they must be removed before DSL service can be enabled.

This chapter begins with a brief, high-level introduction to the "flavors" of DSL that have been defined. The primary focus is on the bit rate and reach capabilities of each flavor. Next, the telephony network is reviewed, and methods to deploy DSL in that network are described. A detailed discussion of the DSL transmission environment follows. The characteristics of twisted-pair lines are examined, and common impairments that have an impact on DSL transmission are described.

Next, the key design options for DSL modems are presented. First, the issue of duplexing — how to partition the available bandwidth between the two transmission directions — is addressed. The modulation scheme (also referred to as the line code) used in a DSL modem is a critical design choice. The chapter provides a detailed description of the line code alternative used in asymmetric digital subscriber line (ADSL) and very high-bit-rate DSL (VDSL) modems: discrete multi-tone (DMT) modulation. Next, the specifics of the flavors of DSL are provided, including the duplexing method and line code used in each. The chapter concludes with a discussion of spectral compatibility, including an explanation of the "near–far" problem that is most problematic on short loops. A brief overview of the promising area of crosstalk cancellation is also presented.

3.2 FLAVORS OF DSL

Several types of DSL have been defined to support transmission over copper in the last mile. The focus in this chapter is on three primary classes of DSL:

- Asymmetric DSL (ADSL) is the most widely deployed type of DSL, providing connectivity primarily to residential subscribers (although about 20% of ADSL lines serve business customers).
- Very high-bit-rate DSL (VDSL) is an emerging DSL that supports higher bit rates than other DSL flavors and may meet the needs of residential as well as business services.
- Symmetrical DSLs, including high-speed DSL (HDSL) and symmetric high-bit-rate DSL (SHDSL), are targeted for business users.

This section provides an overview of the achievable bit rates in the downstream (toward the subscriber) and upstream (away from the subscriber) directions as well as the target loop reaches of ADSL, VDSL, HDSL, and SHDSL. A later section provides details of each type of DSL, such as which line code is used and how the available bandwidth is partitioned between the two transmission directions.

3.2.1 ADSL

As its name suggests, ADSL supports asymmetrical transmission. Typically, the downstream bit rate is up to eight times as large as the upstream bit rate, although the exact ratio depends on the loop length and noise. According to a major North American network operator, network traffic statistics show that the typical ratio of traffic asymmetry is about 2.25:1. ADSL generally operates in a frequency band above POTS or, in some regions of the world, above ISDN. Thus, ADSL allows simultaneous use of the telephone and broadband connection. Originally conceived as a means to provide on-demand video, ADSL has become the world's DSL of choice for broadband access on copper. The vast majority of DSL lines in the world use ADSL. More than 80 million ADSL lines were in operation at the end of 2004.[1]

ADSL was first specified in 1993 for the U.S. in the Committee T1 Standard T1.413,[2] followed by Europe in ETSI TS 101 388,[3] and later in the international ADSL standard, ITU-T Recommendation G.992.1.[4] These specifications are now sometimes referred to as the ADSL1 standards. Today, the ITU-T is the standards organization primarily responsible for continuing ADSL standardization.

ADSL1 was defined to support downstream bit rates of up to 8 Mbps and upstream bit rates of up to 896 kbps. The maximum reach of ADSL1 is approximately 18 kft (1 kft equals 1000 ft); in practice, ADSL1 is seldom deployed on loops longer than 16 kft.

The ADSL2 specification, ITU-T Recommendation G.992.3 (2002),[5] specifies a number of additional modes that expand the service variety and reach of ADSL. The various annexes of ADSL2 allow downstream bit rates as high as approximately 15 Mbps and upstream bit rates as high as 3.8 Mbps, and a mode to extend the maximum reach of ADSL has also been defined. In addition, ADSL2 introduces additional loop diagnostic functions within the modems and modes to save power. Finally, ADSL2 defines all-digital modes of operation, without underlying POTS or ISDN.

At the time this chapter was written, the most recent work in ADSL was in the specification known as ADSL2plus, which approximately doubles the downstream bandwidth, thus increasing the maximum downstream bit rate to 24 Mbps. ADSL2plus also supports all of the same upstream options as ADSL2. ITU-T Recommendation G.992.5,[6] which specifies ADSL2plus operation, was completed in late 2003. The specification is a "delta" standard to ADSL2, meaning that an ADSL2plus modem has all the functionality of an ADSL2 modem. In practical terms, ADSL2plus modems will revert to ADSL2 operation when loop conditions do not allow transmission at the higher frequencies available in ADSL2plus, such as when a loop is long.

Section 3.8.1 provides details of the various ADSL operational modes.

3.2.2 VDSL

Work on VDSL began shortly after the first versions of ADSL (T1.413 and ETS 101 388) were completed. A number of (sometimes conflicting) objectives were originally established for VDSL. One goal was to provide higher downstream and upstream bit rates than ADSL. Another goal was to enable symmetrical as well as asymmetrical transmission. Early bit rate and reach goals for VDSL included 52 Mbps downstream with 6.4 Mbps upstream, and 26 Mbps symmetrical, both on loops up to 1 kft in length. As later sections will explain, achieving both of these objectives with a single system without creating some deployment problems is impossible.

VDSL is intended to provide very high bit rate services on short loops. The maximum reach depends strongly on the selected frequency plan — that is, how the (typically large) available bandwidth is allocated to the downstream and upstream directions. When the frequency plan standardized for use in the U.S., known as plan 998, is used, the reach of VDSL is limited to less than 5 kft unless an optional band is enabled for upstream transmission.

The first version of VDSL, called VDSL1, is defined in the American National Standard T1.424,[7] ITU-T Recommendation G.993.1,[8] and ETSI TS 101 270.[9,10] As does the maximum reach, the downstream and upstream bit rates of VDSL1 depend strongly on the frequency plan used. When plan 998 is used, the maximum downstream bit rate is over 50 Mbps, and the maximum upstream bit rate is over 30 Mbps. These rates are achievable only on very short loops.

At the time this chapter was written, work on VDSL2, the second-generation of VDSL, was just underway. Objectives of VDSL2 are to improve performance relative to VDSL1, meaning higher bit rates and longer reach, as well as to include new features, such as loop diagnostics and fast start-up. Another objective is to define VDSL2 in a manner that facilitates multimode operation, i.e., a modem that can behave as a VDSL2 or ADSL2 modem.

3.2.3 Symmetrical DSLs: HDSL and SHDSL

ADSL and VDSL are primarily designed for residential subscribers, who tend to download more content than they send, and who typically use their telephone lines for POTS. Business users, in contrast, require more symmetrical connections, and their telephony is generally provided through alternate means.

HDSL was developed in parallel with ADSL, primarily as a technology to replace T1 lines, which support 1.544 Mbps symmetrical transmission. T1 lines are very inefficient in their use of the spectrum, but they are very reliable; they provide 1.544 Mbps using 1.544 MHz of bandwidth* using two twisted-pair lines. In contrast, HDSL supports bit rates

---

* The 3-dB bandwidth is 1.544 MHz, but T1 signals also have significant sidelobe energy.

up to 1.552 Mbps (1.544 Mbps plus 8 kbps overhead), but it does so using only 196 kHz of bandwidth.* Like T1, HDSL uses two twisted-pair lines. The reach of HDSL is between 9 and 12 kft without repeaters. Unlike ADSL and VDSL, HDSL supports the use of repeaters, and the reach of HDSL with repeaters can be 30 kft or more. HDSL is defined in ETSI ETR 152[11] and in ITU-T Recommendation G.991.1.[12]

Two new flavors of HDSL, defined in T1 standard T1.418, are now widely deployed in North America. HDSL2 supports the same loop length and bit rate as HDSL, but it uses only a single twisted pair and does not support repeaters. HDSL4 uses two twisted-pair wires, but its reach is 33% longer than HDSL. HDSL4 supports repeaters and is more compatible with other DSL services than HDSL is (see Section 3.9).

SHDSL is the newest symmetrical DSL, specified in ITU-T Recommendation G.991.2. It supports bit rates up to 5.696 Mbps using more bandwidth than HDSL but only one twisted-pair line. SHDSL operates on loops up to about 18 kft in length. Like HDSL, SHDSL supports repeaters so the maximum loop reach can be extended well beyond 18 kft.

Intended for deployment to business customers, who typically do not require provision of POTS on the same line as their data traffic, neither HDSL nor SHDSL preserves POTS.

3.2.4 Choosing a Flavor of DSL

Table 3.1 summarizes the key characteristics of the various DSL flavors, namely, the downstream and upstream bit rates, loop reach, and whether POTS (or ISDN) may be simultaneously supported on the same pair of wires. The table is a simplified presentation, and a number of caveats are documented in footnotes to the table entries. Figure 3.1 provides a simplified graphical interpretation of the loop reaches where the various DSLs operate. The reader is cautioned that the reaches shown are approximate and that the actual reach of any DSL is a function of a number of variables, including loop

---

* For technologies that use single-carrier modulation, the quoted bandwidths are 3-dB bandwidths.

Table 3.1 Summary of Characteristics of DSL Variants

DSL type	Appropriate standards	Maximum reach[a] (kft)	Maximum bit rate (Mbps)		POTS/ISDN simultaneous transport
			Downstream	Upstream	
ADSL1	G.992.1, T1.413, ETS 101 388	18	8	0.896	Always
ADSL2	G.992.3	>18	13[b]	3.5	Yes[c]
ADSL2plus	G.992.5	~9[d]	24	3.5	Yes[c]
VDSL1	G.993.1, T1.424, ETS 101 270	~5[e]	50+[f]	30+[f]	Always
VDSL2[g]	N/A	>5	100?	100?	Yes
HDSL	G.991.1	9	1.552	1.552	No
HDSL2	T1.418	9	1.552	1.552	No
HDSL4	T1.418	12	1.552	1.552	No
SHDSL	G.991.2	18	5.696	5.696	No

[a] Approximate reaches are provided. Actual reach depends on a number of factors, including loop attenuation and noise environment.

[b] Assumes the use of frequency-division duplexed operation (see Section 3.7.1), which is the most popular ADSL duplexing mode as of the time of writing. Overlapped operation would increase the achievable downstream bit rate by up to approximately 2 Mbps.

[c] All-digital modes of operation are also defined in ADSL2 and ADSL2plus.

[d] ADSL2plus reverts to ADSL2 operation on loops longer than about 9 kft.

[e] VDSL1 reach is limited to about 5 kft unless an optional frequency band is enabled for upstream transmission.

[f] Achievable bit rates in VDSL depend on which frequency plan is used. The numbers provided assume use of the frequency plan known in the industry as 998, which is the only frequency plan defined in VDSL1 in the U.S.

[g] Standardization of VDSL2 was in progress at the time this chapter was written.

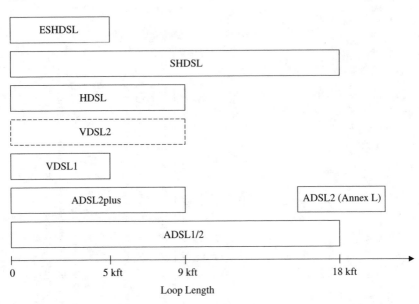

Figure 3.1 Graphical depiction of approximate reaches of various DSLs (assuming 26-AWG lines). At the time this chapter was written, VDSL2 was in the process of being defined, and the loop range shown for VDSL2 is the author's best guess.

attenuation and noise. The factors affecting DSL performance are discussed in Section 3.6.

3.3 THE TELEPHONY NETWORK

In the early days of telephony, copper twisted pairs were the sole means by which connectivity between locations was provided. Today, the telephony network has evolved and grown to support an enormous number of simultaneous connections, as well as long-distance and broadband communications. This section provides an overview of the network architecture.

For the purposes of this chapter, the telephony network can be segmented into two parts: the local loop and the backbone. The local loop is the part of the network between the telephone company's central office (CO) and its customers; the

Figure 3.2 The local loop exists between the telephone company's central office (CO) and subscribers' homes.

rest of the network is lumped together as the backbone. The backbone is generally optical fiber or perhaps satellite links. Although the backbone is clearly a critical component of providing broadband, the primary focus in this chapter is on the local loop.

The local loop is almost exclusively composed of copper twisted-pair lines, primarily for cost reasons. Figure 3.2 depicts the simplest architecture of the local loop. Copper cables emanate from the central office toward subscribers. Individual twisted-pair lines are then tapped from the cable to provide POTS service to subscribers.

Within the CO, POTS lines are connected to a voice switch, which converts customers' analog voice signals to 64-kbps digitized streams and routes them to the public switched telephone network (PSTN). The switch also converts incoming voice signals to analog format and routes them to the appropriate lines. Figure 3.3 illustrates the simplicity of CO-based POTS.

In some networks, POTS is not provided from the CO. Instead, a remote terminal (sometimes called a cabinet) is installed between the CO and subscribers. The remote terminal (RT) contains equipment, such as a digital loop carrier (DLC), that provides POTS service. The DLC essentially performs the digitization and aggregation function of the switch in the CO. A high-speed connection connects the RT to the CO. The connection may be copper-based (for example, a T1), or it might be fiber. Figure 3.4 illustrates the network architecture when POTS is provided from the RT.

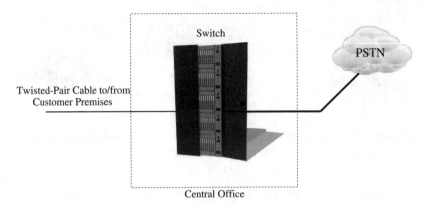

Figure 3.3 The conceptual simplicity of POTS support from the CO.

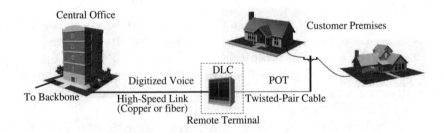

Figure 3.4 Example of the local loop when POTS is provided from the remote terminal.

Installation of an RT shortens the length of the copper portion of the local loop, which facilitates higher bit rates in DSL but also poses some deployment challenges. The next section describes how those challenges are overcome.

3.4 DEPLOYING DSL OVER POTS

DSL can operate over POTS* or without POTS on the same line. This section focuses on the deployment of DSL over

---

* The reader should keep in mind that operation over ISDN is also possible.

POTS, which tends to be more complicated than if POTS preservation is not required. Because POTS can be provided from the CO or from the RT, operators have developed strategies for deploying DSL to their CO-based and RT-based POTS customers.

3.4.1 DSL Deployment from the CO

The central office is a large, climate-controlled building with ample power available for DSL equipment. A single office typically serves 10,000 to 150,000 subscriber lines. The overlay of DSL on the CO-based POTS network is straightforward. It involves the installation of three key components: the customer modem, the POTS splitter, and the digital subscriber line access multiplexer (DSLAM).

3.4.1.1 The Customer Modem

At the customer end of the telephone line, a DSL modem is required to demodulate downstream (incoming) signals and modulate upstream (outgoing) signals. The customer modem is often referred to as the customer premises equipment (CPE) or as the xTU-R (where the appearance of "x" may designate a generic DSL customer modem or "x" may be replaced by another letter to indicate the "flavor" of DSL, and "TU-R" is the acronym for "terminal unit — remote"). Typically, the modem is a stand-alone unit that provides an Ethernet or USB port to facilitate connection to a computer or networking equipment. Depending on the type of DSL and its anticipated application, the CPE may provide additional functionality. For example, an ADSL CPE may also provide a wireless access point to enable 802.11 connectivity. For business services, a CPE may perform some routing or switching functions.

3.4.1.2 The POTS/ISDN Splitter

When DSL operates noninvasively to POTS (or ISDN), as ADSL and VDSL do, then use of a splitter is required at the subscriber side of the line and at the CO to allow simultaneous use of DSL and POTS on the same physical line. In the receive

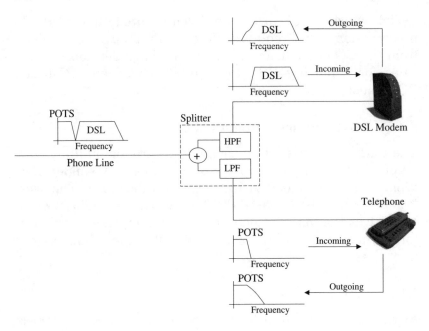

Figure 3.5 Illustration of splitter functionality.

direction, the splitter applies a low-pass filter to the received signal to filter out DSL and presents a "clean" POTS signal to the telephone. In parallel, a high-pass filter is applied to filter out POTS and present a clean signal to the DSL modem. Because the characteristics of the high-pass filter are crucial to DSL operation, the high-pass part is typically designed into the transceiver so that modem designers can control the filter cut-off frequency, roll-off, and out-of-band energy.

In the transmit direction, the low-pass and high-pass filters are applied to the POTS and DSL signals, respectively, to confine them to the required frequency bands, and then the two signals are combined and presented to the local loop. Figure 3.5 illustrates how the splitter isolates POTS and DSL signals. For ease of illustration, the high-pass filter is assumed to reside in the splitter, even though in practice it is almost always part of the DSL modem.

Installation of the splitter at the customer premises requires a skilled technician and often involves the installation

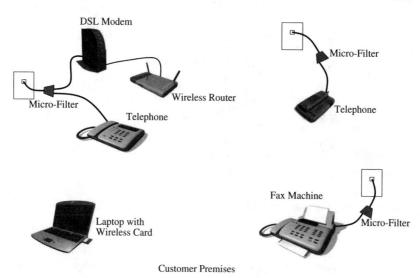

Customer Premises

Figure 3.6 Using microfilters eliminates the need for a techni-
cian-installed splitter at the customer premises.

of a new, dedicated inside wire from the splitter to the DSL
modem. The high cost of labor to install the splitter and inside
wire can be avoided through the so-called "splitterless" customer
premises configuration, shown in Figure 3.6. In a splitterless
installation, the customer inserts an in-line filter (specified in
ANSI T1.421) between each phone and its wall jack. The in-line
filter (also called a microfilter), an example of which is shown
in Figure 3.7, is a low-pass filter that prevents audible noise in
the telephone handset due to DSL signals and isolates DSL
transmission from phone noise and impedance effects. According to a major North American network operator, more than
90% of DSL installations are currently splitterless, allowing the
unskilled customer to complete the DSL installation without a
technician visit to the premises.

3.4.1.3 The Central Office Modem and DSLAM

After passing though the high-pass filter of the CO splitter,
the DSL terminates on the CO side at a modem known as the

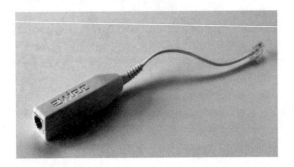

Figure 3.7   Example of an in-line filter used in splitterless DSL installations. (Photo provided by 2Wire, Inc.)

xTU-C (where "xTU" has the same meaning as for the customer and, of course, "C" stands for "CO"). The xTU-C is the complement of the xTU-R, and it demodulates upstream signals and modulates downstream signals. The xTU-C resides on a line card with other xTU-Cs in a digital subscriber line access multiplexer (DSLAM), which interfaces with the backbone of the network. The DSLAM aggregates data from a large number of DSL subscribers and presents a single high-capacity data stream to the Internet. Likewise, it accepts data from the backbone and routes it to the appropriate xTU-C for modulation. Most of today's DSLAMs support ATM traffic, although interest in Ethernet DSLAMs appears to be growing due to the availability of low-cost Ethernet equipment.

The text by Starr et al.[13] contains a collection of photographs of central offices, remote terminals and equipment such as DSLAMs, CPE, and splitters, and an in-line filter.

3.4.1.4   Putting Together the Pieces

Figure 3.8 illustrates the overlay of DSL on CO-based POTS. At the CO, a loop supporting DSL terminates in a POTS splitter, which separates the DSL and POTS signals. The DSL signal is then routed to a DSLAM, where it is demodulated by the xTU-C and aggregated with other users' data streams. The aggregated stream is then routed to the backbone and to

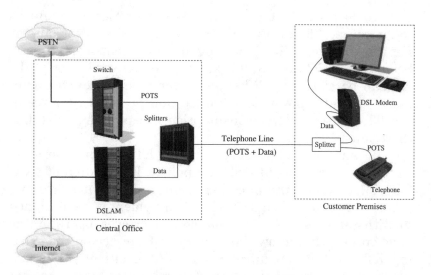

Figure 3.8 Conceptual illustration of an end-to-end connection with DSL provided from the CO as an overlay to POTS.

the Internet. As in the case without DSL, the POTS signal is routed to a time-division multiplexing (TDM) switch, which digitizes it, aggregates it with other digitized POTS streams, and routes the combined stream to the PSTN.

At the customer side, a POTS splitter (or a set of distributed in-line filters) separates the POTS and DSL signals. The DSL signal is then demodulated by the CPE, while the analog POTS signal is available at telephones within the premises.

3.4.2 DSL Deployment from the RT

Primarily due to environmental differences, the strategy for deploying DSL on POTS lines from the RT differs from the CO deployment strategy. Relative to the CO, RTs are physically small and have no climate control. Because of the harsh environment, equipment deployed in the RT must be environmentally hardened. Furthermore, providing power for remotely deployed electronics can be challenging. Although a local utility could be contracted to bring power to the RT, the costs to do so can be high, and back-up batteries are required to guarantee lifeline POTS* in the event of a power outage. For these reasons, line powering, in which some copper pairs between the CO and the RT are used to deliver power to the RT, can be attractive. Whether power is fed through copper pairs or locally, power is a limited and precious resource at the RT, and the power consumption of equipment installed there must be low.

Because of the lack of space in the RT, network operators have limited choices in DSL equipment. Installation of a CO DSLAM is out of the question; even if sufficient space and power were available in the RT, CO equipment is not designed to withstand its harsh environment. However, operators do have options that allow the overlay DSL service on the existing RT-based POTS infrastructure. Alternatively, they can choose to replace existing hardware with new hardware that supports POTS and DSL in a more integrated manner.

3.4.2.1   Overlay of DSL at the RT

In the overlay scenario, operators can choose to install a remote DSLAM, which provides essentially the same functionality as a CO DSLAM (including POTS splitters) in less space and in an environmentally hardened package. The remote DSLAM also provides operators with the ability to configure and manage DSL connections from the CO.[14] An alternative to the remote DSLAM is the remote access multiplexer (RAM), which provides all the functionality of a remote DSLAM but in a fraction of the space. RAMs are so small that they can be installed in spaces in the RT that would otherwise be unoccupied. An individual RAM may support only a few subscribers, or it could serve 48 or more. Not surprisingly, RAMs that support more subscribers are larger than those that support fewer subscribers.[14]

---

* "Lifeline POTS" is guaranteed access to emergency services (911) through any phone line at any time, even when a power outage has occurred.

3.4.2.2   Integrating POTS and DSL

As DSL deployment increases, installing remote DSLAMs or RAMs in the RT will become impractical; an RT only has so much space. Therefore, an alternative deployment option is needed.

Integrating POTS and DSL on a single line card is one way in which DSL can be provided easily from the RT. Integrated line cards give network operators the option to replace existing line cards in a digital loop carrier with line cards that support both POTS and DSL. A key feature of integrated approaches is that they support the same subscriber density at which only POTS was previously supported. Therefore, the integrated line card allows operators to install DSL in existing DLCs without concerning themselves with space availability because the integrated line card is the same size as the POTS-only line card, and it supports the same number of customers. Because an integrated line card provides POTS as well as DSL functionality, the cost is higher than the cost of a DSL-only line card; therefore, a DSL deployment strategy to replace existing POTS cards with integrated line cards is more expensive than simply overlaying DSL. However, due to space limitations in the RT, this is often the only practical option to provide DSL. As the price of integrated POTS and DSL line cards declines, newly installed DLCs will commonly house integrated line cards to meet subscribers' current and future needs for POTS and DSL services.

On the integrated line card, POTS is handled in the same way in which it is by an ordinary DLC. Incoming POTS signals are digitized, aggregated with other digitized voice streams, and passed to the CO via the high-speed link. Therefore, installing integrated line cards in the RT requires no upgrade or change to an operator's POTS infrastructure.

DSL signals are demodulated on the line card, and the data stream is routed to a network interface card incorporated in the chassis housing the integrated line cards.[14] The interface card aggregates the data and places it onto the high-speed link, which carries it to the CO. Within the CO, the voice and data streams are handled in the same manner as when the DSLAM is located at the CO.

3.4.3 Why not Fiber to the Home?

In the ideal network, the copper in the local loop would be replaced by fiber optic cable, thus providing subscribers with broadband capabilities (bit rates) beyond their wildest dreams. Therefore, it is worth considering why operators have not embraced a strategy of replacing existing twisted-pair cables with fiber.

Primarily, operators leverage the existing copper infrastructure for provision of broadband services because the cost of installing fiber all the way from the CO to the home or office is prohibitive. The fiber is a small fraction of the expense; however, the labor required to trench new cable and install the hardware required to convert optical signals in the fiber to/from electrical signals in the twisted pair is quite costly. In the case of fiber to the home, separate trenches need to be dug for each customer, which is a very expensive proposition. Therefore, operators may only install fiber from the CO to remote terminals as a strategy to shorten the local loop. By installing fiber only to the RT, the labor and material costs of the installation can be amortized over many customers. Existing copper pairs are used deeper in the network, where fiber installation costs would be much higher and harder to recuperate through subscriber fees. Using this strategy, operators can justify shortening the length of twisted pair in the local loop, thus increasing the broadband data rates that can be provided to customers.

Although fiber provides an ideal transmission media, copper wire has two vital advantages. First, copper wires are already installed to every customer. Second, these wires can carry ultra-reliable power from the network to operate the equipment at the customer end of the line, thus enabling POTS to continue to function during a local power failure.

3.5 TWISTED-PAIR LINES

The key component of the local loop is the twisted-pair line manufactured by twisting together two insulated copper wires, which reduces interference between adjacent pairs. The insulator in twisted-pair lines is typically polyethylene, although in older cables paper was sometimes used. The twisted pair of wires is known as an unshielded twisted pair (UTP). To form cables, multiple UTPs, usually 10, 25, or 50 pairs, are twisted together tightly into a binder group. Several binder groups are then bound into cables. Cables are deployed from the CO to neighborhoods, and individual twisted pairs are tapped from the cable as necessary to provide service to subscribers, as described in Section 3.3.

In the U.S., twisted pairs are differentiated by their gauges, which for telephony networks range from 26 American wire gauge (AWG) to 19 AWG, with the higher numbers indicating smaller-diameter wires. In other regions of the world, such as Europe and Asia, twisted pairs are defined by the diameter, in millimeters, of the component copper wires. The wire diameters range from 0.4 to 0.91 mm. For comparison purposes, 0.4-mm cables are similar to 26 AWG, and 0.5-mm cables are similar to 24 AWG. In the U.S., 24- and 26-AWG cabling is most common in the local loop.

All physical channels attenuate signals from a transmitter to a receiver, and the UTP is no exception. The amount by which a transmitted signal is attenuated when it reaches the receiver at the end of a twisted-pair line is a function of a number of variables, including frequency, length of the line, dielectric constant, and wire diameter (gauge). On all lines, attenuation increases with frequency. The rate at which attenuation increases with frequency is a function of line length and wire gauge. Signals on long lines are attenuated very rapidly with increasing frequency, whereas short lines cause a more gentle increase in attenuation with frequency. Likewise, a given length of large-diameter wire, such as 19 AWG, attenuates signals less rapidly with frequency than that same length of a smaller-diameter wire, such as 26 AWG. The dependence of attenuation on loop gauge is one reason why a maximum reach value for a particular DSL flavor cannot be stated without caveats. The maximum reach on 24 AWG is always longer than the maximum reach on 26 AWG, assuming the loop noise is the same.

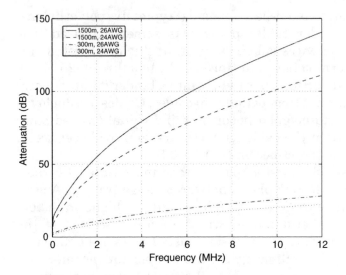

Figure 3.9   Attenuation as a function of frequency of two 26-AWG (0.4 mm) lines and two 24-AWG (0.5 mm) lines.

Figure 3.9 illustrates attenuation as a function of frequency of four lines: 300 and 1500 m of 26-AWG (or 0.4-mm) line, and 300 and 1500 m of 24-AWG (or 0.5-mm) lines. The attenuation curves are smooth because the lines are assumed to be terminated in the appropriate characteristic impedance at both ends. A comparison of the four curves shows clearly the relationships among attenuation, line length and wire gauge: longer loops are more severely attenuated than shorter loops, and "fatter" lines cause less severe attenuation than "skinnier" lines.

Of interest to modem designers is the maximum frequency that can support data transmission, which guides the selection of the system bandwidth. Typically, systems spanning wide bandwidths are more expensive than systems that span smaller bandwidths. If the selected bandwidth is too high, the system cost might be prohibitive, or perhaps components that run at the desired sampling rate with the desired accuracy will not be available. Conversely, if the selected bandwidth is too low, the bit rates on short loops on which a higher bandwidth would have been appropriate will be constrained to artificially low levels. Thus, engineers are faced with a classic design trade-off.

Attenuation curves, such as those in Figure 3.9, indicate how a transmitted signal is attenuated by the channel by the time it reaches the receiver. However, whether a particular frequency can support meaningful data transmission depends not only on the channel attenuation, but also on the level of the transmitted signal and the noise appearing at the receiver at that frequency. In other words, the maximum useful frequency is a function of the transmitter power spectral density (PSD, which is the distribution of transmitted power in the bandwidth) and the channel attenuation and noise. In addition, the maximum useful frequency depends on the target symbol error probability — that is, the probability that a received data symbol is in error. Furthermore, designers often impose a noise margin, which provides a certain number of decibels (dB) of "headroom" to accommodate changes in the channel noise as the system operates. The noise margin ensures that the system can continue to operate at the same bit rate and accommodate increases in noise up to the noise margin before the target symbol error probability is exceeded. If a transmitted PSD, channel noise, and noise margin are the same on all loop lengths, the maximum useful frequency decreases with increasing loop length.

Figure 3.10 plots the maximum useful frequency as a function of loop length for 24- and 26-AWG lines. In generating this figure, the maximum useful frequency was assumed to be the highest frequency at which at least 1 bit of information can be supported at a symbol error probability of 10^{-7}. A uniform transmitter PSD of –60 dBm/Hz was assumed, which is the approximate level used in VDSL, and a noise margin of 6 dB was imposed. The noise was assumed to be Gaussian with a uniform PSD of –140 dBm/Hz. The assumption of additive white Gaussian noise (AWGN) at this level is extremely optimistic with respect to real loops, but it establishes an upper bound on the maximum useful frequency for the selected transmitter PSD (which is low for the longer-reach services such as ADSL).

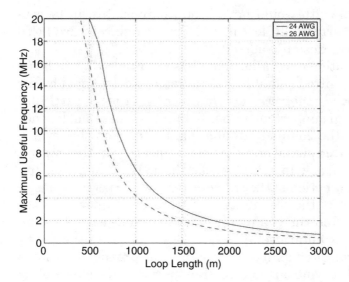

Figure 3.10 Maximum useful frequency as a function of loop length for 24-AWG and 26-AWG lines.

Note that the maximum useful frequency is a decreasing function of loop length, as expected. Of particular interest, however, is the slope of the curves at short loop lengths. The maximum useful frequency decreases rapidly with small increases in loop length. This behavior presents a difficult challenge to designers: if a single system is supposed to operate on a wide range of loop lengths — for example, all loops up to 1 mile (about 1.6 km) in length — what bandwidth should be used? Figure 3.10 indicates that the maximum useful frequency ranges from more than 20 MHz on very short loops to less than 3 MHz on loops that are 1 mile long. If the system is designed to use frequencies up to 20 MHz, much of the available bandwidth will not be useful on many target loops, and the modems might be very expensive. If the system is designed to use frequencies only up to 10 MHz, for example, then the achievable data rates on loops shorter than about 800 m will be restricted by the bandwidth choice.

This dilemma was one of many addressed by the standards bodies specifying VDSL. Eventually, it was agreed that

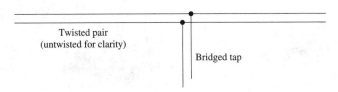

Twisted pair
(untwisted for clarity)

Bridged tap

Figure 3.11 Illustration of a bridged tap.

the maximum bandwidth in VDSL1 would be 12 MHz, which represents a good trade-off between performance and system cost. With a system bandwidth of 12 MHz and a practical noise scenario (which would be characterized by levels significantly higher than −140 dBm/Hz), rates on only the shortest loops (approximately ≤200 m) are restricted by the choice of bandwidth. In real deployments, most loops tend to be long enough that 12-MHz bandwidth is sufficiently large not to restrict data rates. Nevertheless, VDSL2 will allow the use of frequencies above 12 MHz to allow bit rates on the shortest loops to be maximized.

3.5.1 Bridged Taps

Many twisted-pair lines do not exhibit the smooth attenuation of the curves shown in Figure 3.9. For example, bridged tap configurations are common in certain regions of the world, including the U.S., where up to 80% of loops have bridged taps.[15] In a bridged tap configuration (shown in Figure 3.11), an unused twisted-pair line is connected in shunt to a main cable pair. The unused pair is left open-circuited across the main pair.

Bridged taps exist for a variety of reasons. Party lines were common in the early days of telephony, when two or more customers shared a single line in order to share costs. Later, when dedicated lines became more affordable, drops to all but one customer on a party line were disconnected simply by physically cutting the unnecessary lines, leaving bridged taps. Today, loops with bridged taps are the result of cabling installation rules that provide future flexibility in the plant. Cables are deployed to a neighborhood or area, and lines are

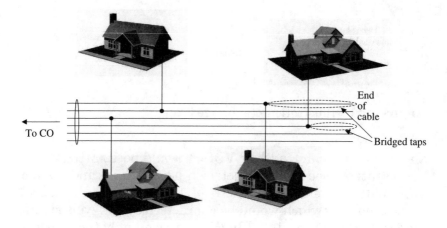

Figure 3.12   Bridged taps are the result of cable deployment rules that provide plant flexibility.

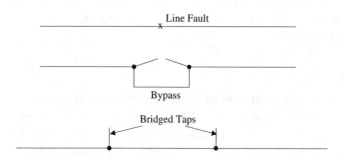

Figure 3.13   Bridged taps caused by line repairs.

tapped as necessary to serve subscribers. This procedure results in unterminated stubs that extend beyond subscriber homes, as shown in Figure 3.12. Repairs to lines can also result in bridged taps. Bingham[16] describes the scenario in which a line fault is simply bypassed with a new segment of twisted pair, leaving two bridged taps, as shown in Figure 3.13. Finally, bridged taps occur inside subscribers' homes. In-home wiring configurations often result in unterminated stubs due to unused telephone jacks.

On lines with bridged taps, the unterminated stub of twisted-pair line reflects signals that, at the receiver, add constructively in some frequency regions and destructively in others. When destructive interference occurs, transmitted signals are attenuated significantly. In the frequency domain, destructive interference appears in the channel's frequency response as notches. In the time domain, bridged taps distort signals, smearing them in time and changing their amplitudes.

The impact on data transmission of a bridged tap depends primarily on its length. The depths and locations of notches in the frequency response of the channel depend on the bridged tap length. Very short bridged taps result in pronounced but fairly shallow notches in the channel frequency response. As the bridged tap length increases, deeper notches appear. Eventually, the bridged tap is long enough so that the notches become less shallow because the reflected signal is attenuated significantly by the bridged tap before it is added to the desired signal at the receiver. Although long bridged taps cause less severe notching than their shorter counterparts, the frequency response of a loop with a long bridged tap still betrays the presence of the tap. At lower frequencies, the bridged tap causes rippling, and at higher frequencies the frequency response of the channel droops by approximately 3 dB relative to what it would have been in the absence of any bridged taps.

Figure 3.14 illustrates four loops, two of which have bridged taps. Loop 1A is a mixed-gauge loop with 400 m of 0.5-mm line followed by 180 m of 0.4-mm line. Loop 1B is the same as Loop 1A, except a 30-m, 0.4-mm bridged tap has been added 30 m from the end of the line. Loop 2A is a mixed-gauge loop with 900 m of 0.4-mm line followed by 100 m of 0.5-mm line. In Loop 2B, two 0.5-mm bridged taps, 50 m in length, have been added 50 m and 100 m from one end of the line. Figure 3.15 shows the insertion gain* transfer function

---

* The insertion gain is similar to the transfer function, except that the insertion gain depends on the source and load impedances. Section 3.6.1.2 discusses the difference in detail.

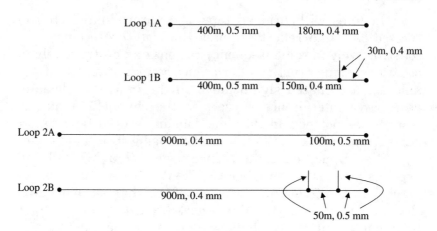

Figure 3.14 Examples of loops without and with bridged taps.

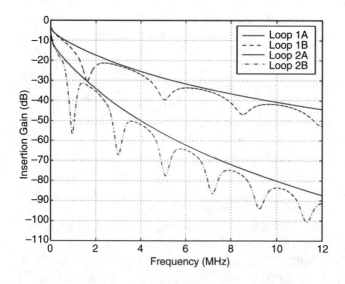

Figure 3.15 Impact of bridged taps on insertion gains of four loops shown in Figure 3.14.

of each of the four loops. Note the difference in the number of notches appearing for the two bridged-tap lines.

Figure 3.16 shows a uniform-gauge loop, Loop 3A, and the same loop with a 200-m, 0.4-mm bridged tap 200 m from

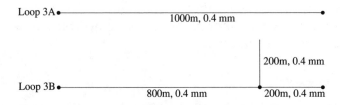

Figure 3.16 Example of a loop without and with a long bridged tap.

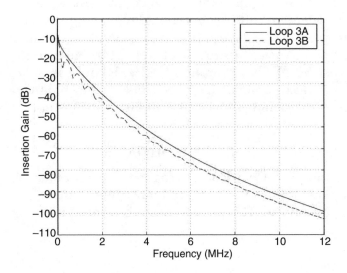

Figure 3.17 Impact of a long bridged tap on the frequency response of a channel.

the end of the line (Loop 3B). Figure 3.17 illustrates the insertion losses of the two lines. Although some ripple appears at low frequencies in the plot of the insertion gain of the loop with the bridged tap, the insertion gain is relatively smooth at higher frequencies. This behavior indicates that high-frequency signals reflected by the bridged tap are attenuated to an almost negligible power level. Therefore, they do not appreciably increase or decrease signal levels on the line, which means the signal arriving at the receiver is approximately half the power it would have been on a line without the bridged tap. As Figure 3.17 shows, the expected overall effect is an average power loss of about 3 dB.

3.5.2   Cable Balance

Ideally, the current in the two directions on a twisted-pair loop is the same, in which case the loop is said to be perfectly balanced. In practice, however, some current can leak into the longitudinal path, resulting in a common-mode component of current. The balance of a line is the ratio of differential mode to common mode current. It is proportional to how tightly the two wires are twisted and generally decreases with increasing frequency. At frequencies up to about 100 kHz, balance is usually at least 50 dB; however, at 10 MHz, balance is only about 35 dB.[16] Section 3.6.2 describes how cable balance relates to radio-frequency interference in last mile systems.

3.6   COMMON LOCAL LOOP IMPAIRMENTS

Like any channel, the local loop is plagued by a number of impairments, including crosstalk, radio-frequency interference, and impulse noise. This section describes these impairments.

3.6.1   Crosstalk

Section 3.5 explained that individual wires composing twisted-pair lines are insulated, and the twisting of these lines into cables limits electromagnetic interference to nearby lines. However, because the shielding between lines is not perfect, signals from one line can and do couple into other lines. As a result, a receiver can detect signals transmitted on other lines, thus increasing the noise power and degrading the received signal quality on that line. The coupling of unwanted signals from one or more lines into a victim line is known as crosstalk and can take two forms in telephone networks: near end and far end. The management of crosstalk is fundamental to ensuring good DSL performance, as Section 3.9 discusses.

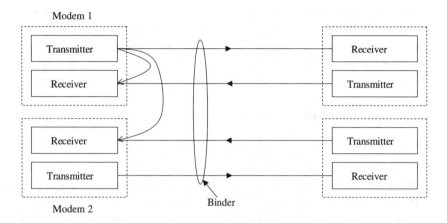

Figure 3.18 Illustration of near-end crosstalk (NEXT).

Crosstalk can occur due to systems that are the same type as the victim system, or of a different type. Use of the term "self-crosstalk" has become common to describe crosstalk due to systems of the same type. However, "self-crosstalk" is somewhat of a misnomer because the victim system is not causing the crosstalk. In Bingham,[16] the terms "kindred" and "alien" are used, respectively, to describe systems of the same and different types. This terminology is adopted here.

3.6.1.1 Near-End Crosstalk (NEXT)

Near-end crosstalk (NEXT) occurs when a local receiver detects signals transmitted on other lines by one or more local transmitters, as shown in Figure 3.18. Signals coupling into the line are transmitted in the direction opposite to the received signal but in an overlapping frequency band. The impact of NEXT depends on the frequency bands used by the receiver.

Kindred NEXT occurs in echo-canceled systems and, to a lesser degree, in frequency-division duplexed (FDD) systems. Echo-canceled systems overlap their transmit and receive bands. On any single line (for example, in the case of Modem 1 in Figure 3.18), because the transmitter and receiver are part of the same modem, transmitted signals can be subtracted from received signals to eliminate interference caused by the band overlap. The component that performs this function is called an echo-canceller. However, transmissions on other lines cannot be canceled without coordination between lines (which does not generally exist in practice today).

Transmissions on other lines therefore couple into victim lines as NEXT, as illustrated by the arrow from the transmitter of Modem 1 to the receiver of Modem 2 in Figure 3.18. In frequency-division duplexed systems, which do not use overlapping transmit and receive bands, kindred NEXT occurs if energy transmitted in a band adjacent to the receive band enters the receive band. In FDD systems employing filters to separate transmit and receive bands, the levels of kindred NEXT are typically far lower than kindred NEXT levels in echo-canceled systems. In FDD systems that do not use filters, kindred NEXT levels at frequencies near band edges can be significant, particularly if the guard band is narrow. However, these levels still are not as high as kindred NEXT levels of echo-canceled systems because the peak energy of the interfering band is, in frequency, far away from the affected band.

Alien NEXT occurs when systems of different types use coincident frequency bands in opposite directions. For example, T1 transmission occurs in both directions in the band up to 1.544 MHz. Two twisted-pair lines are used, one in each direction. Consequently, for any DSL system operating in a frequency band lower than 1.544 MHz, T1 lines are a potential source of (severe) alien NEXT.

The level of NEXT detected at a receiver depends on the line characteristics (such as balance); the number of interferers and their proximity to the line of interest; the relative powers and spectral shapes of the interfering signals; and the frequency band over which NEXT occurs. The power spectral density of NEXT from one particular loop to another is not a smooth function of frequency. However, more than one line typically contributes to the NEXT appearing at a receiver. Consequently, NEXT is characterized in terms of sums of pair-to-pair NEXT coupling powers from other lines.

For the purposes of analysis and simulation, a statistical model of NEXT is used in the industry. This model represents the expected 1% worst-case power sum crosstalk loss as a function of frequency, which means that 99% of pairs tested will be subject to a power sum crosstalk loss no higher than that predicted by the model at a given frequency. Under the assumption that the model is accurate, because the model is a smooth function of frequency measured crosstalk levels of 99% of twisted pairs will lie below the model.[17] Thus, the model provides an upper bound on crosstalk levels for 99% of twisted pairs.

The NEXT coupling model used in the industry for the evaluation of DSL systems is

$$\text{NEXT}(f) = K_{\text{NEXT}} \cdot n^{0.6} \cdot f^{3/2} , \qquad (3.1)$$

where K_{NEXT} is the NEXT coupling constant, n is the number of identical disturbing lines in the binder, and f represents frequency. The value of K_{NEXT} is 8.536×10^{-15}, which was derived from measurements of 22-AWG cables. It has been found to be accurate for other cable types,[17] although cables in some parts of the world are characterized by a larger coupling constant (and higher NEXT). Because K_{NEXT} and n are constants, Equation 3.1 indicates that NEXT coupling is a monotonically increasing function of f. Thus, signals residing higher in frequency are subject to higher NEXT coupling than are lower-frequency signals.

Figure 3.19 shows NEXT coupling due to ten disturbers over the frequency range of 0 to 12 MHz, using the model given by Equation 3.1. The dependence of coupling on frequency is evident in the figure. Note that NEXT coupling is significant for frequencies above a few hundred kilohertz.

NEXT coupling characterizes how lines in a binder are affected by disturbing lines. The level of NEXT detected by a receiver depends also on the PSDs of signals transmitted on the disturbing lines. Using the model of Equation 3.1, the PSD of NEXT at a receiver, due to n disturbers of a given type, is given by

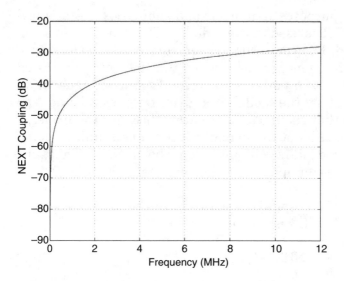

Figure 3.19 NEXT coupling due to ten disturbers, illustrating dependence of coupling on frequency.

$$S_{\text{NEXT}}(f) = \text{NEXT}(f) \cdot S_d(f) = K_{\text{NEXT}} \cdot n^{0.6} \cdot f^{3/2} \cdot S_d(f) \text{ , (3.2)}$$

where $S_d(f)$ is the PSD of signals transmitted by all modems on the disturbing lines. From Equation 3.2, it is clear that NEXT levels increase if the transmit power on the interfering line(s) is increased. For example, if the transmit power of all interferers is doubled, so is the power of NEXT appearing at other receivers.

3.6.1.2 Far-End Crosstalk (FEXT)

Far-end crosstalk (FEXT) occurs when a local receiver detects signals transmitted in its frequency band by one or more remote transmitters. In this case, interfering signals travel in the same direction as the received signal, as illustrated by Figure 3.20. (Note that the positions of the transmitters and receivers have been switched relative to Figure 3.18.) As is the case with NEXT, kindred and alien FEXT can occur.

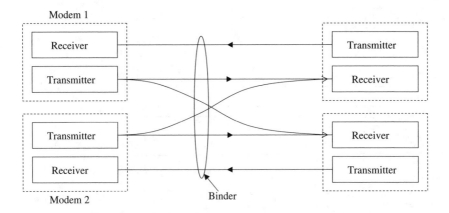

Figure 3.20   Illustration of far-end crosstalk (FEXT).

Like NEXT, the PSD of FEXT from one line to another is not a smooth function, and a worst-case model has been defined. The model for FEXT coupling is

$$\mathrm{FEXT}(f) = K_{\mathrm{FEXT}} \cdot n^{0.6} \cdot f^2 \cdot L \cdot \left|H(f)\right|^2, \qquad (3.3)$$

where, as before, n is the number of identical disturbing lines in the binder, and f represents frequency; L is the length of line over which coupling occurs; and K_{FEXT} is the FEXT coupling constant. When L is measured in feet, the value of K_{FEXT} is 7.744×10^{-21}.[17] As in the case of the NEXT coupling constant, this value of K_{FEXT} has been shown to be accurate for many cables; however, exceptions do occur. $\left|H(f)\right|$ is the magnitude of the "insertion gain transfer function" of the length of line over which disturbing signals travel prior to reaching the disturbed receiver.

It is important to realize that $\left|H(f)\right|$ is *not* simply the magnitude of the classical transfer function of the loop. The insertion gain transfer function depends on the source and load impedances. Figure 3.21 is useful to explain how the insertion gain transfer function is computed. The upper diagram shows a voltage source with matched source and load impedances, which are assumed to be real and equal to R_N.

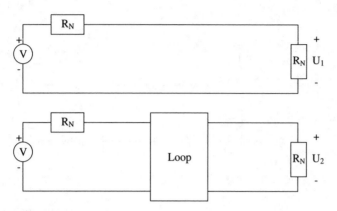

Figure 3.21 Insertion gain transfer function is the ratio of load voltages with and without the loop in the circuit.

The voltage U_1 appears across the load resistance. The lower diagram shows the same network, but with the loop inserted between the source and load resistors. The voltage U_2 now appears across the load resistance. The insertion gain transfer function is defined as U_2/U_1. (Conversely, the insertion loss is defined as U_1/U_2.) The terms "transfer function" and (even worse, although the author must admit to being as guilty of misuse as anyone) "insertion loss" are often used interchangeably. In many cases, what is meant is "insertion gain transfer function."

Figure 3.22 shows three loop configurations to illustrate how to determine the proper values of L and $|H(f)|$ for FEXT calculations. For simplicity, only the transmitters and their corresponding receivers on two lines are shown. The case labeled (a) is trivial: both lines are the same length and the transmitters and receivers are co-located. Thus, to determine FEXT coupling for both lines, the value of L is just l, and $|H(f)|$ is the insertion gain transfer function of a loop of length l.

The case labeled (b) shows a slightly more complicated scenario. In this case, the transmitters are co-located, but the loops are different lengths. Regardless of which loop is considered, the length of line over which signals couple is l_1, meaning that $L = l_1$. For Loop 1, disturbing signals are attenuated by a

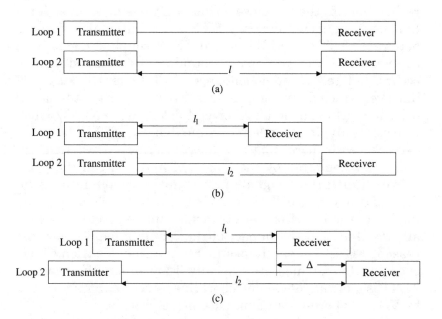

Figure 3.22 Loop configurations to illustrate the determination of L and $|H(f)|$ for FEXT coupling model.

line of length l_1, so the appropriate $|H(f)|$ is the insertion gain transfer function of a loop of length l_1. For Loop 2, the appropriate $|H(f)|$ is that of a loop of length l_2 because the interfering signal is attenuated between its transmitter and the victim receiver. In the case labeled (c), neither the transmitters nor the receivers are co-located, but the coupling length L remains l_1. The insertion gain transfer functions are now a bit more complicated than before. For Loop 2, the appropriate insertion gain transfer function is of a line of length $l_1 + \Delta$, which is the length over which interfering signals are attenuated before reaching the receiver on Loop 2. For the FEXT appearing at the receiver on Loop 1, interfering signals are attenuated by a loop of length $l_2 - \Delta$, so this length should be used to compute $|H(f)|$.

Comparing Equation 3.1 and Equation 3.3, note that the frequency-dependence of FEXT coupling is to the power of 2

rather than to the 3/2 power. Thus, FEXT is a stronger function of frequency than is NEXT. An additional difference
between NEXT and FEXT is that FEXT coupling depends on
the length of line over which coupling of unwanted signals
occurs. Related is the dependence of FEXT on the length of
line over which disturbing signals travel (and are attenuated)
prior to reaching a victim receiver. In contrast to NEXT, which
is essentially independent of line length, FEXT tends to
decrease with increasing line length because unwanted signals are attenuated by the length of line between the point
at which disturbing signals first begin to couple into victim
lines and that at which they enter the receiver. Therefore, the
impact of the insertion gain transfer function factor is generally greater than the impact of the coupling length. For this
reason, FEXT is usually a minor impairment on long lines
such as ADSL, which are typically 3 km in length or longer.
FEXT is more significant on short lines, such as those targeted
by VDSL, where it can dominate noise profiles.

Figure 3.23 shows FEXT coupling due to ten disturbers
as a function of frequency for three lengths of 26-AWG lines.
The solid curve is FEXT coupling, assuming that the length
over which coupling occurs is 300 meters, and the dashed and
dash-dot curves show coupling for loops of length 1 and 1.5
km, respectively. For simplicity, all disturbing and victim lines
are assumed to be the same length, and all transmitters and
receivers are co-located, which corresponds to case (a) in Figure 3.22. The three curves show clearly that FEXT coupling
decreases with increasing line length. They also illustrate the
dependence of FEXT on line attenuation, which, as discussed
previously, increases with loop length and frequency. On
longer loops, the line attenuation is severe enough to coun-
teract the f^2 contribution to the FEXT coupling expression,
resulting in coupling curves that, beyond some frequency,
decrease rapidly with increasing frequency.

As was the case with NEXT, the level of FEXT appearing
at a receiver depends on the PSD of signals transmitted on
disturbing lines. The PSD of FEXT due to n disturbers of a
given type is the coupling given by Equation 3.3 multiplied
by the PSD of signals transmitted on disturbing lines:

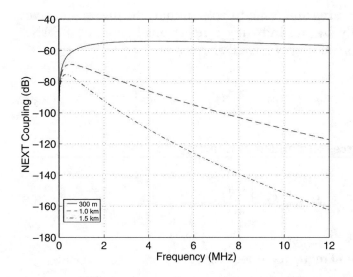

Figure 3.23 FEXT coupling due to ten disturbers, illustrating the dependence of coupling on frequency and line length (26 AWG).

$$S_{\text{FEXT}}(f) = \text{FEXT}(f) \cdot S_d(f)$$
$$= K_{\text{FEXT}} \cdot n^{0.6} \cdot f^2 \cdot L \cdot \left| H(f) \right|^2 \cdot S_d(f), \qquad (3.4)$$

where $S_d(f)$ is the PSD of signals transmitted on the disturbing lines.

3.6.1.2.1 Adding Crosstalk from Different Sources

The NEXT and FEXT models used in the industry are based on worst-case coupling between lines. These models are accurate as long as all disturbers are the same type; that is, they are the same length, they use the same transmit PSD, etc. If two or more different disturber types are present in a binder, clearly not all disturbers can be in worst-case positions relative to a disturbed line. Thus, because the models represent worst-case conditions, direct addition of different crosstalk PSDs results in an overly pessimistic noise PSD.

For this reason, a modified summation rule was derived heuristically by the industry consortium known as FSAN.[18] The rule for adding N FEXT sources is

$$S_{\text{FEXT}}(f) = \left[\sum_{i=1}^{N} \left(S_{\text{FEXT},i}(f) \right)^{\frac{1}{0.6}} \right]^{0.6} . \qquad (3.5)$$

NEXT sources are added similarly:

$$S_{\text{NEXT}}(f) = \left[\sum_{i=1}^{N} \left(S_{\text{NEXT},i}(f) \right)^{\frac{1}{0.6}} \right]^{0.6} . \qquad (3.6)$$

3.6.2   Radio-Frequency Interference

Radio-frequency interference into and from telephone lines is a concern for last mile transmission. Ingress results when over-the-air signals in overlapping frequency bands couple into phone lines. Egress is the opposite process: leakage of signals on the twisted pair into over-the-air antennae. Ingress and egress are caused by imbalance in the twisted pair. Ingress results when unwanted signals couple unequally into the two wires, and egress when the two wires radiate unequally to a receiving antenna.

Section 3.5 explained that loop balance degrades with increasing frequency, and systems transmitting on shorter lines can use a higher maximum frequency than systems transmitting on longer lines. Because VDSL systems transmit at higher frequencies due to the shorter loop lengths, radio-frequency interference becomes more of a problem than in systems that confine signals to lower frequency bands, such as ADSL, HDSL, and SHDSL. Telephone lines near subscriber premises (specifically, overhead distribution cable and wires within the home) are particularly susceptible to ingress from over-the-air radio-frequency signals, including AM radio and amateur radio signals. Likewise, transmitters must be designed to ensure that they do not radiate excessively into vulnerable over-the-air bands. Of particular concern are the

Table 3.2   Amateur Radio Frequency Bands

Start frequency (MHz)	End frequency (MHz)
1.8	2.0
3.5	4.0
7.0	7.3
10.1	10.15
14.0	14.35
18.068	18.168
21.0	21.45
24.89	24.99
28.0	29.7

amateur radio bands, although other bands, such as maritime emergency bands, must also be protected.

3.6.2.1   Ingress

AM radio ingress can be problematic for DSL receivers due to the high power levels used by and density of AM radio stations in the over-the-air spectrum. AM interferers appear in the frequency domain as high-level, 10-kHz-wide noise spikes in the band between 525 kHz and 1.61 MHz. When present, AM interferers tend to remain at the same level for the duration of a DSL connection.

Amateur radio (HAM) signals can be an even larger problem for VDSL transceivers than AM radio because HAM signals are intermittent and may even change carrier frequency. Furthermore, HAM signals can be transmitted at high levels of up to 1.5 kW, although in most situations their power levels are more typically 400 W or less.[15] Generally, HAM signals are less than 4 kHz wide.[16]

The amateur radio bands recognized internationally are given in Table 3.2. In VDSL1 systems, ingress from at least the lowest four amateur bands must be expected. In ADSL2plus systems, ingress from the lowest amateur band can occur. Depending on the maximum allowed frequency, VDSL2 systems may be affected by all of the bands shown in Table 3.2.

3.6.2.2 Egress

To control egress, last mile transceivers must be designed so that the levels of signals in potential victim frequency bands — particularly the amateur radio bands — are low enough not to interfere with operation of the victim system. Studies by British Telecom (now BT) suggest that signals of last mile systems within the frequency bands used for amateur radio operation must be at a level of −80 dBm/Hz or lower to ensure that transmissions on twisted-pair lines are not audible by amateur radio operators.[19] Because DSL transmit PSD levels are generally at levels significantly higher than −80 dBm/Hz, modems must provide notches in the transmitted spectrum to ensure the transmitted PSD does not exceed −80 dBm/Hz in the amateur radio bands.

3.6.3 Impulse Noise

Impulse noise is a temporary, high-power burst of energy that can overwhelm information-bearing signals. As is the case with ingress, cable imbalance is the mechanism that allows impulse noise to enter the local loop. Thus, this noise is actually a type of radio-frequency interference. However, unlike ingress and despite its name, impulse noise tends to be highest at low frequencies.

Impulse noise is caused by electromagnetic events in the vicinity of telephone lines. Although the sources of ingress are not fully identified or understood, some examples are cited in Starr et al.,[15] including control voltages to elevators in apartment buildings, opening of refrigerator doors, and ringing of phones on lines in the same cable binder. Power line discharges and lightning can also cause impulse noise. Typically, these noise events last 10 to 100 μs, although events lasting as long as 3 ms have been recorded. The differential voltages caused by impulse noise can be as high as 100 mV, although levels below 5 mV are more typical.[15] Impulse noise can overwhelm received signal levels.

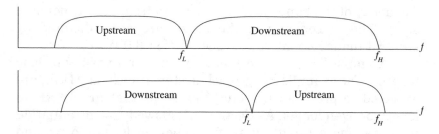

Figure 3.24 Possible placements of the downstream and upstream channels in two-band FDD.

3.7 DSL TRANSMISSION

This section considers transmission alternatives for DSL, including how the available spectrum is allocated to the downstream and upstream directions, and line code options.

3.7.1 Duplexing Alternatives

Duplexing defines how a system uses the available bandwidth to support bidirectional data transmission. Three duplexing approaches are used in DSL transmission: frequency-division duplexing, echo cancellation, and time-division duplexing.

3.7.1.1 Frequency-Division Duplexing (FDD)

Frequency-division duplexed systems define two or more frequency bands, at least one each for upstream and downstream transmission. These bands are disjoint in frequency — thus the name "frequency-division duplexing." Critical to the performance of FDD systems are the bandwidths and placement in frequency of the upstream and downstream bands.

Figure 3.24 illustrates the simplest FDD case, which provides a single upstream band and a single downstream band. As the figure illustrates, the upstream band may reside above or below the downstream band. The lower band edge frequency of the band residing higher in frequency is denoted as f_L. The placements of the bands are often dictated by spectral management rules. (See Section 3.9.)

A viable DSL system must operate on a variety of loop lengths. For example, long-reach systems, such as ADSL, are expected to provide connections on loops that are 18 kft in length or even longer. As Section 3.5 showed, line attenuation increases more rapidly with frequency on longer loops, and the maximum useful frequency decreases as the loop length increases. To allow successful transmission, the upstream and downstream channels must be located below the maximum useful frequency. Referring again to Figure 3.24, if the maximum useful frequency falls below f_L, then the upstream or the downstream channel fails, and bidirectional transmission is not possible.

Another primary consideration when designing an FDD system is the bandwidths of the upstream and downstream channels. The appropriate choices for the bandwidths depend on the desired data rates and downstream-to-upstream data rate ratio. The appropriate bandwidth allocation for asymmetric 8:1 data transport differs substantially from that appropriate to support symmetric data. Choosing the downstream and upstream channel bandwidths is complicated further by the variability of line length and, as a result, the variability of the channel signal-to-noise ratio and useful frequency band. Depending on the number of bands into which the spectrum is divided, the appropriate bandwidth allocation for 8:1 service on a 300-m line may be very different from the appropriate allocation for 8:1 service on a 1.5-km line.

3.7.1.1.1 "Optimal" Frequency Plans

Selecting a viable frequency plan to support a specific data rate combination (for example, 26 Mbps downstream and 3 Mbps upstream) on a loop of specific length and gauge is straightforward. The allowed transmitter PSD, total allowed transmit power, and approximate attenuation of the desired loop length as a function of frequency are known. An expected worst-case (or average) noise scenario can be assumed. In the absence of additional spectral constraints, the useful frequency band (which can be determined by computing the signal-to-noise ratio as a function of frequency and determining the maximum frequency at which transmission at the lowest spectral efficiency can be supported) simply must be partitioned into two bands, one upstream and one downstream, so that the desired bit rates are supported.*

Under these conditions, such a frequency plan is close to optimal for that specific data rate combination, target reach, and loop gauge. However, a two-band frequency plan designed in this manner is suboptimal for any other loop length, even if the goal is to support data rates in the same ratio as the one for which it was designed. For example, if the frequency plan was optimized to support 26 Mbps downstream and 3 Mbps upstream on loops of length L, it is not optimal to support 13 Mbps downstream and 1.5 Mbps upstream on loops longer than L, nor is it optimal to support 52 Mbps downstream and 6 Mbps upstream on loops shorter than L. The problem, of course, is that the frequency range capable of supporting data transmission decreases with increasing loop length.

Figure 3.25 illustrates this effect for an arbitrary frequency plan that allocates the region from 0 to 4 MHz to the downstream direction and 4 to 20 MHz to the upstream direction. Note that as the loop length increases from 300 m to 1.0 km, the data-carrying capability of the upstream channel decreases dramatically, but the downstream channel's data-carrying capability hardly decreases at all. On a 300-m loop, the upstream bit rate far exceeds that of the downstream channel, but on a 1.3-km loop, the upstream channel rate is zero.

Frequency plans with only two bands are suboptimal for nearly all loop lengths and bit rate combinations because the data-carrying capacity of the band residing in the higher

---

* Clearly, the desired bit rate and reach combination must be achievable. If the capacity of the loop is insufficient to support the sum of the desired downstream and upstream bit rates at the desired reach, then no frequency plan will solve the problem.

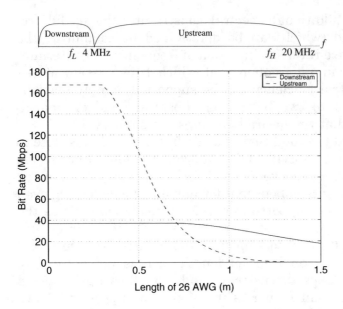

Figure 3.25 Inconsistent downstream-to-upstream bit rate ratio resulting from an example two-band frequency plan.

frequency range diminishes rapidly as the loop length increases. In contrast, the degradation to the band located lower in frequency is much less severe. As a result of these two effects, the bit rates become lopsided relative to the desired data rate ratio on all but a very narrow range of loop lengths, as illustrated in Figure 3.25. A frequency plan that provides proportional increases or decreases in the downstream and upstream bit rates as the loop length changes would be more attractive from a service provisioning standpoint.

One possible definition of the "optimal" frequency plan is the band allocation that provides a specific data rate ratio (for example, symmetrical or 2:1 asymmetrical transmission) on all loop lengths. With this definition, the optimal plan to support symmetrical services is shown in Figure 3.26. The available frequency band is partitioned into M subbands, each of which is Δf Hz wide. Under the assumption that the transmitter PSDs and noise PSDs in the downstream and upstream

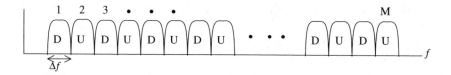

Figure 3.26 Optimal frequency plan for symmetrical services.

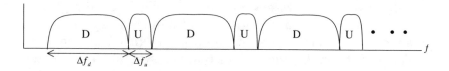

Figure 3.27 Optimal frequency plan for asymmetrical services.

directions are the same, symmetric transmission is supported by assigning odd subbands to the downstream direction and even subbands to the upstream direction (or vice versa).

Ideally, the subbands are infinitely narrow so that the capacity of each is equal to the capacity of its nearest neighbor to the right. In reality, of course, the subband widths will be finite, and the capacities of adjacent subbands will not likely be equal, but they will be close. If the capacities of adjacent subbands are nearly equal, then the downstream and upstream data rates are also nearly equal, irrespective of the loop length. On any loop, half (or perhaps one more than half) of the useful subbands are always assigned to each direction. Even on long loops, when the total available bandwidth is small, symmetrical transmission is still possible because subbands are available in both directions.

The optimal frequency plan for asymmetrical services is shown in Figure 3.27. In this case, the widths of the downstream and upstream subbands differ. For example, if the desired downstream-to-upstream ratio is 2:1, then Δf_d is approximately two times Δf_u (again assuming the downstream and upstream transmit PSDs and noise PSDs are approximately equal). As with the symmetrical allocation, the performance of this allocation (in terms of supporting a particular data rate ratio on all loop lengths) is best for small Δf_u so that the capacity of each downstream band is as close as possible to the desired multiple of the capacity of the upstream channel to its right.

In order to implement the optimal frequency plan for a particular data rate ratio, a system must be capable of transmitting and receiving in the defined narrow subbands. However, because the bands adjacent to the transmit bands are receive bands, the system must not transmit significant out-of-band energy because any out-of-band energy from the transmitter could appear as high-level NEXT at the local receiver. Section 3.8.2 describes how the system standardized for VDSL meets these requirements and can operate using the optimal frequency plan for any arbitrary data rate ratio.

3.7.1.2   Echo Cancellation (EC)

Some systems use a coincident frequency band to support transmission in both directions. When the transmissions occur at the same time, an echo canceller is used to "subtract" the transmitted signal (which is known) from the received signal. However, if other systems (kindred or alien) in a binder transmit in a modem's receive band, NEXT results. NEXT from other lines cannot be canceled without coordination between lines. As Section 3.9.1.1 describes, coordination between lines and the potential to cancel crosstalk are topics attracting the attention of the DSL industry.

Because NEXT is an increasing function of frequency, echo cancellation is used in DSL only below a few hundred kilohertz, where NEXT levels are low enough to allow meaningful transmission to occur. Some systems, such as SHDSL, use fully overlapped downstream and upstream channels with equal downstream and upstream transmitter PSDs. The advantage of this approach is that the bit rates in the two directions are equal, assuming that the noise PSD is the same in both directions, which is often the case in SHDSL because NEXT from other SHDSL systems tends to be the primary component of the noise.

Supporting asymmetrical services with a system that uses EC generally requires one of the bands, usually the downstream, to extend beyond the upstream band. In this case, the asymmetry of the bit rates will depend strongly on loop length. In ADSL systems that use overlapped spectra, the downstream channel extends fully across the upstream channel, but the bandwidths of the two channels differ significantly. Although allowing the downstream band to start at a lower frequency increases the downstream bit rate, NEXT from other EC ADSL in the binder reduces the upstream bit rate. In deployment scenarios in which the downstream channel fails before the upstream channel does, the use of overlapped spectra can provide additional reach. However, when performance is upstream-limited, the use of FDD spectra is preferred to avoid NEXT in the upstream band.

One disadvantage of using overlapped spectra is the need for an echo canceller, which can increase system complexity. Because the complexity is directly proportional to the bandwidth of the system, this is another reason to restrict the use of overlapped spectra to a few hundred kilohertz.

3.7.1.3 Time-Division Duplexing (TDD)

In contrast to FDD solutions, which separate the upstream and downstream channels in frequency, and echo-cancelled systems, which overlap downstream and upstream transmissions, time-division duplexed (TDD) systems support upstream and downstream transmissions within a single frequency band but during different time periods. Figure 3.28

Figure 3.28 A single band supports downstream and upstream transmission in TDD systems.

illustrates the single frequency band used by TDD systems. Use of the time-shared channel bandwidth is coordinated by means of superframes. A superframe consists of a downstream transmission period, a guard time, an upstream transmission period, and another guard time. The durations of the downstream and upstream transmission periods are integer numbers of symbol periods. Superframes are denoted as A-Q-B-Q, where A and B are the number of symbol periods allocated for downstream and upstream transmission, respectively, and the Qs represent quiescent (guard) times that account for the channel's propagation delay and allow its echo response to decay between transmit and receive periods.

As an example, the duration of a superframe might be 20 symbol periods. The sum of A and B could be 18 symbol periods, which would mean the sum of the Qs would be 2 symbol periods. The values of A and B can be chosen by the operator to yield the desired downstream-to-upstream data rate ratio. For example, if the noise profiles in the upstream and downstream directions are assumed to be equivalent, setting A equal to B results in a configuration that supports symmetric transmission. Setting A = 16 and B = 2 yields an 8:1 downstream-to-upstream bit rate ratio. When A = 12 and B = 6, 2:1 transmission is supported. Figure 3.29 illustrates the superframes that support 8:1, 2:1, and symmetrical transmission.

The use of superframes enables TDD systems to compensate for differences in the downstream and upstream noise levels. If the noise in the upstream direction is more severe than in the downstream direction, a TDD system can allocate additional symbols to the upstream direction to compensate. For example, if symmetric transmission is required, an 8-Q-10-Q superframe can be used instead of the nominal 9-Q-9-Q superframe, resulting in increased range at a given data rate.

Use of TDD systems requires that modems on lines within a single binder group be synchronized to a common superframe clock so that all downstream transmissions occur simultaneously on all lines, and all upstream transmissions occur at approximately the same time on all lines. If a common

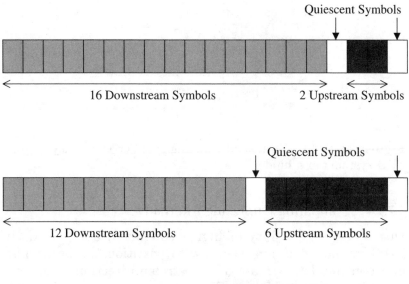

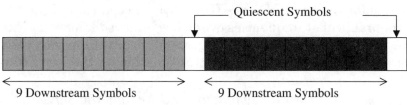

Figure 3.29 TDD superframes enable support of a variety of downstream-to-upstream bit rate ratios.

superframe structure is not used, lines supporting TDD in a binder group can cause NEXT to one another, significantly degrading the data rates they can support. The common clock can be provided by a number of methods; for example, it can be derived from the 8-kHz network clock, sourced by one of the TDD modems, or derived using global positioning satellite (GPS) technology. When the clock is sourced by one of the modems, it must be assumed that all other modems operating the binder have access to the clock signal. Thus, in this case, coordination between modems at the CO (or RT) is a requirement.

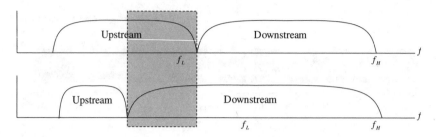

Figure 3.30 NEXT caused by mixing symmetric and asymmetric FDD systems in a binder.

3.7.1.4 Comparing Duplexing Alternatives

Due to the availability of high frequencies, a key issue in VDSL standardization was how to partition the bandwidth between the downstream and upstream directions. The only candidates were TDD and FDD because echo cancellation is not practical at frequencies over a few hundred kilohertz. This section examines the advantages and disadvantages of TDD and FDD for very high speed transmission on the local loop.

3.7.1.4.1 Mixing Symmetric and Asymmetric Services

The ideal VDSL system would be configurable to enable support of asymmetrical and symmetrical services. Given that the best FDD bandwidth allocations for symmetric and asymmetric services differ, as do the appropriate TDD superframe structures, the question of whether symmetric and asymmetric services can reside simultaneously in the same binder arises naturally.

Unfortunately, when the optimal time/frequency allocations are used, spectral incompatibilities result regardless of whether a system is TDD or FDD. To illustrate, Figure 3.30 shows example spectral allocations for an FDD system. The upper allocation supports symmetrical transmission (on some loop length) and the lower allocation supports more asymmetrical transmission. The shaded portion is the frequency band in which NEXT between lines occurs. As the figure illustrates, mixing symmetric and asymmetric FDD VDSL systems

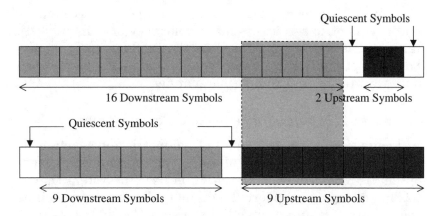

Figure 3.31 NEXT caused when symmetric and asymmetric TDD superframes are mixed in a binder.

causes NEXT in part of the frequency band, but all the time. To ensure spectral compatibility between symmetric and asymmetric services, use of a suboptimal, compromise spectral allocation is necessary. Unfortunately, no single spectral allocation for FDD supports symmetrical and asymmetrical services without a performance degradation to one or both types of service.

TDD systems suffer from a similar degradation when symmetric and asymmetric superframe structures are mixed in a binder. Figure 3.31 illustrates the case when a line supporting 8:1 transmission with a 16-Q-2-Q superframe resides in the same binder as a line supporting symmetrical transmission with a 9-Q-9-Q superframe. Note that the 9-Q-9-Q superframe has been shifted in time by one symbol period to minimize overlap between the downstream symbols on the 8:1 line and the upstream symbols on the symmetric line. However, five symbols on both lines are still corrupted by NEXT. Whereas NEXT in the FDD case spans only part of the bandwidth but all the time, NEXT with TDD spans the entire bandwidth but only part of the time. The severity of the NEXT in either case depends on how different the mixed ratios are and the frequency band(s) affected.

NEXT between lines optimally supporting symmetric and asymmetric services is not a deficiency of FDD or TDD. Rather, it is a problem caused by the fundamental impossibility of optimally supporting symmetrical and asymmetrical data rate ratios in a compatible manner.

3.7.1.4.2 Support of Required Data Rate Ratios

Although simultaneous support of symmetric and asymmetric data rate ratios is impractical unless a compromise (and suboptimal) time/frequency allocation is used, a system that is flexible enough to support the optimal allocations as well as the compromise allocation still has value. Consider, for example, DSL deployment in densely populated regions and large cities. In such places, the loops of business and residential customers may reside in the same binder. Most operators agree that business customers will require symmetrical service, whereas residential customers will need more asymmetrical service to support Internet access, video on demand, and the like.

The needs of the two customer types are conflicting. However, most business customers require service during working hours on weekdays and most residential customers require service in the evening and on weekends. Modems capable of supporting multiple frequency plans or superframe allocations would enable operators to offer symmetrical service during the day and asymmetrical service at night and on weekends. In this way, operators could provide the desired rates to business and residential customers whose lines happen to reside in the same binder. This capability is only useful in regions where network operators have full control over the services in a binder. Some countries have regulations that require local loop unbundling; many service providers may use the pairs in the cable, and the prospect for coordination between the service providers is small. In these environments, the time/frequency flexibility of a DSL implementation cannot easily be exploited.

The superframe structure used in TDD inherently enables support of symmetric and a wide range of asymmetric downstream-to-upstream bit rate ratios with a single transceiver. The desired bit rate ratio is determined for the most part by setting the software-programmable values of A and B to the appropriate values. Furthermore, if the downstream and upstream noise profiles differ substantially, the superframe structure can be modified to compensate for the difference.

To support both multiple frequency plans in a reasonable fashion, FDD modems must be capable of changing the bandwidths of the upstream and downstream channels. If a system uses analog filters to separate the downstream and upstream bands, these filters must be designed to support multiple frequency allocations; this increases the complexity, cost, and power consumption of the transceiver. Thus, a system that uses programmable analog filters is unlikely to be able to support a wide variety of frequency plans.

One example of an FDD implementation that provides flexibility in bandwidth allocations is the discrete multi-tone (DMT) system described in detail in Section 3.8.2. This system partitions the available bandwidth into a large number of subchannels and assigns each subchannel to the downstream or upstream direction. As a result, the bandwidths (and, indeed, the number) of the downstream and upstream channels can be set arbitrarily. In the extreme situation, odd subchannels can be used downstream, and even subchannels upstream (or vice versa), which maximizes the number of downstream and upstream bands. These systems eliminate the need for flexible analog filters and provide tremendous flexibility in FDD bandwidth allocations.

3.7.1.4.3 Complexity, Cost, and Power Consumption of TDD

Relative to FDD systems, TDD systems can provide reduced complexity in digital signal processing and in analog components. For TDD modems that use discrete multi-tone modulation, these complexity reductions result from sharing hardware common to the transmitter and receiver. Sharing of hardware is possible because the transmitter and receiver functions of a DMT modem are essentially equivalent: both require the computation of a discrete Fourier transform (DFT), which is usually accomplished using the fast Fourier transform (FFT) algorithm.

Because TDD is used, a modem can only transmit or receive at any particular time. As a result, hardware to compute only one FFT is required per modem. This FFT spans the entire system bandwidth and is active throughout the superframe except during the guard periods. Additional analog hardware savings are realized in TDD modems because the same band is used to transmit and receive. The path not in use can be turned off, which reduces power consumption. In contrast, modems using FDD always must provide power to the transmit and the receive paths because both are always active.

3.7.1.4.4   Synchronization Requirements

TDD modems must operate using a common superframe so that the overall system performance is not compromised by NEXT from line to line. For this reason, a common superframe clock must be available to all modems at the CO or RT. The common clock is easily provided; however, operators outside Japan* view the distribution of this common clock as a difficult undertaking, particularly when unbundled loops are considered (see Section 3.9). Operators are uncomfortable with assuming the responsibility to provide a common, reliable clock to those who lease lines in their networks. They worry that a common clock failure would render them vulnerable to lawsuits from companies or individuals who are leasing lines and relying on the common clock for their last mile systems. As a consequence, despite all the benefits and flexibility of TDD, FDD was selected as the duplexing scheme for VDSL.

3.7.2   Line Code Alternatives

Two classes of line codes — single-carrier modulation and multicarrier modulation — are used in DSL. Within the

---

* In Japan, TDD has been used successfully for decades to provide ISDN service.

single-carrier class, pulse-amplitude modulation (PAM) and quadrature-amplitude modulation (QAM) line codes are used successfully in DSL. From the multicarrier family, only the discrete multi-tone (DMT) line code is used in DSL.

Well-established transmission techniques, PAM and QAM are described in many references, including Proakis[20] and Lee and Messerschmitt.[21] DMT is a newer line code, and many readers may not be familiar with it. Therefore, in this section, a high-level explanation of DMT is provided. More detailed information about DMT can be found in Cioffi[22] and Golden et al.[23]

3.7.2.1 Discrete Multi-Tone (DMT)

Multicarrier modulation is a class of modulation schemes in which a channel is partitioned into a set of orthogonal, independent subchannels, each of which has an associated subcarrier. Discrete multi-tone (DMT) modulation is a specific type of multicarrier modulation that has been standardized worldwide for ADSL and VDSL.

DMT uses an N-point complex-to-real inverse discrete Fourier transform (IDFT) to partition a transmission channel bandwidth into a set of orthogonal, equal-bandwidth subchannels. In DSL applications, the modulation is baseband, and $\bar{N} - 1$ subchannels are available to support data transmission, where $\bar{N} = N/2$. A DMT system operates at symbol rate $1/T$, with period T.

During each symbol period, a block of B bits from the data stream is mapped to the subchannels, with each subchannel allocated a number of bits that can be supported with the desired bit error probability and noise margin, based on its SNR. (The appropriate value of B is determined during an initialization procedure.) Thus,

$$B = \sum_{k=1}^{\bar{N}-1} b_k, \tag{3.7}$$

where each b_k is determined based on the subchannel SNR. A maximum value for b_k is always imposed; in ADSL it is 15.

For some subchannels, b_k may be zero. The aggregate bit rate of the system is $R = B/T$.

Each subchannel can be considered an independent QAM system. Therefore, bits are mapped to ordinary QAM constellation points. Indexing the subchannels by k, the constellation points are denoted X_k.

The IDFT is defined as

$$x_i = \frac{1}{\sqrt{N}} \sum_{k=0}^{N-1} X_k \cdot e^{j\frac{2\pi}{N}ki} \quad \forall \ i \in [0, N-1] , \qquad (3.8)$$

where i is the sample index. Because the IDFT is applied to the subchannel constellation points (the subsymbols), the bitstream can be considered a frequency-domain variable. During a symbol period, the IDFT transforms the subsymbols, as a block, to the time domain for transmission over the channel.

Ideally, a multicarrier transmission system would partition the channel into perfectly independent subchannels. However, subchannels with "brick wall" characteristics in the frequency domain would require high implementation complexity, and the processing delay would be infinite. In DMT systems, therefore, subchannels overlap, but they remain orthogonal at the subcarrier frequencies.

Each symbol that the DMT transmitter applies to the channel, where each symbol is the result of an IDFT operation, can be considered to be windowed in the time domain by a rectangular pulse, which is caused by the finite duration of each symbol. Denoting the subcarrier spacing as f_0, the nth DMT symbol is the sum of components that can be written in the time domain in the form of

$$x_{n,k}(t) = \left[X_k \cdot e^{j2\pi f_0 kt} + X_k^* \cdot e^{-j2\pi f_0 kt} \right] \cdot w_T(t) , \qquad (3.9)$$

where $x_{n,k}(t)$ represents the components of the nth symbol due to the kth subchannel. The rectangular window is defined as

$$w_T(t) = \begin{cases} 1 & t \in [0,T) \\ 0 & t \notin [0,T) \end{cases} , \qquad (3.10)$$

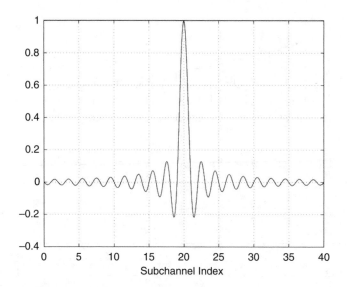

Figure 3.32   The sinc function — the basis of the DMT transmitter.

where $T = 1/f_0$ is the duration of each symbol. The Fourier transform of $w_T(t)$ is

$$w_T(f) = \operatorname{sinc}\left(\frac{f}{f_0}\right), \qquad (3.11)$$

which is a sinc function with its peak at 0 Hz and zeros at multiples of f_0.

Because multiplication in time corresponds to convolution in frequency, and because $e^{j2\pi f_0 kt}$ can be written as $\cos(2\pi f_0 kt) + j \sin(2\pi f_0 kt)$, the Fourier transform of $x_{n,k}(t)$ is the convolution of signals of the form $X_k \cdot \delta(f - f_0)$ and $W_T(f)$. This convolution simply corresponds to copies of $W_T(f)$ that are centered at multiples of f_0 and scaled by the X_k corresponding to the subchannels. Figure 3.32 illustrates the sinc function that appears at the subchannel with index 20. In ADSL and VDSL, the copies of $W_T(f)$ appear at multiples of 4.3125 kHz.

Note that for any selected copy of $W_T(f)$, its value at any other integer multiple of f_0 is zero, due to the properties of the sinc function. Therefore, at any subchannel center frequency,

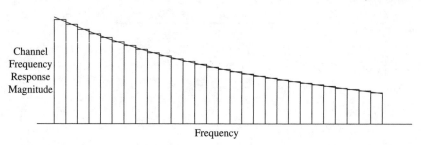

Figure 3.33   The partitioning of a channel into subchannels.

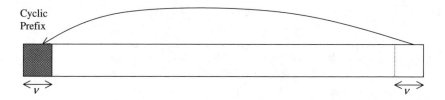

Figure 3.34   The cyclic prefix is a copy of the last samples of the symbol inserted before the start of the symbol.

the value of the aggregate signal — the sum of all the sinc functions corresponding to the subchannels — is due only to the signal on that subchannel. Thus, the signal at any multiple of f_0 is independent of all other subchannels.

The number of subchannels into which the channel is partitioned is selected to be large enough that the frequency response of each subchannel is roughly constant across its bandwidth, as illustrated in Figure 3.33.[23] Under this condition, the resulting subchannels are almost memoryless, and the small amount of intersymbol and intersubchannel interference caused by the channel's nonunity impulse response length can be eliminated by use of a cyclic prefix. As its name implies, the cyclic prefix precedes each symbol. It is a copy of the last ν time-domain (that is, post-IDFT) samples of each DMT symbol, as shown in Figure 3.34.[23] Thus, the cyclic prefix carries redundant information and is transmission overhead. If no more than ν + 1 samples are in the channel impulse response, then intersymbol interference (ISI) caused by each symbol is confined to the cyclic prefix of the following symbol. By discarding the cyclic prefix samples in the receiver, ISI is eliminated completely.

One might wonder why the last v data samples are used as the cyclic prefix rather than zeros, or even random samples, to eliminate ISI. In fact, the cyclic prefix serves another purpose. It is well known that with continuous-time signals, convolution in the time domain corresponds to multiplication in the frequency domain. Thus, the output of a channel can be determined by performing a convolution of the time-domain signal input and the channel impulse response, or by multiplying the Fourier transform of the signal and the channel frequency response and computing the inverse transform of the result. In discrete time, convolution in time corresponds to multiplication of the Fourier transforms only if at least one of the signals in the convolution is periodic, or the size of the DFT is infinite. In practice, the DFT size cannot be infinite, which means that the input signal or the channel impulse response must be periodic. Because the input signal corresponds to real data, it is clearly not periodic. Likewise, the channel is not periodic. Therefore, some manipulation of the signal is required to ensure that the relationship between convolution in time and multiplication in frequency holds.

Denoting the discrete-time input sequence as x, the samples of the channel as h, and the output sequence as y,

$$y = x * h \, , \qquad (3.12)$$

where * denotes convolution. Because practical channels are not infinite in length, the channel is assumed to have constraint length (or memory) of v samples (where v may be arbitrarily large), and the convolution corresponding to the nth symbol can be written as

$$y_n = \sum_{k=0}^{v} h_k x_{n-k} \, . \qquad (3.13)$$

The relationship

$$x * h \Leftrightarrow X \cdot H \qquad (3.14)$$

holds only if the signal x is made to appear periodic during each symbol period. Assume x during a specific symbol period is prefixed by a copy of the last v samples to become the new sequence $[x_{N-v} \ x_{N-v+1} \ \cdots \ x_0 \ x_1 \ \cdots \ x_{N-1}]$. The reader can verify that the convolution of this sequence and $h = [h_0 \ h_1 \ \cdots \ h_v]$ yields an output sequence that, for the last N samples, depends only on samples from the current symbol samples. (Because the previous input symbol was not the same as the current symbol, the first v samples of the output have components of the previous symbol's samples.) Thus, the last N samples of the received signal are exactly what they would have been if the input signal had been truly periodic. Therefore, use of the cyclic prefix ensures that the relationship in Equation 3.14 holds for the period of interest to the receiver.

Indexing the symbols by n,

$$Y_n = X_n H_n , \qquad (3.15)$$

and, in the absence of channel noise, the input sequence can be recovered (conceptually, at least) at the receiver by a simple division by the channel frequency response.

If the cyclic prefix length were zero, the symbol rate of the DMT system would be simply the inverse of the subchannel bandwidth. However, all practical channels have nonzero impulse response lengths, so a nonzero cyclic prefix is necessary. If the sampling rate of the system is f_s, the subcarrier spacing is calculated as $f_0 = f_s / N$. In contrast, the rate at which data-carrying symbols are transmitted is $1/T = f_s/(N + v)$, which excludes the "excess time" required to transmit the cyclic prefix. Thus, $f_0 > 1/T$.

To minimize overhead due to the cyclic prefix, v should be very small relative to N, which is achieved by using a large IDFT. In ADSL and VDSL systems, the cyclic prefix overhead is less than 8%.

To support bidirectional transmission using FDD, different subchannels are used in the downstream and upstream directions. In some systems, such as VDSL, the subchannels span from zero to the maximum downstream or upstream frequency; subchannels allocated to the opposite direction (or overlapping POTS) are simply not used by the transmitter.

During the transceiver initialization procedure, the SNRs of the subchannels are computed using estimations of the channel attenuation and noise profile. Based on these SNRs and the desired bit rate and error rate performance, the number of bits that each subchannel can support, b_k, is computed. The resulting mapping of bits to subchannels, commonly called the bit loading, is used during each DMT symbol period to compute the subsymols for the subchannels. During each symbol period, the subsymbols are input to the IDFT, which converts them to time-domain samples. The cyclic prefix is then prepended to the IDFT output, and the resulting signal is converted from digital to analog format and is (possibly) filtered and transmitted over the channel.

In the receiver, after analog-to-digital conversion, the cyclic prefix is stripped and discarded from the sampled signal, and the time-domain samples are input to a DFT. Each value output by the DFT, which is a noisy, attenuated, and rotated version of the original QAM subsymbol on that subchannel, is then scaled by a single complex number to remove the effects of the magnitude and phase of its subchannel's frequency response. The set of complex multipliers, one per subchannel, is known as the frequency-domain equalizer (FEQ). Changes in the channel magnitude or phase as the system operates are accommodated by updating the FEQ taps. Following the FEQ, a memoryless (that is, symbol-by-symbol) detector decodes the resulting subsymbols. Because a memoryless detector is used, DMT systems do not suffer from error propagation; rather, each subsymbol is decoded independently of all other (previous, current, and future) subsymbols. Figure 3.35 shows a high-level block diagram of a DMT transmitter and receiver pair.

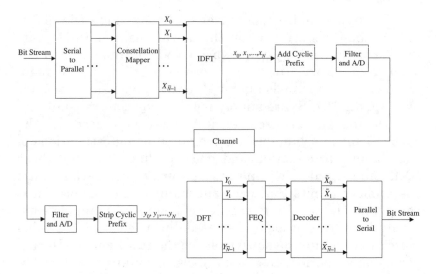

Figure 3.35 DMT transmitter and receiver block diagrams.

3.8 SPECIFICS OF DSL FLAVORS

This section describes the individual variants of DSL in more detail. The key modem aspects are described, including line code, duplexing, and mechanisms to overcome the transmission impairments described in Section 3.6.

3.8.1 ADSL

ADSL uses DMT modulation and FDD or EC duplexing. The subcarrier spacing is always 4.3125 kHz. ADSL1 specifies 256 subchannels in the downstream direction, with 32 in the upstream direction. Therefore, the downstream subchannels span from 0 to 1.104 MHz. The 32 upstream subchannels span from 0 to 138 kHz for ADSL over POTS or 138 to 276 kHz for ADSL over ISDN.

When ADSL operates on the same line as POTS, the lowest subchannels (for example, those below 25 kHz) are not used (and in fact will be filtered out by the POTS splitter) in either transmission direction. When ADSL operates on the same line as ISDN, subchannels below 120 kHz are typically not used. ADSL1 over POTS is defined in Annex A of the ITU-T

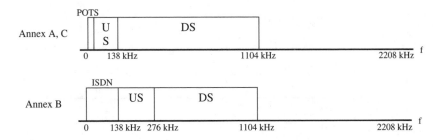

Figure 3.36 Frequency bands used in ADSL1. Annex C, for Japan, uses the same downstream and upstream frequency bands as Annex A.

Recommendation G.992.1. For Japan, ADSL1 over POTS operation is defined in Annex C. ADSL1 over ISDN is defined in Annex B of the recommendation. Figure 3.36 illustrates the frequency band usage corresponding to the three annexes.

ADSL2 specifies all the operational modes of ADSL1, also in Annexes A, B, and C, and some additional modes. In Annexes I and J, ADSL2 defines all-digital modes of operation for use on lines on which simultaneous support of POTS or ISDN is not necessary. In addition, ADSL2 defines a mode to double the upstream bandwidth for over-POTS operation in Annex M. (Annex M is essentially "Annex J over POTS.") In this mode, the upstream band extends to 276 kHz, using 64 subchannels. Lastly, in Annex L, ADSL2 specifies an operational mode to extend the reach of ADSL. In this mode, the downstream and upstream channels are confined to smaller frequency bands, and their PSD levels are boosted slightly to improve performance on long lines with severe attenuation. Figure 3.37 provides a graphical representation of the bandwidths used in the ADSL2 operational modes.

In ADSL2plus, the downstream bandwidth is doubled and up to 512 subchannels are available. Relative to ADSL2, only Annex L is not defined because the objective of ADSL2plus is to improve the bit rates on short loops, which is contradictory to increasing the reach of ADSL. Otherwise, all the operational modes are similar to those in ADSL2, except that the downstream bandwidth extends to 2.208 MHz.

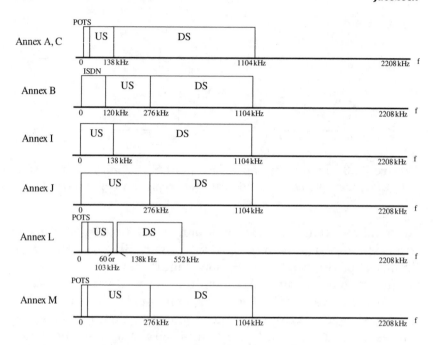

Figure 3.37   Frequency band usage in ADSL2.

Figure 3.38 illustrates the frequency band usage in ADSL2-plus.

The exact frequencies (subchannels) used by any ADSL system depend on a number of factors, including regional requirements, duplexing choice and modem design. Table 3.3 details the various ADSL operational modes and maximum frequency band usage. ADSL systems initialize to operate with 6 dB of noise margin and provide a bit error rate of no higher than 10^{-7} even if the noise margin during a connection degrades to zero.

3.8.2   VDSL

Standardization of VDSL first began in the mid-1990s, well before operators had begun mass deployment of ADSL. In hindsight, the standards effort began prematurely. Operators simply were not prepared to define a new DSL flavor while

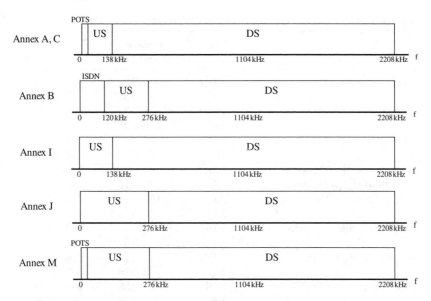

Figure 3.38 Frequency band usage in ADSL2plus.

they were struggling to determine how best to deploy the one they already had. Therefore, operators did not initially take a strong position on fundamental aspects of VDSL, such as line code and duplexing scheme. Consequently, vendors aligned behind their favorite line codes and duplexing schemes, discussions became acrimonious, and, because DSL standards organizations operate by consensus, virtually no progress could be made.

Three distinct proposals, each different in some fundamental aspect, were proposed: TDD-based DMT, FDD-based DMT, and FDD-based QAM. Although the decision to use FDD was made in 2000, the line code remained unresolved. Eventually, to allow some progress to be made, T1E1.4 agreed to work on the two line codes in parallel, and ETSI TM6 decided to standardize both line codes. The ITU-T spent its energy on ADSL and other standards unencumbered by line code uncertainty.

In 2003, a VDSL line code decision was finally made, in large part due to pressure from the IEEE 802.3ah task force,

Table 3.3 Summary of ADSL Band Usage

Type of system	Type of duplexing	Upstream band (kHz)	Downstream band (kHz)
ADSL1 over POTS	FDD	25–138	138–1104
	EC	25–138	25–1104
ADSL1 over ISDN	FDD	138–276	276–1104
	EC	138–276	138–1104
ADSL2 over POTS	FDD	25–138	138–1104
	EC	25–138	25–1104
ADSL2 over POTS with	FDD	25–276	138–1104
extended upstream	EC	25–276	25–1104
ADSL2 over ISDN	FDD	120–276	276–1104[a]
	EC	120–276	120–1104
ADSL2 all-digital mode 1	FDD	3–138	138–1104
	EC	3–138	3–1104
ADSL2 all-digital mode 2	FDD	3–276	276–1104
	EC	3–276	3–1104
ADSL2plus over POTS	FDD	25–138	138–2208
	EC	25–138	25–2208
ADSL2plus over POTS with	FDD	25–276	276–2208
extended upstream	EC	25–276	25–2208
ADSL2plus over ISDN	FDD	120–276	276–2208
	EC	120–276	120–2208
ADSL2plus all-digital mode 1	FDD	3–138	138–2208
	EC	3–276	3–2208
ADSL2plus all-digital mode 2	FDD	3–276	276–2208
	EC	3–276	3–2208

[a] In ADSL2 over ISDN with FDD, the crossover from upstream to downstream transmissions has some flexibility. The transition frequency can be as low as 254 kHz.

which was working to specify Ethernet in the first mile (EFM). The EFM task force had established two performance objectives: 2 Mbps symmetrical on 2700 m of 26-AWG line (the "long-reach" goal) and 10 Mbps symmetrical on 750 m of 26-AWG line (the "short-reach" goal). SHDSL was selected as the physical layer for the long-reach goal, and VDSL was chosen for the short-reach goal. However, to ensure interoperability, EFM needed to specify a single line code and asked T1E1.4 for assistance.

In response to the request from EFM, T1E1.4 established a process by which one of the line codes would be selected as "the" VDSL line code. Two independent laboratories, BT Exact and Telcordia, would conduct tests of VDSL1 modems in what became known as the "VDSL Olympics." It was hoped that one approach would prove to have better real-world performance; that line code would then be the obvious choice for VDSL.

Two DMT modems (from Ikanos and STMicroelectronics) and two QAM modems (from Infineon and Metalink) were submitted for the VDSL Olympics. The results showed that the performance of the DMT-based systems was significantly better than the performance of QAM-based systems. (See References 24 through 37.) As a result, T1E1.4 decided to elevate DMT to the status of an American National Standard. To ensure that work on QAM was not lost, the QAM VDSL specification was captured in a technical requirements document. T1E1.4 also agreed to begin work on VDSL2, with an agreement from the start that DMT would be the one and only line code. Based on the decision in T1E1.4, EFM then adopted DMT-based VDSL as the short-reach physical layer in its specification.

Shortly after the T1E1.4 decision had been made, ETSI TM6, which had no VDSL1 line code dilemma, also decided to begin work on VDSL2 — also with DMT as the only line code. In early 2004, the ITU-T reached a compromise for VDSL1. Recommendation G.993.1, which is the international version of VDSL1, specifies DMT in the main body of the standard and captures QAM in an annex. In line with the other standards bodies, the ITU also agreed to start work on VDSL2, based only on DMT.

At the time of writing, work on VDSL2 was underway in T1E1.4, ETSI TM6 and the ITU. One key objective in VDSL2 is to facilitate multimode implementations that can operate as ADSL or VDSL. Another objective is better performance than VDSL1, which means not only higher bit rates on short loops, but also longer reach.

Relative to ADSL, several enhancements are incorporated in DMT-based VDSL to improve performance. These enhancements were first described in Isaksson et al.[38,40,41] and Isaksson and Mestdagh.[39] The standardized version of DMT is known as "digitally duplexed" DMT because analog filters are not required to segregate transmit and receive bands.

In each transmission direction in VDSL, DMT provides a full set of subchannels that span the entire defined bandwidth. As in ADSL, the subcarrier spacing is 4.3125 kHz, although an option is defined in VDSL1 to allow the use of 8.625 kHz to allow more bandwidth to be spanned with fewer subchannels. This is one strategy to provide higher bit rates on short loops.

With all subchannels available for downstream or upstream transmission, the two modems in a VDSL connection can agree which subchannels will be used downstream and which will be used upstream.* However, allocating different subchannels to the two transmission directions does not entirely eliminate the need for filters to separate received signals from transmitted signals. Although the transmitted subchannels will exhibit the necessary orthogonality because the subcarriers will align in frequency with the zeros of the other transmitted subcarriers, echo from the transmitted to received subchannels is still possible because the two signals are not likely to be in phase.

To eliminate the echo, the transmitted and received symbol boundaries must be aligned so that the zeros of all subchannels, transmitted and received, lie precisely at the subcarrier frequencies. In this case, bidirectional transmission can be supported without the use of band-splitting filters. Furthermore, adjacent subchannels ideally can be allocated to different transmission directions. Because the zero crossings and subcarrier frequencies are aligned, use of a guard band between downstream and upstream bands is, at least in theory, unnecessary.

To allow the required alignment of the transmitted and received symbol boundaries, a cyclic suffix is defined. The required length of the cyclic suffix can be minimized through

---

* Regional spectrum management rules generally impose a frequency plan, thus limiting the flexibility of the modems to negotiate the best frequency plan.

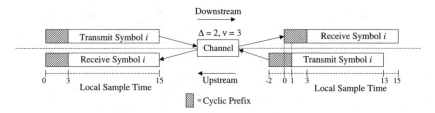

Figure 3.39 The transmitted and received symbol boundaries cannot be aligned at both ends of the line.

the use of a timing advance. Finally, windowing is applied to reduce susceptibility to ingress.

3.8.2.1 Cyclic Suffix

The cyclic suffix is the same idea as the cyclic prefix. However, as its name implies, the cyclic suffix is a copy of some number of samples from the beginning of the symbol (before the prefix has been added), and it is appended to the end of the symbol. Let the length of the cyclic suffix be denoted as 2Δ, where Δ is the phase delay in the channel. The sum of the cyclic prefix and cyclic suffix (usually called the cyclic extension) must be large enough that the transmitted and received symbol boundaries can be aligned at both ends of the line. It is assumed the cyclic prefix is not overdimensioned and is thus entirely corrupted by ISI. Therefore, the symbols are properly aligned if, at the same time, there are sets of N samples in the transmitted and received data streams that do not include any part of the cyclic prefix.

Alignment at one end of the line is easily achieved simply by observing the symbol boundaries of received symbols and transmitting symbols so that they are synchronized to the received symbol boundaries. However, achieving alignment at one end of the line almost always results in misalignment at the other. Figure 3.39[23] illustrates the simple case of a system with $N = 12$ (i.e., six subchannels) and a channel with phase delay Δ of two samples and a constraint length (memory) ν of three samples. (Obviously, this system is not very efficient,

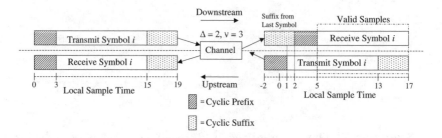

Figure 3.40 Adding a cyclic suffix allows the transmitted and received symbols to be aligned at both ends of the line.

but it is only for illustration purposes.) Including the cyclic prefix, the length of each transmitted symbol is 15 samples. Note that to achieve alignment of the transmitted and received symbols on the left-hand side of the figure, the transmitter on the right must advance transmission of its symbols by two samples. This action results in significant misalignment on the right-hand side because the valid transmitted symbol now overlaps the cyclic prefix of the received symbol, and vice versa.

Now assume a cyclic suffix of length 4 samples has been appended to all symbols so that each transmitted symbol is 19 samples in duration. Figure 3.40[23] illustrates the impact of the cyclic suffix. With the cyclic suffix, 12 samples in the transmitted and received symbols on the right-hand side are not corrupted by ISI. Thus, valid transmit and receive symbols exist from samples 5 through 16, inclusive. Note that the transmitted symbol on the right-hand side of the figure is a shifted version of the actual symbol. However, because a shift in time corresponds to a rotation in frequency, the receiver on the left-hand side will simply use FEQ taps that are correspondingly rotated to demodulate the signal. The rotation of the FEQ taps happens naturally during the initialization process because the cyclic suffix is always appended.

3.8.2.2 Timing Advance

Dimensioning the cyclic suffix to be twice the phase delay of the channel results in the desired condition of time-synchronized valid transmit and receive symbols at both ends of the

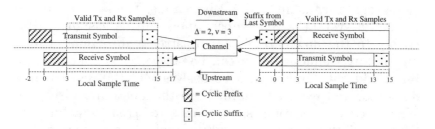

Figure 3.41 Timing advance reduces the required duration of the cyclic suffix by half, improving system efficiency.

line. However, like the cyclic prefix, the cyclic suffix is redundant information and thus results in a bit rate penalty. (In reality, the penalty will not be as severe as the example would suggest.) Thus, if possible, it is desirable to reduce the size of the cyclic suffix.

Referring again to Figure 3.40, the reader can verify that, with the cyclic suffix appended, there are actually several sets of valid transmit and receive symbols at the modem on the left-hand side. Any continuous set of 12 samples starting with sample 3, 4, 5, 6, 7, or 8 constitutes a valid symbol in the transmit and receive directions. To minimize the overhead, it is desirable to reduce the number of sets of valid symbols at both ends of the line to exactly one.

The timing advance is a method to achieve this goal. Rather than precisely synchronizing the transmitted and received symbol boundaries on the left-hand side, the transmitted symbol is advanced by a number of samples equal to the phase delay of the channel, Δ. The cyclic suffix length can then be halved, as Figure 3.41[23] illustrates. When the transmitted symbol on the left-hand side is advanced by 2 samples, a single set of valid transmit and receive samples, from 3 through 14 inclusive, results on the left-hand side. On the right-hand side, because the cyclic suffix length has been halved, the valid samples are now also 3 through 14.

By using the cyclic suffix and timing advance, the desired condition of synchronous transmitted and received symbols is achieved, thus yielding a system that does not require filters to segregate transmit and receive bands. The key advantage to such a system is that subchannels can be assigned arbitrarily to the downstream and upstream directions, which means any frequency plan can be supported. In fact, in the limit, odd subchannels can be assigned to the downstream direction, and even subchannels to the upstream direction. This mode of operation is sometimes referred to as "zipper" because of the alternating subchannel usage, which corresponds to the optimal symmetrical frequency plan described in Section 3.7.1.1.1.

3.8.2.3   Windowing

Windowing is used in VDSL to mitigate the effects of ingress from amateur and AM radio signals. Although the subchannels in DMT do not interfere with each other at the subcarrier frequencies due to alignment of the zero crossings of the sinc functions and the subcarrier frequencies, the subchannels have side lobes that can pick up noise outside the main lobe of the subchannel. This noise then reduces the capacity of the system, or it could cause performance problems during a connection if the noise source is of the on/off type, such as amateur radio.

The purpose of windowing is to reduce the subchannel side lobe levels such that the susceptibility of the system to ingress is decreased. In VDSL1, the first part of the cyclic prefix and the last part of the cyclic suffix are windowed by the transmitter. Subsequent symbols are then partially overlapped so that the windowed samples of the cyclic suffix of symbol M are added to the windowed samples of the cyclic prefix of symbol $M + 1$. The side lobes can be reduced significantly by windowing only a small number of samples.

3.8.2.4   Egress Suppression

VDSL systems must ensure that they do not interfere with the operations of amateur radio enthusiasts. Therefore, VDSL systems are required to reduce their transmit PSDs to −80 dBm/Hz within the amateur radio bands. By dividing the channel into subchannels, DMT systems are inherently well equipped to meet strict egress requirements in the amateur

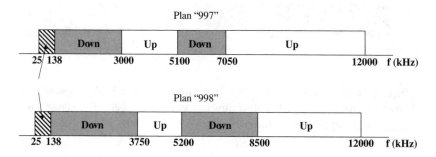

Figure 3.42 VDSL1 frequency plans.

radio bands. Subchannels overlapping the amateur radio bands can be disabled, which immediately reduces the transmit PSD within these bands by 13 db, without windowing, and to even lower levels with windowing. Additional simple digital techniques can be used to reduce the PSD further to the required level of –80 dBm/Hz. Thus, systems can meet egress suppression requirements with little additional complexity.

3.8.2.5 VDSL Frequency Plans

As mentioned in Section 3.7.1, standard compliant VDSL uses FDD; thus, definition of a frequency plan is necessary. The frequency plan definition was a difficult chore in standardization. Operators in different regions preferred different bit rate combinations and levels of asymmetry. Therefore, the frequency plans that resulted are true compromises.

In Europe, the various operators were unable to agree on a single frequency plan. As a result, the two frequency plans shown in Figure 3.42 were standardized. Both feature four primary bands — two for the upstream direction and two for the downstream direction. Both plans also define an optional band that overlays the upstream band used in ADSL over POTS. This band can be used in the downstream or upstream direction. The plan known as "998" was optimized to support 22 Mbps downstream with 3 Mbps on 1 km of 0.4-mm line; plan "997" was designed for use when less asymmetrical

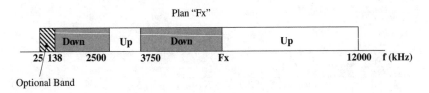

Plan "Fx"

Optional Band

Figure 3.43 The "Fx" VDSL1 frequency plan, which allows flexibility in the band-split frequency between the second downstream and second upstream bands.

transport is desired. In the U.S., the operators were able to agree on "998" as the only frequency plan. In the ITU, a third frequency plan, known as "Fx," was also standardized (Figure 3.43). The transition frequency between the second downstream and second upstream bands is not specified. Instead, operators or regulators can choose an appropriate value for Fx based on their regional requirements and objectives.

Note that, due to spectral overlap, each of the frequency plans causes NEXT to the other two. Thus, to avoid NEXT, a single frequency plan must be used in a binder.

3.8.2.6   VDSL System Parameters

Table 3.4 contains the key parameters for VDSL1 systems. The VDSL1 standards allow great flexibility in the number of subchannels, which gives designers the option to design systems that meet particular needs (such as very high bit rates or more moderate bit rates). When only 256 subchannels are used, the optional band below 138 kHz must be enabled, and it must be used in the upstream direction; otherwise, there would be no upstream band with any of the standardized frequency plans. In this mode, a VDSL modem could potentially interoperate with an ADSL modem. In fact, one goal in VDSL2 is to facilitate implementations that can support ADSL and VDSL modes.

All of the 12-MHz bandwidth specified by the standardized frequency plans in VDSL1 can be spanned in two ways. A system could use 4096 subchannels at a subcarrier spacing of 4.3125 kHz. In this case, the upper (about) one third of the

Table 3.4 VDSL System Parameters

System parameter	Valid values
IDFT/DFT size	512, 1024, 2048, 4096, or 8192
Number of subchannels	256, 512, 1024, 2048, or 4096
Symbol rate	4 kHz (other values optional)
Cyclic extension length	40, 80, 160, 320, or 640 (other values optional)
Subcarrier spacing	4.3125 kHz mandatory; 8.625 kHz optional
Sampling rate	2.208, 4.416, 8.832, 17.664, or 35.328 MHz
Bandwidth	1.104, 2.208, 4.416, 8.832, or 17.664 MHz
Overhead due to cyclic extension	7.8% with mandatory values

subchannels would not be used because they extend beyond 12 MHz. Alternatively, a system could use 2048 subchannels with 8.625-kHz subcarrier spacing. Again, the upper one third of the subchannels would extend beyond 12 MHz and would need to be disabled.

Note that any of the mandatory cyclic extension lengths with the appropriate sampling rate corresponds to a duration of about 18 µs (assuming 4.3125-kHz subcarrier spacing). An analysis in Ginis and Cioffi[42] indicates that a cyclic prefix duration of 9 µs is typically sufficient for VDSL lines, and that the phase delay of channels up to 1 mile in length is less than 8 µs, which means the cyclic extension needs to be at least 17 µs in duration. Thus, the cyclic extension is appropriately dimensioned for loops up to about 1 mile long.

Like ADSL systems, VDSL systems operate with 6 dB of noise margin. In practice, they provide a bit error rate no greater than 10^{-7} when the actual noise margin is non-negative.

3.8.3 Symmetric DSLs

Unlike ADSL and VDSL, the symmetric DSLs almost exclusively use single-carrier modulation (SCM). SCM is used for symmetric DSLs primarily due to the low latency requirements for some of the applications. To achieve the desired condition of equal downstream and upstream bit rates, symmetric DSLs use overlapped spectra with echo cancellation.

Most HDSL systems use 2B1Q baseband transmission, although DMT and carrierless amplitude/phase (CAP) modulation are used in some parts of Europe. Fully overlapped transmission occurs on two wire pairs, with each pair supporting 784 kbps, including overhead. The 3-dB bandwidth of HDSL signals is 196 kHz.

SHDSL improves on HDSL by supporting symmetrical bit rates from 192 kbps to 5.696 Mbps on a single twisted-pair line. SHDSL uses trellis-coded pulse amplitude modulation (TC-PAM), which is a one-dimensional baseband scheme. Each trellis-coded symbol can support 4 bits (referred to as 16-TCPAM) or 5 bits (32-TCPAM). The 3-dB bandwidth of SHDSL is 387 kHz when the bit rate is 2.312 Mbps using 16-TCPAM, and 712 kHz when the bit rate is 5.696 Mbps using 32-TCPAM. Fully overlapped spectra are used. SHDSL also provides means to support lower bit rates using correspondingly smaller bandwidths. For example, the 3-dB bandwidth of the signal that supports 192 kbps is only 33 kHz.

HDSL2 and HDSL4 use 16-TCPAM modulation.

3.9 UNBUNDLING AND SPECTRAL COMPATIBILITY

Phone companies — originally Bell Telephone and later the regional Bell operating companies (RBOCs) — installed the telephone lines in the U.S. and continue to own these lines today. Prior to the mid-1990s, network operators provided all services on their lines. However, in the mid-1990s, the RBOCs wanted to offer long-distance service, which previously was prohibited. In exchange for allowing the RBOCs to offer this service, the Federal Communications Commission (FCC) required the RBOCs to "unbundle" their loops, which meant allowing other companies to lease telephone lines to provide services such as telephony and xDSL. The objective was to increase competition in the local loop.

Today, competition has indeed increased, but some technical difficulties have also arisen. When the phone companies provided all services on all loops in a particular region, they could ensure those services would not interfere with each other: they simply avoided offering combinations of services that resulted in excessive crosstalk and bit rate or reach degradations. With unbundling, however, RBOCs could no longer be assured that the services provided on all loops would be compatible. Even if they knew the services that they were providing were all compatible, there was no assurance that services being provided by companies leasing lines would be compatible with the RBOC's services. Likewise, the companies leasing lines had no assurance that services offered by the RBOC would not interfere with the services that they wished to provide.

Spectral compatibility is the term used to describe, qualitatively, the impacts of systems on each other. If two systems are spectrally incompatible, then at least one causes high levels of crosstalk noise to the other. Conversely, spectrally compatible systems do not cause excessive levels of NEXT or FEXT to each other. However, the amount of crosstalk considered acceptable as opposed to excessive has been and continues to be the subject of much debate in standards organizations. In one loop plant, a 10% penalty in reach due to crosstalk from another system might not cause any problem, but in another that same penalty in reach could result in an operator not being able to offer a service because too low a percentage of the customer base can be served. How much of a penalty is acceptable? Clearly, achieving agreement on the conditions for spectral compatibility is important.

It is an understatement to say that the area of spectral compatibility is fuzzy, and few blanket statements can be made concerning the spectral compatibilities of two systems. However, it is possible to state at least one set of conditions resulting in spectrally compatible systems: two systems are spectrally compatible if

- They use the same frequency bands in the same directions — that is, they cause little or no NEXT to each other if they are nonoverlapped and they cause reciprocal NEXT to each other if they are echo canceled.
- They transmit at power levels that result in signal levels along the cable that are roughly the same.

This second condition means that the FEXT between lines is approximately reciprocal, i.e., FEXT from line A to line B is roughly the same as FEXT from line B to line A. These conditions imply that systems of the same type are spectrally compatible.

Certainly, forcing all systems to use the same frequency bands in the same directions at the appropriate power levels would result in global spectral compatibility. However, today such an approach is impractical for several reasons. First, early xDSL systems (such as ADSL and HDSL, for example) evolved independently before unbundling was a reality. Thus, assuming a cable could be mandated to contain only systems of one type was not unreasonable, which meant spectral incompatibilities between different systems might never occur in practice. Furthermore, although the goal of achieving spectral compatibility between different systems that might reside in the same cable was recognized, no official mechanism was in place to evaluate the spectral (in)compatibilities of different systems. Even when spectral incompatibilities were noted, they could not be resolved easily because the objectives of the different systems were conflicting: a spectral allocation necessary to support symmetrical service optimally is different from that required to support asymmetrical service optimally. Additionally, the appropriate spectral allocation for a short-reach asymmetrical system is quite different from the appropriate spectral allocation for a long-reach symmetrical system. Thus, if all systems are dimensioned nearly optimally (and in xDSL development, many systems have been), spectral incompatibilities will occur.

In practice, a vague notion of spectral compatibility is not terribly useful. A quantitative measure of spectral compatibility is needed, particularly following unbundling, so that operators can be confident in their deployments, and new systems can be developed under a well-defined set of rules that protects existing systems. For these reasons, T1E1.4, the standards body responsible for xDSL standardization in North America, has generated a technical specification, T1.417,[43] to provide definitions of spectral compatibility and methodologies to determine whether two systems are spectrally compatible.

The T1.417 standard defines "spectrum management classes," which correspond to existing systems that have been deployed in volume or emerging systems expected to be deployed in volume. Many of the classes correspond to DSL technologies. Each class is characterized by a PSD, total average transmit power, transverse balance, and longitudinal output voltage. Deployment guidelines, which specify the maximum length of loop on which a class may be deployed while ensuring spectral compatibility with other classes, are also defined for each class. A system that meets all the requirements of a spectrum management class is in compliance with the spectrum management and spectrum compatibility requirements of T1.417. Therefore, if a system meets all the requirements of, say, Class 5 (ADSL), it is compliant with T1.417, even if the system is not actually ADSL.

Because the spectrum management classes were derived from existing or emerging systems, they are not immediately useful to evaluate the spectral compatibility of a new system that may not exactly meet the requirements of a specific class. For example, the PSD of a new system might not lie entirely under the defined PSD template, or the transmit power might be higher than that allowed by the class. For this reason, T1.417 also provides a generic analytical method to evaluate the spectrum compatibility of a system that does not qualify for one of the spectrum management classes. This method, known as Method B, is defined in (unfortunately) Annex A of the standard. The analytical method provides a means for a system proponent to show through simulations that the data rate reduction in each of the existing classes due to the introduction of the proposed system in the network will be acceptable as defined in T1.417. A system satisfying the requirements of Annex A is said to meet the spectrum compatibility requirements of the T1.417 standard.

Although T1.417 is perhaps the best known spectrum management standard, the reader should be aware that different spectral compatibility rules apply in other countries due to differences in network characteristics and regulatory policies. For example, the U.K. requires compliance with the access network frequency plan (ANFP), which specifies strict PSD limits based on the line length.[44]

For a more detailed discussion of spectrum management and T1.417, see Chapter 10 of Starr et al.[13]

3.9.1   Near–Far Problem and Upstream Power Back-Off

The previous section gave two criteria that guarantee systems are spectrally compatible: they transmit in the same direction(s) in all frequency bands, and they transmit at power levels that result in equal signal levels along the cable length. This section describes the near–far problem, which is spectral incompatibility between kindred systems that results when the second criterion — roughly equal power at any point along the cable — is not achieved.

In VDSL, FEXT is a dominant impairment due to the short lengths of the loops and, consequently, the availability of high-frequency bands to support transmission. In the upstream direction, FEXT appearing at a receiver is (some type of) a sum of FEXT due to transmitters on other lines that are almost certainly at different distances from the receiver. Therefore, if all upstream transmitters launch signals at the maximum allowed PSD level, then on any line, FEXT contributions from transmitters on shorter loops are higher than contributions from transmitters on longer loops, which causes spectral incompatibility between kindred systems. The result is a dramatic decrease in the capacities of longer loops due to this "near–far" effect. Figure 3.44 illustrates that degradations in upstream data rates due to near–far FEXT can be extreme.

To mitigate the near–far problem, upstream VDSL transmitters must adjust (generally reduce) their transmit PSDs so that upstream transmissions do not unfairly increase the received FEXT levels on longer lines supporting kindred systems. The process of reducing the upstream transmit PSDs is known as upstream power back-off.

Ideally, the upstream bit rate on each line following application of upstream power back-off would be the same as if all

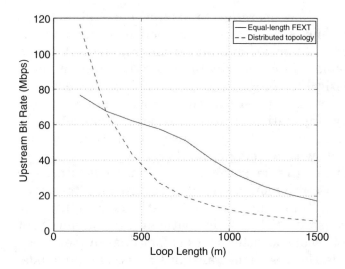

Figure 3.44 Degradations of upstream bit rates due to the "near–far" effect.

lines were the same length as the line under consideration and all remote transceivers transmitted at the maximum PSD level. Practically, however, degradation from this "equal-FEXT" performance level is unavoidable on at least some loops. Projecting data rate losses with a particular power back-off algorithm is complicated and cumbersome because the achievable data rates are a function of the loop plant topology and characteristics as well as the chosen values of the power back-off parameters. Consequently, provisioning of VDSL services is difficult because achievable upstream rates cannot be determined easily in advance.

Application of upstream power back-off should meet a number of goals. First and foremost, upstream power back-off must improve spectral compatibility between VDSL systems by forcing upstream transmit PSDs on short lines to be at lower levels than the PSDs on longer lines. Within this constraint, upstream power back-off ideally should allow support of higher bit rates on short loops and lower rates on longer loops, proportional to the loop capacities. This property is desirable to allow operators to maximize overall network performance and also because the service mix cannot be known in advance in an unbundled environment. A power back-off scheme that arbitrarily limits service alternatives is highly undesirable. Furthermore, for unbundled environments, power back-off must not require coordination between lines. Finally, application of upstream power back-off on an initializing line must not compromise the upstream performance on lines with ongoing connections.

Typically, upstream power back-off methods use some criterion to spectrally shape the PSD of upstream transmissions on each line. The simplest methods compute the upstream transmit PSD on a line without information about the loop topology or characteristics of other lines in the binder. In other words, the upstream transmit PSD is computed independently for each line by each remote modem, possibly using some information sent by the downstream transmitter. Jacobsen,[45,48] Jacobsen and Wiese,[52] Wiese and Jacobsen,[46] Pollett and Timmermans,[49] Sjoberg et al.,[50] Schelstraete,[51] and the FSAN VDSL Working Group[47] provide information about the specific methods of upstream power back-off discussed in the context of VDSL1 standardization.

Upstream power back-off methods that use information about all the lines in a binder to compute globally optimized upstream transmit PSDs can provide better performance than the simpler methods. However, such methods are typically computationally expensive. Furthermore, coordination between lines is not possible in today's networks, particularly with unbundling. These methods have nonetheless sparked significant interest in recent years and undoubtedly will play a role in increasing the overall capacity of the local loop in the future.

3.9.1.1 Crosstalk Cancellation

Because crosstalk is a significant impairment in last mile networks, the natural question arises as to whether anything can be done about it. Replacing all the cables in the network with higher quality cables (for example, Cat-5) is not a practical approach, so can crosstalk be canceled? The answer is yes. With additional computational complexity and knowledge of the coupling characteristics between lines, improved performance is possible, particularly for VDSL lines that tend to be dominated by FEXT. This section describes promising work on a particular method that utilizes coordination between lines.

Vectored transmission, described in detail in Ginis and Cioffi,[42] takes advantage of the co-location of transceivers at the CO or RT to cancel FEXT and optimize transmit spectra in the downstream and upstream directions. Upstream FEXT is canceled using multiple-input–multiple-output (MIMO) decision feedback at the upstream receivers, which are presumed to be co-located at the CO or RT. Downstream FEXT is mitigated through preprocessing at the downstream transmitters.

Vectored transmission relies on knowledge of the binder crosstalk coupling functions, which describe precisely how transmissions on each line couple into every other line. The technique presumes use of DMT modulation, with two additional constraints. First, the length of the cyclic prefix must be at least the duration of the maximum memory of the channel transfer function and the crosstalk coupling functions. Assuming a propagation delay of 1.5 μs/kft for twisted-pair lines, the memory of the channel transfer function is approximately 8 μs for a line 1 mile long. Measurements of FEXT[42] indicate that the maximum expected memory of the crosstalk coupling functions is approximately 9 μs. Thus, the cyclic prefix duration must be at least 9 μs for vectored transmission. Section 3.8.2 confirms that standard DMT-based VDSL meets this requirement.

The second requirement for vectored transmission is that of block-synchronized transmission and reception. Assuming co-located, coordinated transceivers at the central site, synchronized downstream transmission is easily achieved. All the downstream transmitters are simply synchronized to a common symbol clock. Synchronized upstream reception can be achieved through use of the cyclic suffix in addition to the cyclic prefix, as used in VDSL. The cyclic suffix is dimensioned so that the combined duration of the cyclic prefix and cyclic suffix is at least twice the maximum propagation delay of the channel.

Ginis and Cioffi[42] provide simulation results of vectoring and Chapter 11 of Starr et al.[13] offers an in-depth discussion of crosstalk cancellation.

ACKNOWLEDGMENTS

The author would like to thank the various reviewers of this chapter for their insightful comments and suggestions. I am particularly indebted to Tom Starr (SBC) and Richard Goodson (Adtran), who agreed to be the final reviewers of the chapter and, as experts in the field, helped me ensure that all the material was accurate and up to date (at least for the next hour or so). I must also acknowledge my former employer, Texas Instruments, for donating the spiffy graphics in some of the figures. Finally, as always, thank you to Jim for putting up with all the weekends I've spent working on chapters that I always think will take a lot less time than they actually do.

GLOSSARY

ADSL	asymmetric digital subscriber line
ANFP	access network frequency plan
AWG	American wire gauge
AWGN	additive white Gaussian noise
CAP	carrierless amplitude/phase
CO	central office
CPE	customer premises equipment
db	decibel
DFT	discrete Fourier transform
DLC	digital loop carrier
DMT	discrete multi-tone
DSL	digital subscriber line
DSLAM	digital subscriber line access multiplexer
EC	echo canceled (or cancellation)
FCC	Federal Communications Commission
FDD	frequency-division duplexing (or duplexed)
FEQ	frequency-domain equalizer

FFT	fast Fourier transform
FSAN	full-service access network
GPS	global positioning system
HAM	amateur radio
HDSL	high-speed digital subscriber line
IDFT	inverse discrete Fourier transform
ISDN	integrated services digital network
ISI	intersymbol interference
MIMO	multiple input–multiple output
PAM	pulse amplitude modulation
POTS	plain old telephone service
PSD	power spectral density
PSTN	public switched telephone network
QAM	quadrature amplitude modulated
RAM	remote access multiplexer
RBOC	regional Bell operating company
RT	remote terminal
SCM	single-carrier modulation
SHDSL	single-pair high-speed digital subscriber line
TC-PAM	trellis coded-pulse amplitude modulation
TDD	time-division duplexing (or duplexed)
UTP	unshielded twisted pair
VDSL	very high bit-rate digital subscriber line

REFERENCES

1. DSL Forum press release. DSL hits 85 million global subscribers as half a million choose DSL every week. Available at http://www.dslforum.org/pressroom.htm.

2. Network and customer installation interfaces — asymmetric digital subscriber line (ADSL) metallic interface. ANSI Standard T1.413-1993.

3. Asymmetric digital subscriber line (ADSL) — European specific requirements [ITU-T G.992.1 modified]. ETSI TS 101 388 (2002).

4. Asymmetric digital subscriber line (ADSL) transceivers. ITU-T Recommendation G.992.1 (1999).

5. Asymmetric digital subscriber line (ADSL) transceivers — 2 (ADSL2). ITU-T Recommendation G.992.3 (2002).

6. Asymmetric digital subscriber line (ADSL) transceivers — extended bandwidth ADSL2 (ADSL2plus). ITU-T Recommendation G.992.5 (2003).

7. Very-high-bit-rate digital subscriber lines (VDSL) metallic interface (DMT based). ANSI Standard T1.424 (2003).

8. Very high speed digital subscriber line. ITU-T Recommendation G.993.1 (2004).

9. Very high speed digital subscriber line (VDSL); part 1: functional requirements. ETSI TS 101 270-1 (2003).

10. Very high speed digital subscriber line (VDSL); part 2: transceiver specification. ETSI TS 101 270-2 (2003).

11. Transmission and multiplexing (TM); high bit-rate digital subscriber line (HDSL) transmission system on metallic local lines. ETSI ETR 152 (1996-12).

12. High bit rate digital subscriber line (HDSL) transceivers. ITU-T Recommendation G.991.1 (1998).

13. T. Starr, M. Sorbara, J.M. Cioffi, and P.J. Silverman. *DSL Advances*. Prentice Hall, Upper Saddle River, NJ, 2002.

14. DSL Forum. DSL anywhere, white paper. Available at http://www.dslforum.org/about_dsl.htm?page=aboutdsl/tech_info.html. 2001.

15. T. Starr, J.M. Cioffi, and P.J. Silverman. *Understanding Digital Subscriber Line Technology*. Prentice Hall. Upper Saddle River, NJ, 1999.

16. J.A.C. Bingham. *ADSL, VDSL, and Multicarrier Modulation*. John Wiley & Sons, New York. 2000.

17. C. Valenti. NEXT and FEXT models for twisted-pair North American loop plant. *IEEE J. Selected Areas Commun.*, 20(5), 893–900, June 2002.

18. R. Heron et al., Proposal for crosstalk combination method. ETSI TM6 contribution 985t23, Sophia Antipolis, France, November 1998.

19. K.T. Foster and D.L. Standley. A preliminary experimental study of the RF emissions from dropwires carrying pseudo-VDSL signals and the subjective effect on a nearby amateur radio listener. ANSI T1E1.4 contribution 96–165, April 1996.

20. J.G. Proakis. *Digital Communications*. McGraw-Hill, New York, 1989.

21. E.A. Lee and D.G. Messerschmitt. *Digital Communication*. Kluwer Academic Publishers, 1993.

22. J.M. Cioffi. EE379A/C course notes. Stanford University, Stanford, CA. Available at http://www.stanford.edu/class/ee379c/.

23. P. Golden, H. Dedieu, and K.S. Jacobsen, Eds. *Fundamentals of DSL Technology*. CRC Press, Boca Raton, FL, 2005.

24. BTexact. VDSL line code analysis of Ikanos — mandatory tests. T1E1.4 contribution 2003-600, Anaheim, June 2003.

25. BTexact. VDSL line code analysis of Ikanos — optional tests. T1E1.4 contribution 2003-601, Anaheim, June 2003.

26. BTexact. VDSL line code analysis of Infineon — mandatory tests. T1E1.4 contribution 2003-602, Anaheim, June 2003.

27. BTexact. VDSL line code analysis of Metalink — mandatory tests. T1E1.4 contribution 2003-604, Anaheim, June 2003.

28. BTexact. VDSL line code analysis of Metalink — optional tests. T1E1.4 contribution 2003-605, Anaheim, June 2003.

29. BTexact. VDSL line code analysis of STMicroelectronics — mandatory tests. T1E1.4 contribution 2003-606, Anaheim, June 2003.

30. BTexact. VDSL Line code analysis of STMicroelectronics — optional tests. T1E1.4 contribution 2003-607R1, Anaheim, June 2003.

31. Telcordia Technologies. Mandatory VDSL transceiver test results for Infineon. T1E1.4 contribution 2003-608, Anaheim, June 2003.

32. Telcordia Technologies. Mandatory VDSL transceiver test results for STMicroelectronics. T1E1.4 contribution 2003-609, Anaheim, June 2003.

33. Telcordia Technologies. Mandatory VDSL transceiver test results for Metalink. T1E1.4 contribution 2003-610, Anaheim, June 2003.

34. Telcordia Technologies. Mandatory VDSL transceiver test results for Ikanos. T1E1.4 contribution 2003-611, Anaheim, June 2003.

35. Telcordia Technologies. Optional VDSL transceiver test results for STMicroelectronics. T1E1.4 contribution 2003-612, Anaheim, June 2003.

36. Telcordia Technologies. Optional VDSL transceiver test results for Metalink. T1E1.4 contribution 2003-613, Anaheim, June 2003.

37. Telcordia Technologies. Optional VDSL transceiver test results for Ikanos. T1E1.4 contribution 2003-614, Anaheim, June 2003.

38. M. Isaksson, D. Bengtsson, P. Deutgen, M. Sandell, F. Sjoberg, P. Odling, and H. Ohman. Zipper: a duplex scheme for VDSL based on DMT. T1E1.4 contribution 97-016, February 1997.

39. M. Isaksson and D. Mestdagh. Pulse shaping with zipper: spectral compatibility and asynchrony. T1E1.4 contribution 98-041, March 1998.

40. M. Isaksson et al. Asynchronous zipper mode. ETSI TM6 contribution 982t16, April 1998.

41. M. Isaksson et al. Zipper: a duplex scheme for VDSL based on DMT. *Proc. Int. Conf. Commun.* S29.7, June 1998.

42. G. Ginis and J.M. Cioffi. Vectored transmission for digital subscriber line systems. *IEEE J. Selected Areas Commun.*, 20(5), 1085–1104, June 2002.

43. Spectrum Management for Loop Transmission Systems. American National Standard T1.417, 2003.

44. Ofcom. UK access network frequency plan. Available at http://www.ofcom.org.uk/static/archive/oftel/publications/broadband/llu/2003/anfp0103.htm.

45. K.S. Jacobsen. Methods of upstream power back-off on very high-speed digital subscriber lines (VDSL). *IEEE Commun. Mag.*, 39(3), 210–216, March 2001.

46. B. Wiese and K.S. Jacobsen. Use of the reference noise method bounds the performance loss due to upstream power backoff. *IEEE J. Selected Areas Commun.*, 20(5), 1075–1084, June 2002.

47. FSAN VDSL Working Group. Power-backoff methods for VDSL. ETSI TM6 contribution 983t17a0. Lulea, Sweden, June 1998.

48. K.S. Jacobsen. The equalized-FEXT upstream power cutback method to mitigate the near–far FEXT problem in VDSL. ETSI TM6 contribution 985t05r0. Sophia Antipolis, France, November 1998.

49. T. Pollet and P. Timmermans. Power back-off strategies for VDSL: TDD vs. FDD, performance comparison. ETSI TM6 contribution 985t24a0. Sophia Antipolis, France, November 1998.

50. F. Sjoberg et. al. Power back-off for multiple target rates. ETSI TM6 contribution 985t25a0. Sophia Antipolis, France, November 1998.

51. S. Schelstraete. Defining power backoff for VDSL. *IEEE J. Selected Areas Commun.*, 20(5), 1064–1074, June 2002.

52. K.S. Jacobsen and B. Wiese. Use of the reference noise method bounds the performance loss due to upstream power back-off. ETSI TM6 contribution 994t16a0. Amsterdam, Holland, November 1999.

53. Network and customer installation interfaces — asymmetric digital subscriber line (ADSL) metallic interface. ANSI Standard T1.413-1998.

4

Last Mile HFC Access

DANIEL HOWARD, BRUCE CURRIVAN,
THOMAS KOLZE, JONATHAN MIN, AND
HENRY SAMUELI

4.1 INTRODUCTION

This chapter deals with coaxial and hybrid fiber-coaxial (HFC) last mile networks. HFC networks started as purely coaxial networks and were developed initially for the distribution of network television signals where they were unavailable; thus, they were designed with a shared physical medium that in most cases did not support upstream transmission. However, the desire to increase reliability and to deliver new, interactive services such as high-speed data and telephony led to upgrading the original architecture to support wider bandwidths on the downstream, two-way signal amplification for interactive services, and higher reliability via fiber trunk lines that significantly reduced the number of radio frequency (RF) amplifiers between the home and the headend. Today's modern HFC networks can deliver over 500 video channels that can be fully interactive and on demand, provide circuit-switched and IP telephony, and of course provide high-speed data services.

Coaxial networks began in 1948 as community antenna television (CATV) networks in small towns where the off-air broadcast signals were unavailable or of very poor quality due to geography or distance from transmitters. Large antennas were placed on towers or hilltops and coax networks run to homes for distribution. Channel lineups grew from 3, initially, to 110 conventional analog channels (6 MHz each in the U.S.) on today's 750-MHz HFC networks. The rapid growth during the 1970s was primarily due to the advent of satellite distribution to the headend; subsequently, new channels became available to consumers that were not available from off the air broadcasts. Thus, although the early days of television involved three major networks that provided content designed for all citizens, the last three decades have seen the development of cable channels targeted at progressively smaller audiences. With IP streaming of video and video-on-demand systems growing rapidly in HFC networks, it will soon be possible to have, in effect, personal channels targeted at or programmed by individuals.

Interactive services were made possible by the introduction of two-way RF amplifiers, in which diplexers split the upstream and downstream signals, amplified each separately, and then recombined them. Then, the addition of fiber was a cost-effective way to increase network reliability because most of the network outages common to early HFC networks were due to RF amplifiers failing. The fiber runs reduced what was often up to 20 amplifiers in series to 5 or less. As of this writing, some cable plants run the fiber so deeply into the neighborhood that the coax portion of the plant is entirely passive, which means there are no RF amplifiers, only taps. From deep fiber to all coax networks, the HFC architecture provides a scalable broadband solution in which cost vs. performance trade-offs can be made with more flexibility than in most other broadband architectures.

A typical HFC architecture is shown in Figure 4.1. In the headend (or master headend), satellite signals, off-air signals, and local TV content are collected and modulated onto a single RF downstream signal. Receivers for the upstream signals are located in the headend or the hubs. The RF signals in the headend or hubs are converted to optical for transmission (vice versa for reception) and transported between the

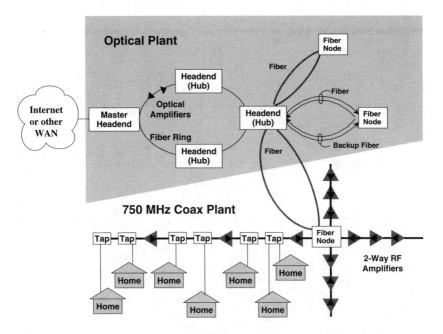

Figure 4.1 Typical architecture of a cable network.

headend and hubs. From there they are sent to optical nodes via single mode optical fiber, often with redundant backup paths using a different route in order to recover quickly from damage due to construction or inclement weather. The master headend can also be connected directly to fiber nodes in smaller networks.

In the optical node, which is typically mounted on poles or, increasingly, in curbside enclosures, the optical downstream signals are converted to RF signals, and the upstream RF signals to optical signals so that the transport over the coax portion of the network is purely via RF. In isolated cases, the fiber node can also contain traditional telephony signals that are then transported to businesses or homes via fiber or twisted pair. However, the focus in this chapter is on transport between the fiber node and the customer premises via coaxial cable. This coaxial cable is usually deployed in a "tree and branch" network architecture (shown in Figure 4.1), with the

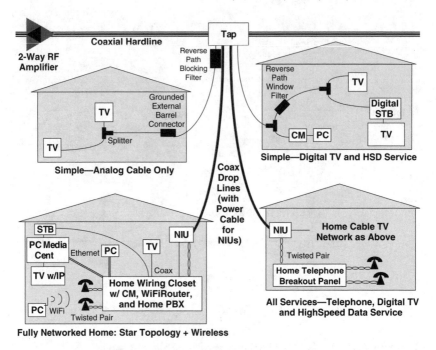

Figure 4.2 Example home network architectures and filtering.

number of cascaded RF amplifiers in current plant designs anywhere from zero to the more typical three to five.

The last hundred feet of HFC networks are represented by the coaxial drop line from the taps to the home and the coax and other networks within the home. Several different home network architectures are possible, as depicted in Figure 4.2. Depending on the services to which a particular customer subscribes, filters may be used at the customer premises or at the tap to block upstream interference from entering the main part of the HFC network. The HFC network thus depicted can support standard- and high-definition video, high-speed data, and circuit switched or voice over Internet protocol (VoIP) telephony.

This chapter begins with an overview of the HFC network and then details the physical nature of the hybrid fiber-coax transmission system, including RF impairments that can affect communication system performance over the cable plant. Then, details of the physical layer for DOCSIS 1.0, 1.1, and 2.0 standards are described, followed by a comparison of the two basic modulation schemes in DOCSIS 2.0: time division multiple access (TDMA) and synchronous code division multiple access (S-CDMA).

Because the media access control (MAC) layer and protocol can affect network performance as much as physical (PHY) layer technologies, a detailed discussion of the MAC protocol of DOCSIS is provided after the PHY discussions. Next, system level considerations are discussed — in particular those that have an impact on the performance of the system. Finally, new technologies for HFC networks are briefly described as a method of demonstrating the flexibility of the network.

4.2 OVERVIEW OF HFC NETWORKS

We begin with an overview of a basic cable data communications system, depicted in Figure 4.1. The HFC network is essentially a frequency division multiplex (FDM) communications system with repeaters (amplifiers) distributed throughout the RF or coaxial portion of the network. The amplifiers include diplexers for splitting the upstream and downstream frequency bands for separate amplification and then recombination. They also contain equalizers for compensation of the coax cable's greater attenuation at high carrier frequencies than at low carrier frequencies. In order to minimize the number of amplifiers required in the network, the amplifiers are operated at close to or even slightly into the nonlinear amplification region of the amplifier transfer curve, which leads to the creation of intermodulation distortion products between the downstream carriers, which land in the upstream and the downstream bands.

Taps are used at various points along the cable to provide the composite cable waveform to the customer premises. In order to present the downstream signal to the home at a prescribed power level, taps closer to the amplifier downstream output must attenuate the signal more than taps

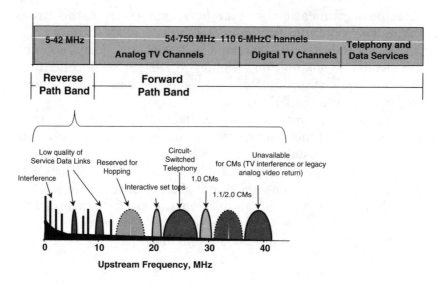

Figure 4.3   Cable network frequency allocations.

located near the end of a coax run. Because the tap attenuations are selected to compensate for losses at downstream frequencies (54 to 860 MHz in North American DOCSIS), they can overcompensate in the upstream band (5 to 42 MHz), where coaxial cable has lower signal loss. This can cause a power variation effect in which customers closer to the headend have less upstream path loss from the CM to the headend than from remotely located CMs. Taps can also be a source of intermodulation distortion if their connectors become corroded or oxidized.

The FDM nature of an HFC network is shown in greater detail in Figure 4.3, which depicts a typical frequency plan for the services transported over the HFC network. These services include analog and digital broadcast and pay-per-view video, video on demand, high-speed Internet service, and cable telephony. In addition, status monitoring systems and set top box return path signals are also transported and, in less common cases, other services such as personal communications services (PCS) over cable, return path analog video, or legacy proprietary cable modem systems.

In North America, the downstream signals on cable plants are constrained to reside in 6 MHz RF channels, while in Europe the downstream signal resides in 8 MHz RF channels. Thus, in North America, the downstream RF carriers are typically spaced apart by 6 MHz. On the upstream, the RF signaling bandwidth is variable, depending on the symbol rate used for the upstream signal. In the case of current DOCSIS cable modems, the upstream symbol rate varies from 160 kilosymbols per second to 5.12 megasymbols per second.

The DOCSIS cable data system used to provide high-speed Internet service shown in Figure 4.4 comprises primarily a cable modem termination system (CMTS), which resides in the cable operator's headend facility, and the cable modem (CM) that resides in the customer premises. In addition, servers for time of day (ToD), trivial file transfer protocol (TFTP), and dynamic host control protocol (DHCP) services are required for initialization and registration of the cable modem when it first boots up on the network.

Modem connectivity for a downstream channel is one-to-many, allowing a continuous broadcast in the downstream using a time division multiplex (TDM) format. On the other hand, the upstream modem connectivity is typically many-to-one, so the upstream data transmissions are bursted in a time division multiple access (TDMA) format; they present a more challenging communications problem than the continuous downstream for that reason and also because the RF interference on the upstream channel is much more severe than that present on the downstream channel.

For these reasons, signal processing in the upstream receiver is generally more complex than in the downstream receiver in a cable modem system. Upstream impairments to be mitigated include passive channel impairments, such as group delay variation and microreflections (a type of multipath distortion), and active channel impairments such as ingress of off-air communication, radar, and navigation signals into the cable plant, intermodulation distortion products, and conducted impulse and burst noise from appliances in the home. The signal processing blocks that mitigate these impairments include the interleaver, forward error correction

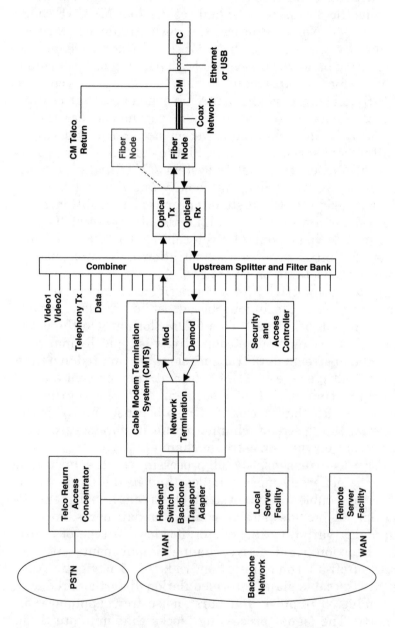

Figure 4.4 Data-over-cable reference architecture (after DOCSIS 2.0 RFI specification).

(FEC), equalization, and ingress cancellation blocks. However, to understand the purpose and function of these blocks, it is important first to understand the passive and active channel impairments. These are presented in the next subsection.

4.3 HFC NETWORK PROPAGATION CHANNEL DESCRIPTION

In this subsection, the passive and active channel effects of the hybrid fiber-coax (HFC) network will be described. Effects such as group delay variation and microreflections are described, as are impairments that arise in the media such as intermodulation distortion, ingress, and conducted interference. Because these impairments are important to modem design, they are described in some detail. When possible, models of the interference are provided to aid future system designers.

4.3.1 Passive Channel Effects

The coaxial cable transmission line shown in Figure 4.5 is a fundamental element of HFC networks; it is a skin-effect-dominated metallic transmission line with transfer function of the form e^{-jkL}, where k is the wave factor for the transmission line and L is the length of the transmission line. For coaxial cable, k has real and imaginary parts and is a function of frequency; thus, the transfer function of coaxial cable has an attenuation that varies with frequency. Because the total loss in the cable increases proportionally with the cable length, L, it is common to characterize coaxial cable by the loss per unit length of the cable, e.g., the loss in decibels per 100 ft of cable. This loss and its behavior vs. frequency will depend on the materials used and dimensions of the coax.

Coaxial cable loss can be primarily ascribed to conductive losses and dielectric losses in the cable. The conductive losses are usually the largest and go as the square root of the frequency in the cable, while the dielectric losses go linearly with frequency. Figure 4.6 shows a typical loss in decibels per 100

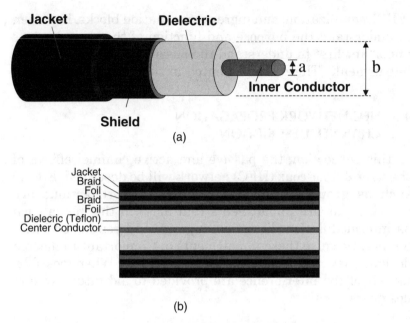

Figure 4.5 Elements of coax cable (a) and detailed construction of RG-6 quad shield typically used for drop lines and in homes (b).

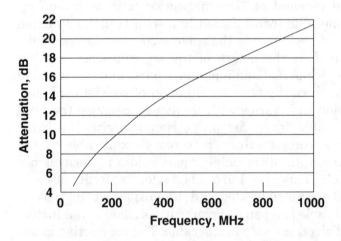

Figure 4.6 Typical loss vs. frequency curve for coaxial cable.

ft vs. frequency characteristic of a common coaxial cable used in HFC networks. The larger the cable is, the lower the resistivity and thus the lower the loss per 100 ft of cable.

The minimum attenuation vs. frequency in coax cable using a dielectric constant close to that of air occurs at a characteristic impedance of approximately 80 ohms.[1] Close to this figure is 75 ohms, which has the additional benefits of being close to the antenna input impedance for a half wave dipole as well as having a simple 1:4 relationship with the impedance of twin lead cable of impedance 300 ohms (previously used for connecting antennas to televisions); therefore, 75-ohm coax cable has become the universal standard for HFC networks and devices such as televisions, set top boxes, and cable modems connected to these networks.

In order to achieve this impedance, the dimensions of the coax must be set properly. From Figure 4.5A, the basic dimensions of a coaxial cable transmission line are the center conductor diameter, a, and the diameter of the inner portion of the outer conductor, b. The characteristic impedance of coaxial cable is then given by[2]

$$Z_{0,\text{coax}} = (377/2\pi\sqrt{\varepsilon_r}) \ln(b/a)$$
$$= (138/\sqrt{\varepsilon_r}) \log(b/a)$$
$$= 75 \text{ ohms for HFC networks} \qquad (4.1)$$

where ε_r is the relative permittivity of the dielectric constant used between the center and outer conductors of the coax.

As long as all connector and amplifier impedances are also 75 ohms, and all pieces of the coax network are properly terminated, no reflections from impedance mismatches occur in the network. In reality, imperfections are always present and lead to microreflections on the coaxial transmission line that lead to excessive attenuation at certain frequencies in the cable plant that may require equalization to mitigate, depending on the order of modulation used. The most common causes of microreflections are

- Impedance mismatches at the connection between the coax and the amplifier
- Not high enough return loss on devices such as amplifiers, taps, power inserters, fiber nodes, and connectors
- Corrosion in any connector on the plant
- Breaks in the coax due to damage by animals or falling objects
- Unterminated drop lines or tap ports

Frequently, multiple low-level reflections or echoes are present, with strong dominant echoes resulting from damage or imperfections in the cable plant. Echoes on the downstream cable plant are attenuated by a larger amount due to the attenuation increasing at higher frequencies; however, on the upstream, echoes can be passed back to the headend with relatively large amplitudes. Furthermore, because the fiber and coax attenuate signals more when they are more distant, the amplitude of echoes is lower as the echo becomes more distant. The DOCSIS specification gives the following guidelines for the upper bound on dominant echoes/microreflections on the downstream, where the decibels relative to the carrier level (dBc) power level indicates how much weaker the echo must be than the original signal:

–30 dBc for echoes greater than 1.5-μsec delay
–20 dBc for echoes between 1.0- and 1.5-μsec delay
–15 dBc for echoes between 0.5- and 1.0-μsec delay
–10 dBc for echoes less than 0.5-μsec delay

The guidelines for the upstream are:

–30 dBc for echoes greater than 1.0-μsec delay
–20 dBc for echoes between 0.5- and 1.0-μsec delay
–10 dBc for echoes less than 0.5-μsec delay

However, note that in these DOCSIS guidelines for bounding dominant echoes, the bounds on upstream echoes are less severe than in the downstream, counter to the general situation found in cable plants. Comprehensive upstream and downstream channel models for echo profiles, as well as other impairments described in this section, are presented by Kolze.[3] These detailed channel models have been widely accepted and applied within the industry to guide advances in cable technology from the first generation of DOCSIS onward. The models were geared toward providing utilitarian channel models which when combined with the theoretical and practical discussions and the illustrative examples in this chapter, should provide the reader with a complete picture of physical channel modeling for HFC networks.

The frequency spectrum of microreflections can be a periodic notch in the frequency response when a single strong

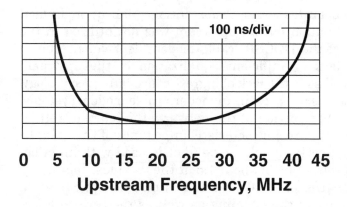

0 5 10 15 20 25 30 35 40 45
Upstream Frequency, MHz

Figure 4.7 Typical group delay variation on the cable upstream.

echo is present, with the period of the notches equal to the reciprocal of the echo's delay. Alternately, a more randomly varying frequency amplitude response will be produced when multiple echoes are present.

The preceding guidelines can be used to determine partially the equalization requirements for cable modem designs that use single-carrier modulation. The other plant impairment that drives equalization requirements is the group delay variation curve, an example of which is shown for the upstream in Figure 4.7. The increase in group delay variation at the lowest frequencies is due to DC filters and lightning protection filters; the diplex filters that separate the downstream and upstream bands cause the group delay variation to increase at the highest upstream frequencies. The group delay variation worsens as the number of amplifiers in cascade increases because each additional amplifier adds more filters to the chain. In some cases, the group delay variation can be hundreds of nanoseconds within the signal bandwidth, and values this high can limit the order of modulation and/or symbol rate that can be used on the upstream near the band edges.

QPSK modulation is robust enough to operate on practically all plants with little or no equalization. However, the impact of microreflections and group delay variation is such that in many plants and in many portions of the upstream band, the higher orders of modulation (16 QAM through 256 QAM on the upstream) will not operate properly at the higher

DOCSIS symbol rates without substantial equalization. In fact, 16 QAM modulation on the upstream was not utilized in the early stages of the DOCSIS cable modem deployment due in part to the lack of sufficient equalization in the upstream receiver, as well as to the lack of ingress cancellation technology. Both technologies are present in modern cable modem systems.

For the fiber portion of the network, the index of refraction is the analogy to characteristic impedance of the coax. In most HFC networks, single mode fiber is used with an effective index of refraction of about 1.5 at the two most commonly used wavelengths of 1310 and 1550 nm. The optical wave travels partially in the cladding as well as the core, and the index of refraction can also vary within the core in order to reduce dispersion in the fiber. Attenuation in single mode fibers is quite low compared to the coax, with typical attenuations of 0.35 dB/km at 1310 nm and 0.25 dB/km at 1550 nm. More detail on optical fiber transport can be found in Chapter 5 of this text.

Having discussed attenuation in the coax and the fiber portion of the network, another passive channel effect is the electrical delay through the HFC network — an effect with an impact on the protocol for HFC networks. The electrical (or propagation) delay in coaxial cable is given by the physical length divided by the velocity of propagation in the cable. The velocity of propagation in coaxial cable with relative dielectric constant ε_r is

$$v = c / \sqrt{\varepsilon_r} \qquad (4.2)$$

where c is the speed of light in a vacuum.

Modern coaxial cables have velocities of propagation in the range 85 to 95% of the speed of light,[1] which means that the relative dielectric constants of the foam used in modern cable are in the range of 1.1 to 1.4. As an example delay calculation in an older all-coax plant, consider a plant with a maximum coax trunk/feeder run of 20 RF amplifiers, each of which supplied 25 dB of gain and extended to a maximum frequency of 550 MHz. One-inch rigid coax has a loss of about 0.6 dB/100 ft at 550 MHz, so a 25-dB loss between amps corresponds to about 4160 ft between amplifiers, neglecting insertion loss of components in the plant. A total length of about 83,000 ft or 25.4 km would be yielded by 20 such amplifiers. If the velocity of propagation is 0.9c, the one-way time delay on the plant would be 25.4 km/(0.9*3 × 10⁵ km/sec) or about 0.1 msec.

In older cable plants, the delay was almost entirely due to the coax portion of the network; however, in modern plants, the fiber generally contributes the largest component to delay. The propagation velocity in fiber is c/n, where n is the index of refraction in the fiber. Because the DOCSIS standard specifies a maximum fiber run of 161 km, the maximum one-way delay in the fiber portion of the network using an effective index of refraction of about 1.5 for single mode fiber is thus 161 km/(2 × 10⁵ km/sec) or about 0.8 msec. When fiber is used, the maximum number of amplifiers is typically only three to five, so the delay in the coax portion is now less than 3% of the total delay in an HFC network.

Wind-loading and temperature variations can alter the electrical length of the fiber and coax and thereby alter the timing and synchronization of signals on the plant. For quasisynchronous modulation schemes such as that used in the TDMA portion of the DOCSIS specification, minor variations in this timing will have a negligible impact on modem operation. However, for synchronous timing schemes such as S-CDMA, timing variations in the cable plant can have a significant impact. For example, Appendix VIII of the DOCSIS 2.0 specification[4] shows that the quasisynchronous method of timing used for the TDMA mode results in an allowable plant timing variation of 800 ns for all orders of modulation, and the synchronous requirements of S-CDMA mode require 90 nsec of timing accuracy for QPSK and 2 nsec of timing accuracy for uncoded 64 QAM/S-CDMA mode.

Temperature changes and wind loading on the fiber can cause such plant timing changes; for maximum fiber runs of 161 km, a temperature change of only 0.3°C will cause a 2-nsec timing variation on the plant. Likewise, wind-loading effects are shown to cause a timing variation of about 6 nsec for moderate wind and relatively short cable runs. Thus, aerial HFC networks have the potential for the greatest timing variations on the network; this can be a limiting factor in the maximum order of modulation and/or modulation rate that can be transported over the HFC network.

One other timing variation that HFC networks can suffer happens whenever a cable operator uses more than one fiber run to a fiber node (shown in Figure 4.1) in order to provide a back-up path in case the primary run is damaged by storm or construction. In this case, the timing on the plant takes a discrete jump when the switchover is made. The impact of this switchover depends on the type of modulation used and system technologies designed to survive momentary discrete jumps in timing on the plant.

4.3.2   Intermodulation Distortion Effects

Intermodulation distortion affects the downstream and the upstream of HFC networks. In the downstream, intermodulation of downstream carriers in the amplifiers leads to composite second order (CSO) and composite triple beat (CTB) distortion products. In the upstream, corrosion on the connectors to and from amplifiers and taps leads to a diode-like behavior on the plant that produces common path distortion (CPD) in the upstream spectrum. Both effects can be modeled similarly via a transfer function that gives the output voltage, V_o, from the plant as a power series in terms of the input voltage, V_i:

$$V_o = GV_i + \alpha V_i^2 + \beta V_i^3 + \ldots \qquad (4.3)$$

where G is the gain of the plant and α and β are coefficients of higher order terms in the power series. CSO distortion can be developed by considering the even order terms, while CTB is developed via the odd order terms. For CPD, even and odd terms are considered.

The most common even order intermodulation distortion products are at difference frequencies of the downstream video carriers: 6 MHz, 12 MHz, and so on for North American HFC networks. Odd order intermodulation distortion products yield difference frequencies of two times one RF carrier frequency plus or minus another RF carrier frequency. However, a complete characterization of intermodulation distortion products reveals many more frequencies, and also the different bandwidths of these frequencies. It can be shown that there are three scales of intermodulation distortion spectral structure[5]:

- Coarse
  - Main intermodulation distortion frequencies that depend on whether the plant is set up for harmonically related carriers (HRCs), incrementally related carriers (IRCs), or standard carriers (STDs). (Listings of these frequency plans can be found on the Internet — for example, http://www.jneuhaus.com/fccindex/cablech.html.) These include the well-known beats at $n \times 6$ MHz.
- Medium
  - Sidebands around each coarse intermodulation frequency that result from the use of offset carriers in certain cable channels per FCC regulations for avoiding aeronautical radio communications. These offsets are 12.5 or 25 kHz away from the nominal downstream frequencies.
- Fine
  - Spreading of intermodulation coarse and medium frequencies with occasional tone-like peaks that result from carrier frequency inaccuracy in downstream modulators. Typical carrier frequency accuracy of cable modulators is on the order of ±5 to ±8 kHz.

In order to determine the frequencies and relative amplitudes of the intermodulation distortion products, an approximate frequency domain representation of the downstream analog carriers can be used:

$$S(f) = \sum_{n=-N_c, n \neq 0}^{N_c} \delta(f - f_n) + \alpha\delta(f - [f_n - f_a]) \qquad (4.4)$$

where
δ = the Dirac delta function
N_c = the number of downstream cable channels
f_n = the video carrier frequency of each channel (generally spaced by 6 MHz in North American HFC networks)

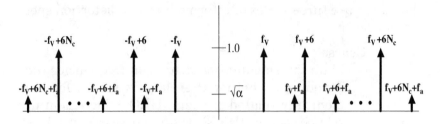

Figure 4.8   Simplified downstream spectrum model.

α = the amplitude of the audio carrier relative to the video carrier (typically in the range of –8.5 to –15.0 dB)

f_a = the spacing between the audio carrier and the video carrier (4.5 MHz in North American HFC networks)

This spectrum is depicted in Figure 4.8.

The second-order mixing products can then be determined from

$$S_2(f) = S(f) \otimes S(f) \qquad (4.5)$$

where $\otimes$ denotes convolution. A similar approach is used to derive the third-order mixing products:

$$S_3(f) = S_2(f) \otimes S(f) = S(f) \otimes S(f) \otimes S(f) \qquad (4.6)$$

Additional intermodulation frequencies are produced, for example, at $f_k + f_j - f_i$, and also at $2f_j - f_i$ and $f_j - 2f_i$. For HRC systems, these additional frequencies are at multiples of 1.5 MHz because the original carriers are at multiples of 6 MHz + (0 or 4.5 MHz).

Note that STD and IRC plans have carrier frequencies that are offset by 0.25 MHz from those of HRC plans. Although this does not affect the location of the second-order mixing products, it will affect the location of third-order products. For example, in an IRC or standard plant, the audio carrier of channel 19 will be at 151.25 + 4.5 = 155.75 MHz. Twice the video carrier of Channel 4 is 2·67.25 = 134.5 MHz. The difference between the two is 21.25 MHz.

Figure 4.9 shows the CPD spectrum from second- and third-order distortion products for an STD frequency plan

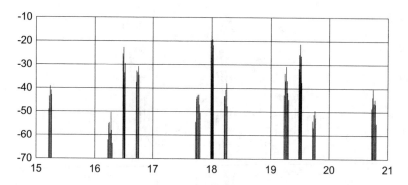

Figure 4.9 Modeled CPD coarse structure for STD plant.

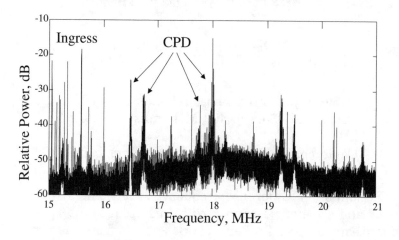

Figure 4.10 Measured CPD coarse structure for STD plant.

using the preceding model, and Figure 4.10 shows measured intermodulation tones (along with ingress) from a plant that uses the STD plan. The medium scale intermodulation frequencies are produced because the FCC requires cable operators to offset the carriers in certain bands by 25 or 12.5 kHz to prevent any leakage signals from interfering with aeronautical radio communications in those bands. The rules are as follows[6]:

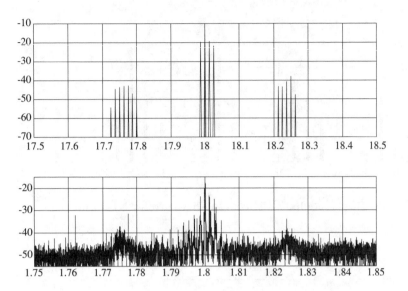

Figure 4.11 Modeled (top) and measured (bottom) medium CPD structure for STD plant.

- Cable in the aero radiocom bands 118 to 137, 225 to 328.6, and 335.4 to 400 MHz must be offset by 12.5 kHz.
- Cable channels in the aero radiocom bands 108 to 118 and 328.6 to 335.4 MHz must be offset by 25 kHz.

Second-order difference frequencies between an offset carrier and a nonoffset carrier will thus produce intermodulation frequencies at 12.5 and 25 kHz offsets from the previously predicted frequencies. Third-order offset products will produce additional intermodulation frequencies at 37.5, 50, 62.5 kHz, etc., from the nonoffset products, which will be lower in amplitude because the number of cable channels that must be offset is less than the number that are not offset. Figure 4.11 shows modeled and measured medium scale intermodulation structure near 18 MHz from an STD plant. Note that the FCC offsets produce a widening of the intermodulation products via additional tones from the offset frequencies. The resulting bandwidth can approach 100 kHz, as seen below in a magnified view near 18 MHz:

Finally, in standard cable plants, the modulators are not locked to a comb generator and thus do not always produce carriers of exactly the specified frequency. The frequency accuracy specifications for typical modulators are $f_{acc} = \pm 5$ kHz or ± 8 kHz.[7] Thus, the actual carrier frequency of any particular modulator will be that specified by the STD frequency plan, plus the specified FCC offsets, if applicable, and, finally, plus a very slowly varying random frequency offset selected from a probability distribution with rough limits of ± 5 or ± 8 kHz. Because this is a significant fraction of the medium CPD frequency structure (at increments of 12.5 kHz), the result is a spreading of CPD frequencies about the nominally predicted frequencies by about half the spacing between CPD medium frequency structure tones. This characteristic is visible in the strongest central tones in the bottom half of Figure 4.11.

Note that as cable operators replace analog carriers with digital carriers, the intermodulation distortion spectrum will no longer have tones at certain frequencies, but rather will have a thermal noise-like spectrum. This is because instead of convolving narrowband carriers with each other, the digital QAM pedestal spectra will be convolved and because the gaps between QAM carriers are much smaller than the carrier width; the result will be due to the convolution of two broad rectangular spectra.

A final effect on intermodulation distortion is the cresting effect reported by Katznelson, in which the downstream CTB and CSO products can suddenly increase in amplitude by 15 to 30 dB for a duration of up to hundreds of milliseconds.[8] The effect is theorized to occur when a sufficient number of downstream RF carriers become phase aligned and thus the intermodulation distortion beats add more coherently for a short period of time. This burst of much stronger intermodulation distortion was estimated to occur on the order of every 10 to 30 sec. A similar effect has been described for CPD on the upstream.[9]

Recent measurements of the upstream and downstream intermodulation distortion products by the lead author of this chapter have confirmed a very rapidly varying amplitude of distortion products of up to 15 dB, which would be accounted for in standard MSO tests for CTB and CSO; however, the 30-dB cresting that occurs several times a minute was not seen in these measurements. Nonetheless, the key point is that the amplitude of the intermodulation distortion (IMD) products is variable, as are other impairments on the cable plant, and thus characterization of IMD and signal processing to mitigate it must take this variation into account.

4.3.3 Ingress

Ingress of off-air communications has previously been modeled using stationary carriers with Gaussian noise modulation.[3] However, actual ingress comes in a variety of forms:

- Strong, stationary HF broadcast sources such as Voice of America
- Data signals with bursty characteristics
- Intermittent push-to-talk voice communications such as HAM and citizen's band (CB) radio signals
- Slow Scan Amateur TV, allowed anywhere amateur voice is permitted, but usually found at these US frequencies: 7.171 MHz, 14.230 MHz, 14.233 MHz, 21.340 MHz, and 28.680 MHz
- Other, less frequent ingress signals such as radar and other military signals

It is relatively straightforward to generate models of all of the preceding communications using commercially available tools for generating communication waveforms; however, the time variation of the signals' power level must be developed. This variation comes from three main sources: fluctuations in atmospheric propagation (multipath, ducting, etc.); fluctuations from vehicular movement (in the case of ham and CB radio signals); and fluctuations because the ingress typically enters the plant in multiple locations. From the evidence that significant reductions in ingress levels occurred after high pass filters were installed throughout the plant, it may be conjectured that ingress typically enters the plant via the subscriber's house or the drop line.

Fluctuations in the signal power of at least 20 dB have been frequently seen with the time scale of fluctuations on the order of tens of milliseconds. Thus, a time-varying, random envelope with power variation of up to 20 dB can be impressed on the preceding signals to generate realistic ingress models for testing new technologies.

For the Morse code communications, captured traces show the on–off cycles of such signals to be on the order of tens of milliseconds ("dots") to hundreds of milliseconds ("dashes").[5] Therefore, a simple model involves gating a CW signal on and off with a 10-Hz rate to emulate such signals. A more complex model includes specific durations for dots and dashes, as well as variations.

Models of voice conversations abound in the telecommunications literature and can be used for detailed modeling of voice signals. Spaces between words (tens of milliseconds) as well as larger silence intervals (seconds) can be applied to the signal models for single sideband and other common ham and CB voice signals. The bandwidths of ingress signals range from extremely narrowband on–off keyed Morse code signals, to voice and slow scan TV signals of bandwidth on the order of 20 kHz, to specialized data signals (e.g., radar waveforms) with bandwidths of hundreds of kHz.

The last point in modeling ingress signals is to determine how many ingress signals are likely to occur in band along with the cable modem signal. This depends highly on whether certain bands are avoided. The DOCSIS specification recommends avoiding the broadcast bands listed in Table 4.1 on the upstream. However, this list was developed before advanced PHY technologies such as ingress cancellation were available. When measurements of upstream ingress are compared with advanced signal processing capabilities such as ingress cancellation, it turns out that many of the bands in the table have few enough ingressors that modems could operate in the bands previously avoided.

On a typical plant, for example, bands 2, 5, and 7 to 9 in the table can turn out to have relatively few strong ingressors, and thus are candidates. A scan of measurement data

Table 4.1 Broadcasting Allocations in 5 to 42 MHz

No.	From (MHz)	To (MHz)	Bandwidth (kHz)
1	5.95	6.20	250
2	7.10	7.30	200
3	9.50	9.90	400
4	1.65	12.05	400
5	13.60	13.80	200
6	15.10	15.60	500
7	17.55	17.90	350
8	21.45	21.85	400
9	25.67	26.10	470

indicates that, with judicious placement of DOCSIS carriers, a typical maximum number of ingress zones to be canceled is about four to six. The term "zone" is used because ingressors frequently occur in groups and must be canceled as a group rather than individually. It is for this reason that the bandwidth to be canceled can often exceed 100 kHz.

Thus, a reasonable model for ingress to use for developing and characterizing advanced signal processing for cable modems is:

- Four to six ingressor zones in band
- Power levels with up to 20 dB fluctuations over tens to hundreds of milliseconds
- Bandwidths in three ranges:
  - 100s of hertz for on–off keyed continuous wave (OOK-CW) signals
  - 3 kHz for amplitude modulated, single side-band (AM-SSB) voice, data
  - 10 (most common) to 20 kHz for frequency-modulated (FM) voice and slow-scan TV (SSTV), and others
- 100 kHz for special signals or for groups of ingress signals to be canceled as a group

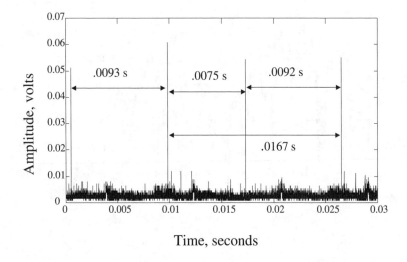

Time, seconds

Figure 4.12 Periodic impulse noise measured on cable plant.

4.3.4 Impulse/Burst Noise

Random impulse noise has been extensively studied in the past, with the following representing current thinking on the subject[10]:

> 1 μsec and less impulse duration (dominant case)
> 10- to 50-μsec burst duration (infrequent case)
> 100 μsec and above (rare)
> Average interarrival time: 10 msec

As for the spectral characteristics, Kolze[3] gives the following model for random impulse noise: Each burst event is AM modulated, zero mean, Gaussian noise with a 1 MHz RF bandwidth, carrier frequency of 5 to 15 MHz (according to measurements) and amplitude ranging from 0 to 60 dBmV.

However, periodic impulse noise such as that depicted in Figure 4.12 is also frequently found on at least one node per headend; to date, no specific models for this phenomenon have been given. These impulses are often quite large in amplitude and appear to occur with different pulse recurrence frequencies that are usually harmonics of the power line frequency

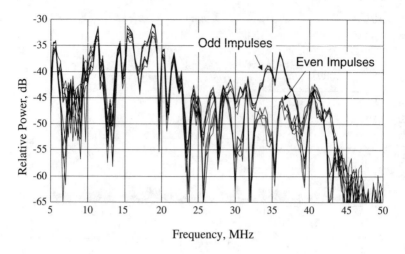

Figure 4.13 FFTs of successive periodic impulses.

of 60 Hz. Examining Figure 4.12 in detail, it appears that the trace is actually two interleaved 60-Hz waveforms.

To check, the fast Fourier transforms (FFTs) of ten successive impulses were captured and plotted in Figure 4.13. Clearly, the even impulses are from one element, and the odd impulses from another different element. Spectra such as the preceding are representative of such events captured from multiple nodes and headends. Therefore, in addition to the current models for random impulse noise and periodic noise with period of 60 Hz, 120 Hz, and so on, we should add to the model interleaved periodic trains of 60-Hz periodic impulse trains with varying offset intervals between them.

4.3.5 Hum Modulation

Hum modulation is amplitude modulation of transmitted signals at the power line frequency or harmonics thereof. Its effect on analog video in North America appears as a "horizontal bar or brightness variation whose pattern slowly moves upward with time (about 16 seconds to move the entire height of the picture)."[1] The period comes from the difference frequency of the power line, 60 Hz, and the frame repetition rate of 59.94 Hz for North America.

Another type of hum observed on cable plants is called "transient hum modulation" on upstream and downstream cable plant.[9] Observations indicate that the RF carrier level was reduced briefly by over 60% for several milliseconds on upstream and downstream cable signals.

4.3.6 Laser Clipping

When the upstream laser clips due to saturation at the input, a burst of clipping noise is generated; this can span the entire upstream spectrum due to harmonics of upstream carriers (or ingress) being generated by nonlinear transfer much as the CSO, CTB, and CPD already described. Causes of input saturation include ingress noise, impulse noise, or simultaneous signal transmission from multiple modems due to collisions of request packets. The result is brief harmonic distortion of upstream carriers as well as the presence of intermodulation products in the bands of other upstream carriers. The burst will be brief in the time domain as well as localized in the frequency domain if impulse noise or intermodulation cresting events are the cause. If ingress is the cause, the effect may last longer, and it can be due to ingress and signal power going into the laser.

4.4 MODULATION SCHEMES FOR CABLE MODEMS

4.4.1 Downstream Modulation Schemes

The DOCSIS downstream physical layer specification[4] uses the ITU J.83B specification for TDM-based single carrier QAM, with support for 64 QAM and 256 QAM. The forward error correction scheme used in J.83B is a concatenation of trellis coding and Reed–Solomon (RS) coding, with interleaving and randomization in between the concatenated codes, as shown in Figure 4.14. The Reed–Solomon block coding can correct up to three symbols within an RS block; the interleaver prevents a burst of symbol errors from being sent to the RS decoder; the randomizer allows effective QAM demodulator synchronization by preventing long runs of ones or zeros; and the trellis coding provides convolutional encoding with the

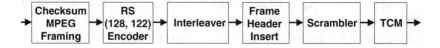

Figure 4.14   DOCSIS downstream error control coding processing.

possibility of using soft decision trellis decoding of random channel errors.

Symbol rates possible on the downstream are as follows:

Modulation type	Symbol rate
64 QAM	5.056941 Msym/sec
265 QAM	5.360537 Msym/sec

Including a modest amount of FEC coding gain, target SNR requirements for a postprocessing bit error rate (BER) of 10^{-8} for the downstream DOCSIS PHY of deployed systems are about 23.5 dB SNR for 64 QAM and 30 dB SNR for 256 QAM. Because both of these SNR values are at least 10 dB below the FCC requirement for analog video SNR on cable downstream channels (43 dB), most operators transmit digital downstream signals at 6 to 10 dB lower than analog signals. This permits improvement in C/CTB and C/CSO performance for the analog channels without compromising C/N performance, as would normally occur on the cable plant, and leaves at least 3 dB margin for customers at the end of the plant as well.

The convolutional interleaving used in DOCSIS downstream signaling delays I groups of J FEC codewords in order to shuffle the transmission of codewords so that a burst of consecutive errors in the channel is transformed into distributed errors with lower individual durations more amenable to FEC receiver processing. In DOCSIS, the downstream packet size is composed of 128 FEC codewords (122 data words plus 6 Reed–Solomon checksum words). To process a single packet, $I \times J$ must thus equal 128. As the value of I increases, the distance between deinterleaved errors increases, which increases the tolerable duration of a burst

Table 4.2 Downstream Latency vs. Interleaving Parameters

I (# of taps)	J (increment)	Burst protection 64 QAM/256 QAM (sec) μ	Latency 64 QAM/256 QAM (msec)
8	16	5.9/4.1	0.22/0.15
16	8	12/8.2	0.48/0.33
32	4	24/16	0.98/0.68
64	2	47/33	2.0/1.4
128	1	95/66	4.0/2.8

noise event, but at the cost of increased latency in transmitting the packet. The variable depth interleaver thus allows a trade-off between impulse noise protection and latency on the downstream, as shown in Table 4.2 from the DOCSIS spec.

As the table shows, at the highest interleaver depth using 64 QAM modulation, burst noise of up to 95-μsec duration can be tolerated at the cost of 4 msec of latency in the downstream. This latency would be too great for VoIP or other high quality of service applications, but would be tolerable for video and best effort Internet data.

4.4.2 Upstream Modulation Schemes

Several upstream modulation schemes were initially proposed for cable modem networks in the IEEE 802.14 committee, including spread spectrum CDMA, S-CDMA, frequency hopping, multitone (several versions), and of course TDMA.[11] For the full story, see the 802.14 committee documents.[12] However, TDMA-QAM finally won the battle via the cable operators' multimedia cable network system (MCNS) consortium, which later became the CableLabs DOCSIS standard. The first version, DOCSIS 1.0, was designed to deliver best effort data for Web browsing. The second version, DOCSIS 1.1, added quality of service (QoS) features and encryption to support VoIP and other QoS intensive services. With DOCSIS 2.0, higher order modulation and a new upstream modulation type, synchronous code domain multiple access (S-CDMA), were introduced to increase capacity and robustness on the upstream.

Table 4.3 Symbol Rates Used in DOCSIS 1.0/1.1

Modulation rate	RF channel bandwidth
160 kHz	200 kHz
320 kHz	400 kHz
640 kHz	800 kHz
1.28 MHz	1.6 MHz
2.56 MHz	3.2 MHz

4.4.3 DOCSIS 1.0 TDMA

The DOCSIS 1.0 and 1.1 upstream specifications provide a TDMA single carrier QAM scheme that includes QPSK and 16 QAM modulations. Symbol rates from 160 kbaud to 2.56 Mbaud are supported, with variable block size Reed–Solomon forward error correction (FEC), and variable frame and preamble structures possible. Table 4.3 shows the different symbol rates and associated RF bandwidths permitted for DOCSIS 1.0 and 1.1 specification-compliant signaling.

The modulation scheme is quasisynchronous, using time stamps for timing and guard symbols to prevent timing errors from degrading performance. Timing is divided into time ticks that are used to specify the size of minislots (the most elemental allocation time on the upstream) in terms of time ticks. Five basic upstream burst types are commonly used: initial maintenance (used for ranging and acquisition); station maintenance (used for periodic ranging); short data transmissions; long data transmissions; and requests for upstream bandwidth. A sixth burst type, request and/or contended data, is specified, but rarely used. These are described in more detail in the following section on MAC protocols.

Although the HFC network is designed to provide downstream signals within a tight range of power levels to each home, the gain/attenuation on the upstream from the homes to the headend varies due to different drop line lengths and tap values and the fact that these values are optimized for downstream, not upstream, signal normalization. Therefore, to get all modem upstream signals to arrive at the headend

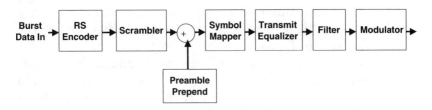

Figure 4.15 Upstream signal processing in DOCSIS 1.0/1.1.

at approximately the same power level, each modem is instructed during the ranging process to adjust its power level up or down so that the desired level is seen at the headend. A transmit range of +8 to +58 dBmV is thus required on DOCSIS cable modems to permit this normalization of power levels at the headend. That the RS block size may be varied permits RS codewords to be optimized for short and long data packets, and the variable symbol rates allow upstream channels to be placed between other upstream RF signals such as legacy proprietary modems, circuit switched telephony, interactive set-top boxes, and status monitoring systems.

Figure 4.15 shows the upstream processing for DOCSIS 1.0/1.1 modems. The data packets are grouped into information blocks that are then converted into RS codewords for FEC encoding. The bits are then scrambled to prevent any long runs of ones or zeroes that might otherwise be transmitted. Scrambling aids the QAM burst demodulator in its symbol synchronization and prevents unwanted peaks in the spectrum. Then, a variable length preamble is prepended, with longer preambles used for noisier conditions or for higher orders of modulation. The preamble is used by the headend demodulator to synchronize to the upstream burst. Next, the bit stream is mapped to symbols to be transmitted — that is, in-phase (I) and quadrature (Q) coordinates in the QAM constellation — and the resulting complex-valued symbol is square-root-Nyquist filtered to reduce frequency side-lobes and intersymbol interference (ISI). Finally, the resulting filtered symbol is modulated up to the RF carrier frequency specified for the channel, which in North America is in the range of 5 to 42 MHz and in Europe is 5 to 65 MHz.

DOCSIS 1.1 implements upstream equalization using predistortion, rather than relying on equalization at the receiver, because if each data burst were preceded by a preamble long enough to train an equalizer, significant amounts of upstream bandwidth would be wasted, especially for small packets. Because thousands of modems can theoretically be connected to a single DOCSIS upstream channel and the RF channel between each modem and the headend is potentially unique, the upstream receiver must equalize the bursts from each modem uniquely to improve the performance.

Given that storing all equalizer coefficients of all modems at the CMTS would be memory intensive and require large processing overhead, the solution used is to send the coefficients to each modem and have the modems predistort the signal prior to transmission. The HFC network then reverses the effect of the predistortion. Thus, during an initialization process, the upstream receiver must measure the distortion in the received signal from each modem, calculate the necessary coefficients for the cable modem based on ranging bursts, and send the coefficients to each modem for use on transmissions.

Figure 4.16 shows an example upstream receiver. For each burst that arrives, the receiver must acquire the timing, gain, and phase of the waveform in order to demodulate it properly. This is the job of the ranging, preamble processing, and tracking loop blocks in combination with the M-QAM

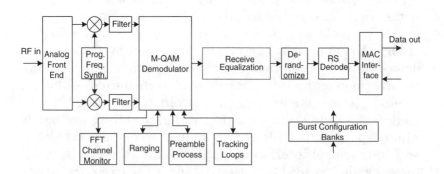

Figure 4.16   Example upstream burst receiver for DOCSIS 1.0/1.1.

demodulator, which, for DOCSIS 1.0 (and 1.1), refers to QPSK and 16 QAM. Following demodulation, the decoded symbols are then equalized, descrambled, and Reed–Solomon decoded. The decoded bits are then sent to the media access control (MAC) processing block for reassembly into IP packets or MAC messages.

4.4.4 DOCSIS 1.1 TDMA MAC Additions

With DOCSIS 1.1, several significant additions to the MAC were made in order to enable VoIP and other high quality of service applications, as well as increase efficiency on the upstream by mandating concatenation, and fragmentation support. Details of the DOCSIS MAC protocol are provided in Section 4.5.

4.4.5 DOCSIS 2.0 Advanced TDMA and S-CDMA

DOCSIS 2.0 added higher order modulation, better coding, RS byte interleaving, a higher symbol rate, synchronous operation, and spread spectrum modulation via synchronized code domain multiple access (S-CDMA). The RS FEC was increased to $T = 16$, which increases the number of correctable errored symbols, and trellis coded modulation (TCM) is added to the S-CDMA mode of transmission. By expanding the constellations to 64 QAM and128 QAM/TCM in S-CDMA mode (which gives the same spectral efficiency as 64 QAM without TCM), the spectral efficiency of large packets can be increased by 50% over DOCSIS 1.0/1.1 systems, which only support a maximum constellation of 16 QAM.

Synchronous operation is required to maintain the orthogonality of codes in S-CDMA mode. It provides the additional benefit of reduced guard times between bursts and reduced preamble for burst demodulation; these improve the efficiency of small packets in which the preamble can occupy a significant portion of the burst. By expanding the symbol rate to 5.12 Mbaud, improved statistical multiplexing, which also provides more efficient use of RF bandwidth, can be obtained.

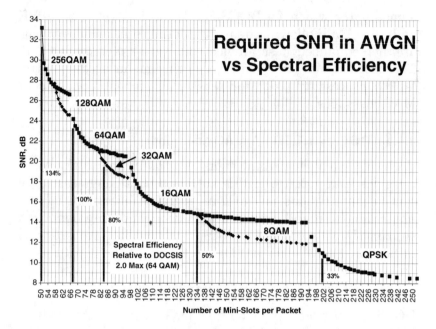

Figure 4.17   Required SNR for 1% packet error rate as a function of modulation order, FEC parameters, and associated number of mini-slots per packet.

DOCSIS 2.0 adds higher orders of modulation as well as more intermediate orders of modulation to the possible burst profiles used on cable upstream transmissions. Increasing the size of the constellation increases the spectral efficiency of the channel via more bits per hertz transmitted, at the cost of requiring higher SNR. Adding more intermediate constellations allows a better trade-off between spectral efficiency and robustness to be made.

Figure 4.17 shows the impact on spectral efficiency from using decreasing orders of modulation and FEC parameters with increased robustness and overhead. The required SNR corresponds to an average packet error rate (PER) of less than or equal to 1%. The abscissa on the chart is the number of minislots required to transmit the same packet as the modulation order and FEC parameters are varied. As the chart shows, the FEC coding can provide more than 5 dB of coding gain for a given order of modulation. However, in some conditions, it is better from a spectral efficiency viewpoint to use a lower order of modulation with less coding gain than a higher order of modulation with maximum coding gain. This conclusion has also been reached by other DOCSIS researchers.[13]

To support the higher orders of modulation, the following are generally required: higher SNR, better equalization in the channel, and longer preambles. It is expected that over time, cable channels will have higher available SNRs due to reduction in the number of homes passed per fiber node and improved maintenance. Equalization is improved via more taps to the pre-equalizer (24 taps in DOCSIS 2.0 instead of 8 previously). The maximum preamble length was also increased from 1024 b in DOCSIS 1.0 to 1536 b in DOCSIS 2.0 in order to provide more robust estimation of synchronization parameters (including gain, carrier frequency and phase, symbol clock frequency and phase, and equalizer coefficients) for the received burst. Note that the necessity of longer preambles for the higher order modulations means that their use on the smallest packets may not be as beneficial because the preamble and packet overhead are significant portions of the overall packet duration on the channel.

For the smallest packets, other channel efficiency technologies such as payload header suppression (described in Section 4.5.2.1) can be more effective. DOCSIS 1.0 specifies QPSK and 16 QAM preambles (the preamble has the same modulation as the data portion of the packet); however, DOCSIS 2.0 uses QPSK preambles for all burst types to save preamble random access memory (RAM) storage space in the CM. It also adds an optional higher power preamble so that short packets and higher orders of modulation may be more effectively acquired.

The use of higher order modulation also leads to larger peak-to-average power ratios in the transmitted waveform. The S-CDMA waveform is the sum of multiple transmitted codes and thus resembles a Gaussian signal, which has a large effective "peak" to average ratio. Therefore, DOCSIS 2.0

requires a reduction in the maximum transmitted power for S-CDMA and the higher order TDMA modulations to control distortion or clipping of the transmitted waveform. For example, A-TDMA, 64 QAM reduces the maximum transmit power to +54 dBmV, but S-CDMA reduces it to +53 dBmV. Clipping of the upstream laser is also a concern, although it is caused by peak excursions of the aggregate upstream waveform, which is the sum of all the channels that share the upstream spectrum.

Trellis coded modulation (TCM) is provided for the S-CDMA mode in DOCSIS 2.0. Unlike the FEC previously described, TCM provides coding gain without reducing channel capacity. The constellation size is expanded such that additional constellation points are used as parity bits for coding; thus, no loss in information capacity occurs. However, this means that 128 QAM/TCM gives the same capacity as 64 QAM without TCM. Theoretically, TCM can provide an additional 4.1 dB of coding gain, but only when the FEC coding gain is minimal; when the Reed–Solomon coding gain is high, the additional TCM coding gain is diminished and can even be negative.[13]

Synchronous operation is currently part of the S-CDMA portion of the 2.0 specification, but was proposed for the advanced TDMA (A-TDMA) specification proposal as well as in the IEEE 802.14a committee discussions. In synchronous schemes, the upstream symbol clock of each modem is locked to the downstream symbol clock so that packets that arrive at the headend from different modems are aligned to within a few nanoseconds of each other. Because the headend receiver needs three items to demodulate a burst (gain, phase, and timing) and the preamble is often used to determine the timing of a burst, the synchronous mode of operation permits shorter preambles by ensuring that timing is already guaranteed. It also reduces the required guard time between bursts. Figure 4.18 shows how synchronous timing is used in DOCSIS 2.0 S-CDMA mode to minimize the guard time and phase error for upstream bursts.

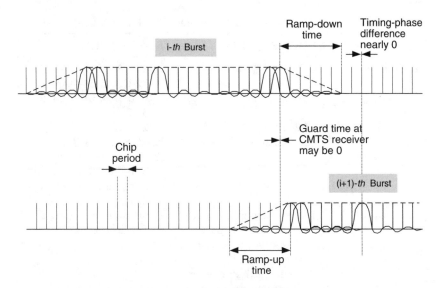

Figure 4.18 Depiction of synchronous timing in S-CDMA.

4.4.5.1 Advanced TDMA in DOCSIS 2.0

Higher order modulation, enhanced equalization, longer FEC, and RS byte interleaving in 2.0 have already been described and are part of the A-TDMA specification. Of these features, RS byte interleaving is the only one not utilized in the S-CDMA portion of the 2.0 specification. Table 4.4 summarizes the additions to the A-TDMA mode of DOCSIS 2.0 and Table 4.5 shows the performance of advanced TDMA interleaving/FEC in the presence of burst noise for Reed–Solomon $T = 16$. The analysis assumes the corruption of all symbols coincident with the noise burst plus an additional symbol before and after the noise burst.

One of the key advantages of A-TDMA is the lack of degradation to existing legacy modem networks running DOCSIS 1.0 and/or 1.1 (termed 1.x networks). A-TDMA transmissions operate in the same logical channel as 1.x transmissions and thus there is no repetition of MAC overhead nor statistical multiplexing loss as occurs when S-CDMA and

Table 4.4   Key Parameters of Advanced TDMA Physical Layer

Category	DOCSIS 1.0/1.1	Advanced TDMA
Modulation	QPSK, 16-QAM	QPSK, 8-QAM, 16-QAM, 32-QAM, 64-QAM
Symbol rates (Mbaud)	0.16, 0.32, 0.64, 1.28, 2.56	0.16, 0.32, 0.64, 1.28, 2.56, 5.12
Bit rates (Mbps)	0.32–10.24	0.32–30.72
FEC	Reed–Solomon, $T = 0$ to 10	Reed–Solomon, $T = 0$ to 16
Interleaving	None	RS byte; block length may be adjusted dynamically to equalize interleaving depths
Equalization	Transmit equalizer with eight symbol-spaced (T-spaced) taps (using symbol spacing obviates the need for fractional equalization schemes)	Transmit equalizer with 24 T-spaced taps
Ingress mitigation	Vendor specific	Receiver ingress cancellation
Preamble	QPSK or 16-QAM; length ≤1024 bits	QPSK-0 (normal power) and QPSK-1 (high power); length ≤1536 bits (≤768 T)
Spurious emissions	Sufficient for 16-QAM	Generally 6 dB tighter to support 64-QAM

TDMA transmission must coexist on the same RF upstream channel (see the section on MAC issues).

Furthermore, although new modulation schemes cannot provide benefits to existing DOCSIS 1.x modems on the network, four robustness improvements in A-TDMA systems apply to legacy DOCSIS cable modems:

- Ingress cancellation processing
- Improved receive equalization
- Improved burst acquisition
- Improved error correction for impulses

Ingress cancellation is not part of the DOCSIS 2.0 specification, but is found in some form in most modern upstream

Table 4.5 Interleaver/FEC Performance in Periodic Burst Noise

Modulation format	Symbol rate (Msps)	Packet length (bytes)	Number of interleaved RS codewords	Maximum correctable noise burst length (usec)	Maximum correctable noise burst repetition rate (kHz)	FEC code rate (%)
64-QAM	5.12	74	No interleaving	3.3	35.7	69.8
64-QAM	5.12	74	2	7.4	27.7	53.6
64-QAM	5.12	74	4	15.2	18.9	22.4
64-QAM	5.12	1528	8	32	2.16	85.7
64-QAM	5.12	1528	16	63	1.88	74.9
64-QAM	5.12	1528	32	64	3.01	59.9
64-QAM	5.12	1528	64	131	2.15	42.7
16-QAM	1.28	74	No interleaving	21	6.06	69.8
16-QAM	1.28	74	2	44	4.69	53.6
16-QAM	1.28	74	4	93	3.19	22.4
16-QAM	1.28	1528	8	196	0.359	85.7
16-QAM	1.28	1528	16	380	0.314	74.9
16-QAM	1.28	1528	32	390	0.502	59.9
16-QAM	1.28	1528	64	790	0.358	42.7

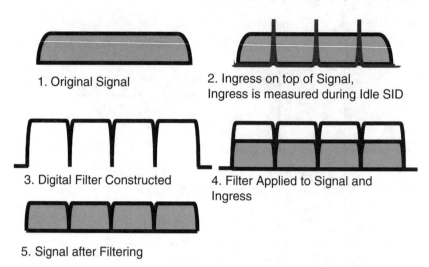

1. Original Signal

2. Ingress on top of Signal,
Ingress is measured during Idle SID

3. Digital Filter Constructed

4. Filter Applied to Signal and
Ingress

5. Signal after Filtering

Figure 4.19   Depiction of ingress cancellation processing.

receivers in order to support the higher order modulations and operation below 20 MHz, where most man-made ingress noise is found. Essentially, the ingress canceller is a digital filter that adaptively responds to narrow-band and wide-band ingress or CPD and filters it out, as depicted in Figure 4.19.

In a typical implementation, the ingress canceller analyzes the channel during the preamble or between packet bursts and computes cancellation coefficients via DSP. These coefficients may be updated at a user-specified rate so that, if slowly varying ingress is present, the overhead burden to the channel capacity is minimized. On the other hand, if rapidly varying ingress is present, the updates may be made much more frequently in order to adapt quickly enough to the ingress.

An example of multiple continuous wave (CW) ingress that has been effectively cancelled by an ATDMA burst receiver is shown in Figure 4.20 along with the 16-QAM constellation after filtering. The signal to interference ratio (SIR) can be improved by tens of decibels using such techniques. In general, a single CW interferer can be canceled even at relatively high CW levels relative to the signal; multiple

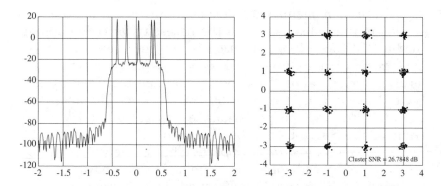

Figure 4.20 Performance of ingress cancellation for 16-QAM constellation.

CW interferers, or wideband modulated interferers, are more difficult to cancel effectively.

Modern burst receivers are able to receive 64-QAM upstream bursts with implementation loss on the order of 0.5 dB. This performance requires precise estimation of synchronization parameters using short preambles and advanced digital receiver design techniques.[14] The low implementation loss translates directly into a net SNR improvement for new and legacy DOCSIS cable modems.

4.4.5.2 S-CDMA in DOCSIS 2.0

In this section, the details of the synchronous code-division multiple access (S-CDMA) mode provided in DOCSIS 2.0 will be presented. The S-CDMA mode provides increased robustness to impulse noise and increased granularity in minislot utilization. The S-CDMA specification also provides improved efficiency in terms of reduced guard time and preambles via synchronous operation and further provides enhanced coding via trellis-coded modulation (TCM). Note that TCM was also proposed for the TDMA mode of operation in the IEEE 802.14a committee's draft specification and is a logical candidate for future versions of the DOCSIS specification. Synchronous TDMA is also a candidate for future DOCSIS upgrades.

Discussion begins with the spreading operation, the derivation of the codes used, and the necessity for code hopping in the DOCSIS 2.0 specification. Then, the requirements of synchronization are presented, as well as MAC concepts such as 2D minislot mapping and logical channels required to make S-CDMA and TDMA compatible on the same RF upstream channel. Finally, the performance aspects of and various trade-offs in S-CDMA will be discussed.

Frequency-spreading property. The frequency-spreading operation of spread-spectrum systems is well known to communications designers. In a direct sequence spread-spectrum system, a low data rate signal has its bandwidth increased by multiplying each data symbol by a sequence of high-rate code chips, e.g., a binary sequence of ± ones. The signal bandwidth is increased or spread in the frequency domain by this operation because the chip rate normally exceeds the symbol rate by a significant factor, typically several orders of magnitude. At the receiver, after synchronizing to the spreading code, the received code sequence corresponding to one data symbol is multiplied, chip by chip, by the same spreading code and summed over the sequence. Because ± ones multiplied by themselves always yield one, the original data symbol is recovered. Any narrowband interfering signal, however, is scrambled by the despreading operation and reduced in power relative to the desired signal by a factor called the "spreading gain," which equals the number of chips per data symbol.

Frequency spreading is only useful if a reduced number of codes is considered. For example, the spreading gain for a single code is the number of binary bits or chips in the code sequence, 128. The data rate carried by a single code is 128 times less than the channel data rate. In the widest DOCSIS 2.0 channel with a chip rate of 5.12 MHz, the symbol rate carried by a single code is 5.12 MHz/128 = 40,000 QAM symbols per second; after spreading, the code is transmitted at 5.12M chips per second over the upstream channel and after despreading in the receiver, the single code yields back the original 40,000 symbols per second of QAM data.

In an actual 2.0 system, a single code is never transmitted alone. The DOCSIS 2.0 spec permits from 64 to 128 codes to be active on the channel — that is, in use and shared at any given time by all the modems on the planet. The spec also permits from two to the entire number of active codes to be allocated to a given modem at a given time; that is, the minimum number of codes that an individual modem can transmit is two. When a given modem is transmitting a subset of the codes, other modems are simultaneously transmitting on the other active codes. The headend receiver must despread all the active codes simultaneously.

If, for example, the number of active codes is 128 (i.e., all codes are in use), then the spreading gain is unity (0 dB), which is to say that there is no spreading gain. In the preceding 5.12 Msps example, the data symbol rate seen by the despreader is 40 ksps per code × 128 codes in use = 5.12 Msps, which is the same as the chip rate. Thus, no frequency-spreading gain occurs when all codes are active. The maximum frequency-spreading gain in DOCSIS 2.0 is only 3 dB and is obtained when 64 codes are active.

Time-spreading property. DOCSIS 2.0 S-CDMA can be visualized as a spread-frequency and spread-time system. The symbol duration used for conventional TDMA transmission is stretched out in time to be 128 times the original symbol period. Then, the spreading code chips are applied to the stretched time symbol at a chip rate equal to the symbol rate used originally for TDMA mode transmission. The added benefit is that the S-CDMA symbols are now 128 times longer and thus the same RS error correction code will be able to correct much longer impulses. Because impulse noise is a problem in cable plants, it can be argued that time spreading has more practical value in S-CDMA than frequency spreading. For S-CDMA ingress mitigation, ingress cancellation processing specific to S-CDMA operation will yield better performance.

Orthogonality property. For the modulation scheme to work optimally, all codes must be orthogonal so that users do not interfere with each other. To accomplish this, the system uses synchronized orthogonal code sequences that can be developed as follows. Start with a maximal length sequence of length 127. These codes have the property that the cyclic

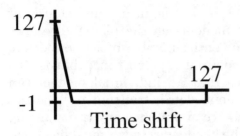

Figure 4.21 Autocorrelation property of maximal length sequence used to derive S-CDMA codes.

autocorrelation of the code is 127 at a shift value of 0, 127, 254, etc., and is –1 for all other cyclic shift values, as shown in Figure 4.21. Therefore, it is possible to envision 127 different codes that are merely time-shifted versions of the original maximal length sequence.

To make the codes truly orthogonal (i.e., to have zero cross-correlation with each other), it is only necessary to append a constant bit to the end of each shifted version of the original code so that the correlation with each other is now 0 instead of –1. The bit that makes each sequence balanced (equal number of plus and minus ones) is used, so the codes will have a cross-correlation of zero with one another and with the all-ones sequence, termed code 0 in the 2.0 specification. Now, 128 codes orthogonal to each other exist, 127 of which are cyclically shifted versions of a single maximal length sequence with an appended bit; the final code is all ones. Because the all-ones code does not have the spreading and whitening property of the codes based on the maximal length PN sequence, it is not often used in actual system operation.

Although the spectrum of the original 127-b PN sequence was white, the spectrum of the modified sequences is no longer flat with frequency, as shown in Figure 4.22. Thus, some codes will be more sensitive to narrowband ingress at certain frequencies than others. As a consequence, code hopping becomes part of the specification so that a particular modem will not be disadvantaged relative to other modems by using a subset of the available codes. Essentially, the active codes are

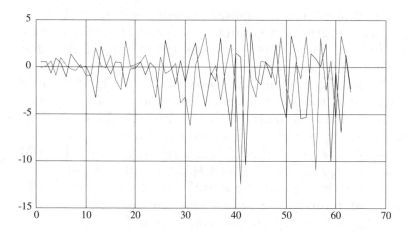

Figure 4.22   Frequency spectrum of two S-CDMA codes.

assigned in a cyclical scheme so that each modem uses a different set of codes on each spreading interval and thus has performance equal to the average across all codes.

To ensure that the total power on the HFC network during an S-CDMA burst does not grow as the number of users/codes increases, the power per code transmitted is equal to the power that would have been used in an equivalent TDMA burst divided by the number of active codes per frame. The number of active codes per S-CDMA burst can be reduced to any nonprime integer from 128 to 64 inclusive in the 2.0 current specification. If the number of active codes is reduced by 3 dB (to 64), the power per code can be increased by 3 dB for an SNR improvement in the channel at a cost of 50% reduction in channel capacity.

S-CDMA operation with increased power per code. A cable modem experiencing abnormally large attenuation in the return path to the headend can be assigned increased power per code and instructed to use less than the full number of active codes. As of this writing, this feature is under discussion as an addition to the DOCSIS 2.0 spec. As an illustrative example, assume a CM experiences 9 dB more attenuation than other CMs and is having trouble reaching the headend with sufficient SNR to operate with 64-QAM modulation. The

CM is told to increase its power per code by 9 dB, and at the same time, its grants are limited to one eighth the total active codes (e.g., a maximum of 16 codes out of 128 active codes; note that 10 log 8 = 9 dB).

The maximum transmit power of the CM is unchanged because eight times fewer codes are transmitted, each code with eight times the power. However, because the slicer SNR at the receiver depends on the power per code and not on the total power, the received SNR is boosted by 9 dB and the CM can now use the network with 64-QAM modulation. Moreover, full network throughput is maintained; although a given CM is assigned fewer codes, the MAC has the flexibility to assign other CMs to transmit on the remaining codes.

A similar trade-off in capacity for SNR could in principle be made in TDMA by reducing the symbol rate by a factor of eight and maintaining the CM transmit power originally used. This would increase the power per symbol by eightfold or 9 dB, while reducing the peak data rate of the CM in question by the same 9 dB. However, because the current DOCSIS TDMA system does not have the granularity to permit the bandwidth thus sacrificed to be used by other CMs during that time period, the total TDMA network throughput would be reduced.

Although the number of active codes can be reduced in S-CDMA mode to achieve an SNR boost, no corresponding mechanism can increase the SNR during spreader-off bursts. If an increase in SNR were required, the system would fail during acquisition or during subsequent spreader-off bursts during maintenance. To limit this concern, the DOCSIS 2.0 specification permits a reduction in active codes to only 64 rather than down to 2 active codes. This attribute of the acquisition phase may be addressed in future releases of the specification.

Because codes are transmitted simultaneously, S-CDMA uses frames as the fundamental timing element as opposed to time ticks used in TDMA mode. The framing structure of S-CDMA in 2.0 is shown in Figure 4.23. The capacity of a frame is found as follows: start with the number of spreading intervals, K, per frame, where a spreading interval is the time

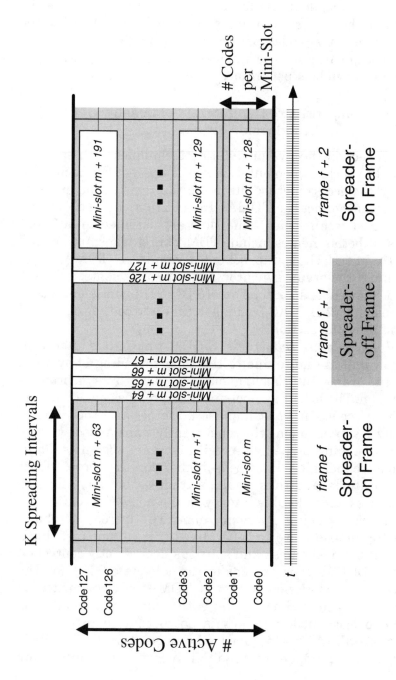

Figure 4.23 Frame structure of S-CDMA.

required to transmit one full code of 128 chips. This sets the time length of a frame for a given symbol rate. Then use the number of active codes per frame and number of codes per minislot to calculate the number of minislots per frame or number of symbols per frame.

4.4.5.3 Comparison of DOCSIS 2.0 TDMA and S-CDMA Modulation Schemes

In this section the two modulation techniques with specifics related to each implementation are compared. For each technology, the access method during initial registration and station maintenance is TDMA. In addition, in each of the considered technologies, time slots are assigned to different users, so both schemes include TDMA burst transmission and TDMA MAC. Thus, the schemes require a TDMA burst modem, with varying synchronization requirements. S-CDMA can be viewed as an extension of TDMA transmission to two-dimensional (2-D) MAC framing in time and code as discussed in the previous section.

Impulse noise robustness. S-CDMA mode has an advantage in impulse noise due to the time spreading of symbols and the frame interleaving made possible by synchronous operation. The longer symbol duration provides an advantage in the presence of weak impulse noise because the impulse energy is spread among the concurrently transmitted symbols (time spreading property). As with most impulse noise mitigation technologies, latency is higher for the more robust configurations, and thus, again, a trade-off between impulse robustness and tolerable latency in the network occurs.

Transmit power dynamic range. The increased robustness to impulse noise in S-CDMA is not without cost. Each transmitter has a noise and spurious floor strictly controlled by the DOCSIS spec to ensure an adequate SNR for the upstream transmission system. In TDMA, only one modem is transmitting at a time, so the entire spurious allocation is given to each modem, which is not permitted to transmit below +8 dBmV to stay above the noise floor. In S-CDMA, time spreading implies that multiple modems are transmitting simultaneously, so the spurious floor must be proportionally lower for each transmitter. The worst-case example occurs if two codes make up a minislot. Then up to 64 S-CDMA modems could in theory be transmitting at one time, if all were granted one minislot in a given frame. However, when reducing from 128 codes to 2 codes, if the spurious output of the modem does not reduce by a factor of 64, but exhibits an (implementation dependent) noise floor, then when the noise floors from the 64 modems sum, the aggregate signal may no longer meet the DOCSIS spec.

To avoid this problem, the spec requires that an S-CDMA modem never transmit below +8 dBmV, even when only one minislot is transmitted. This implies that with all codes allocated, the modem cannot transmit below 8 + 10 log 64 = 26 dBmV, a reduction in low-end dynamic range of 18 dB compared to equal throughput in TDMA for this worst-case example. The spec also limits S-CDMA at the upper end of the range to +53 dBmV, compared to a maximum for TDMA QPSK of +58 dBmV, so the net is 23 dB less range for S-CDMA than TDMA in the worst case.

In practice, this effect is mitigated by the fact that S-CDMA does not use the lowest modulation rates (below 1.28 MHz), which normally would utilize the low transmit power range, and by using minislot sizes larger than 2 codes. Also, the transmit power range of TDMA is limited by the spec when it mixes modulation types other than QPSK. Further enhancements of the spec may relax the power limitations on both modulation types. As discussed earlier, S-CDMA has a potential advantage at the high end of the transmit power range in that it can assign increased power per code. This increase comes at a throughput expense per CM because the maximum number of codes that can be transmitted by a given CM is then limited.

Narrowband ingress robustness. With TDMA and S-CDMA, agility in modulation bandwidth and carrier frequency can be used to avoid RF bands with known severe interference. Remaining narrowband interference can be mitigated by adaptive ingress cancellation, which has been demonstrated effectively for both modulation types. With

S-CDMA, decision-directed ingress cancellation techniques are a challenge at the chip level because the despreading delay prevents the timely availability of reliable decisions. Other estimation and subtraction techniques with typically greater complexity may be employed.[15,16] Also, as already discussed, S-CDMA can use frequency or spectrum spreading as a form of signal repetition, trading bandwidth efficiency for robustness against narrowband noise, although the robustness increase is small compared with a large capacity reduction.

Synchronization sensitivity. S-CDMA requires much tighter timing synchronization compared to single-carrier TDMA, due to the requirement to maintain code orthogonality. For instance, the synchronization accuracy requirement for uncoded 64-QAM is ± 2 nsec (0.01 of a symbol) at the highest symbol rate of 5.12 Mbaud for S-CDMA, compared to ± 250 nsec for TDMA, a factor of over 80. Timing errors can translate into loss of code orthogonality, the result of which can be a self-inflicted noise floor due to intercode interference. An example of this effect for a non-DOCSIS S-CDMA system is shown in Figure 4.24.[17]

4.5. MAC PROTOCOLS FOR HFC NETWORKS

4.5.1 Introduction

In this section, MAC protocols for HFC networks will be discussed, with an emphasis on the DOCSIS MAC protocol because it is the current international standard and likely to continue as the cable modem standard until fiber reaches the home. Initially, the HFC network MAC protocols used by proprietary cable modems were quite diverse. The Zenith cable modems used essentially Ethernet over RF, and the LANcity and Motorola modems used proprietary protocols more appropriate to the cable modem network, but still IP based. The proprietary modems from Com21 used protocols based on ATM. In 1994 the IEEE 802.14 committee was tasked with developing a PHY specification and a MAC protocol for cable modem networks. Many different protocols were proposed, among which were centralized priority reservation (CPR)[18] and

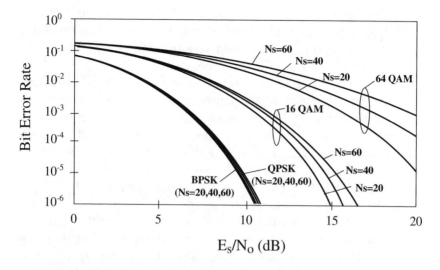

Figure 4.24 Effect of timing error and resulting code noise on S-CDMA performance. (After de Jong, Y. et al., *IEEE Trans. Broadcasting*, 43(2), June 1997.)

specialized protocols required for 2-D modulation techniques such as S-CDMA and multitone schemes such as variable constellation multitone (VCMT).[12]

In the end, the authors of the original DOCSIS standard chose a single carrier PHY specification and a MAC protocol that has become the international standard for cable modem networks; in fact, the DOCSIS MAC protocol is finding its way into fixed broadband wireless networks as well as satellite data networks. For more detail on satellite networks that employ the DOCSIS protocol, see Chapter 7 in this text.

The DOCSIS MAC protocol is a centralized scheme with equal fairness to all modems on the network regardless of their distance from the headend. This is in contrast to protocols such as CPR in which the advantage of modems closer to the headend is used to create a slight increase in capacity for the overall network. At the time of the development of the DOCSIS protocol, HFC networks already supported the MPEG II protocol for transmission of downstream digital video to set top boxes; therefore, the DOCSIS protocol specifies that the downstream data transport be in the same 188-byte MPEG packets as digital video via an MPEG transmission convergence sublayer. The downstream is thus a TDM scheme and the upstream is a TDMA scheme.

Also, the upstream is much more prone to RF interference, so the DOCSIS MAC protocol specifies mechanisms for setting MAC and PHY upstream parameters in order to vary the robustness of the network depending on the actual cable plant conditions. This is done via upstream channel descriptor messages sent on the downstream.

The DOCSIS MAC protocol further provides link-layer security with authentication in order to prevent theft of service as well as providing data security at a level similar to telephone networks. The security mechanism, called Baseline Privacy Plus, provides in-line 56-b DES encryption and decryption, but is not an end-to-end security solution; only the cable modem network is protected. IP security (IPSEC) protocol and other security protocols are used for end-to-end security over HFC networks.

The initial version of the DOCSIS MAC protocol, version 1.0, was designed for best effort service. Version 1.1 added support for multiple QoS and flow types per modem, thereby enabling VoIP and other low-latency/jitter applications to cable modem networks. Version 1.1 also added fragmentation, concatenation, and payload header suppression (for transport of voice packets with headers that are fixed during voice calls) in order to support VoIP and to increase efficiency of small packet transport. Version 1.1 also added security enhancements such as Baseline Privacy Plus, which added authentication to the in-line data encryption standard (DES) encryption/decryption, and encryption support for multicast signaling.

DOCSIS version 2.0 added S-CDMA and ATDMA physical layer technologies (previously described), synchronization to the downstream symbol timing for S-CDMA operation, logical channels for supporting mixed channels with S-CDMA and TDMA, and other modifications to support the advanced physical layer technologies in 2.0. A two-dimensional MAC was also added to support S-CDMA; the logical channel approach was specifically added because, unlike the 802.14

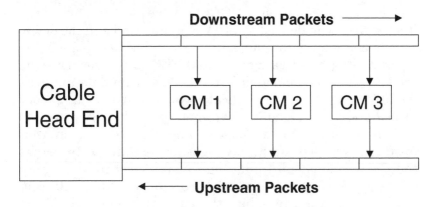

Figure 4.25 MAC model of a cable data network.

committee proposal in which S-CDMA did not need to inter-operate with TDMA on the same RF carrier, in DOCSIS 2.0 the mixing of S-CDMA and TDMA on the same RF channel was required. This permits cable operators to transition to 2.0 systems gradually while maintaining legacy modems on the same network.

4.5.2 Detailed MAC Protocol Description

The usual model for a MAC protocol for HFC networks is shown in Figure 4.25. In the DOCSIS MAC protocol, all stations (cable modems or CMs) on the plant must delay their transmissions so that the farthest station on the network (CM 3) has the same opportunity to request bandwidth as CM 1 and CM 2, which are much closer to the headend. Among other functions, a MAC protocol must specify mechanisms for stations to:

- Initialize and log on to the network
- Configure and update their transmission and reception settings
- Request data bandwidth and to transmit and receive their data
- Initialize, engage, and update encryption of data for privacy

For example, when a DOCSIS cable modem powers up, it must first scan downstream TV channels and find the downstream channel containing data for the cable modem network. Upon finding and locking sync and FEC in the downstream, the CM can then demodulate the downstream data and look for MAC messages that indicate which upstream frequency to use and what signaling parameters on that frequency to use in order to begin ranging and registration on that upstream channel. First, messages are exchanged to permit the CM to adjust its timing and power level in accordance with the CMTS. Next, messages are sent to establish IP connectivity and then to obtain the current time of day. Then, the CM registers with the CMTS and the CMTS assigns the CM a service ID (SID) and begins allocating time slots on the upstream for the modem to use in transmitting data. Lastly, the CM and CMTS establish the encryption and decryption keys to be used on the network via Baseline Privacy in DOCSIS.

During operation, the overall channel parameters to be used by all modems on the channel are specified in the downstream and include the minislot size in units of timebase ticks, the symbol rate to be used, the upstream channel frequency, and the preamble pattern, which may be as long as 1024 b. Individual burst profiles are specified by interval usage codes for each burst profile and include the following parameters:

- Modulation type
- Differential encoding use
- Preamble length
- Preamble starting point
- Scrambler seed
- Scramble enabling
- Number of Reed–Solomon forward error correction (FEC) parity bytes to use ("T")
- FEC codeword length
- Guard time to use for bursts
- Last codeword mode to use (fixed, or shortened if necessary to save overhead)
- Maximum burst length permitted for the burst profile

Individual modems (delineated by service ID) are also instructed as to the RF power level to use, frequency offset to use, ranging (timing) offset, burst length, and, finally, the transmit equalizer coefficients to use.

This process requires well-defined messaging in the MAC protocol so that, in the initialization phases when modems are still acquiring the network, they do not interfere with the operation of modems already on the network. This is especially true for synchronous CDMA (S-CDMA) operation in DOCSIS 2.0. The key elements of the current DOCSIS 2.0 MAC protocol are:

- Data bandwidth allocation is controlled entirely by the cable modem termination system (CMTS).
- Minislots are used as the fundamental data bandwidth element on the upstream.
- Future growth of the protocol definition is provided by reserved bits/bytes in the headers as well as support for extended headers, which permit addition of new header structures into the protocol.
- A variable mix of contention- and reservation-based upstream transmit opportunities is present; for contention-based upstream transmission, a binary exponential back-off scheme is used when collisions occur in contention minislots.
- MAC management messages are provided for registration, periodic maintenance such as power and timing adjustments, and changes to modulation order, FEC, or other channel parameters.
- Data bandwidth efficiency is improved over ATM and similar protocols through the support of variable-length packets (although extensions are provided for future support of ATM or other data types if required).
- Quality of service is provided via support for data bandwidth and latency guarantees, packet classification, and dynamic service establishment.
- Extensions are provided for security at the data link layer.

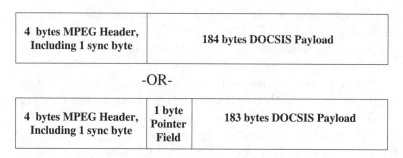

Figure 4.26 Downstream DOCSIS packet structure.

- Robustness vs. throughput can be traded off via support for a wide range of QAM orders and symbol rates, which means a wide range of data rates for the protocol.
- Support for multiple logical channels within a single physical channel so that multiple modulation schemes may be intermixed on the same upstream RF frequency.

4.5.2.1 Downstream Description

As was previously described, the DOCSIS downstream specification follows the ITU-T J.83B specification for digital video, which includes a fixed Reed–Solomon forward error correction (FEC) scheme and a variable interleaving scheme for robustness to burst noise. An example downstream packet is shown in Figure 4.26. The 4- or 5-byte header is composed of a sync byte; various flags, indicators, counters, and control bits; the DOCSIS packet ID (0x1FFFE); and an optional pointer field of 1 byte. When the pointer field is present, the DOCSIS payload is reduced to 183 bytes so that the total is still 188 bytes for a downstream MPEG packet. The downstream is synchronous to all modems connected to the network, and thus is a TDM scheme, as opposed to the upstream, which is a burst scheme using TDMA and a combination of TDMA/S-CDMA in the case of DOCSIS 2.0.

The downstream contains media access partitioning (MAP) messages, which specify how modems get access to the upstream via assignment of upstream minislots to each modem that has requested access. Latency on the upstream often depends on how frequently MAP messages are sent and can determine throughput. For example, if MAP messages are sent every 5 msec, given that a modem must request bandwidth, wait for acknowledgment and assignment of bandwidth, then use the granted opportunity, this process can take at least two MAP cycles. This means that one packet per 10 msec is transmitted on the upstream; if small packets (64 bytes) only are sent, this translates to a throughput for small packets of 51.2 kbps per modem. Note that the physical layer maximum raw burst rate for DOCSIS modems on the upstream is much higher, up to 30.72 Mbps for DOCSIS 2.0 modems. Thus, in this case, the protocol, rather than the physical layer, limits the throughput.

In reality, modems can transmit multiple flows (best-effort data, VoIP, and others), so the modems do get to use more of capacity provided by the physical layer; however, it is important to note that MAC and PHY work together to provide (or limit) the speed of a cable modem network. Although it is possible to increase the frequency of downstream MAP messages and increase throughput on the upstream, there are practical limits on how frequent MAPs can be due to downstream overhead. Furthermore, because the MAP lead time must account for the longest possible interleaver delay, modification of the MAP rate in general will not improve upstream latency.

The downstream MAC protocol also provides specification of the upstream burst parameters to use on a suite of different burst profiles matched to different packet sizes and applications. Configuration files in the CMTS specify the breakpoint between "large" and "small" packets — a decision made solely on the packet size rather than the function of the packets. These interval usage codes (IUCs) for each burst profile are specified in an upstream channel descriptor (UCD) message on the downstream, which also contains channel parameters applicable to all burst profiles, such as the size of minislots used, symbol rate, RF frequency, and the preamble pattern to be used. The different burst types associated with the different IUCs in DOCSIS 2.0 include:

- Initial maintenance (IUC 3) — used for initial contact between remote station and controller (broadcast or multicast)
- Station maintenance (IUC 4) — used to maintain transmission alignment between remote station and controller (unicast) and to adjust power level of CM
- Request (IUC 1) — used by remote station to request bandwidth for "upstream" transmission (broadcast, multicast, or unicast)
- Request/data (IUC 2) — used by remote station for sending a request OR sending "immediate" data (broadcast or multicast); currently not used by most system designers
- Short grant (IUC 5) — used for transmission of data using smaller FEC codeword size (unicast) for DOCSIS 1.0 and 1.1 CMs; applies to TCP ACKs, VoIP packets, and other small packets
- Long grant (IUC 6) — used for transmission of data using larger FEC codeword size (unicast) for DOCSIS 1.0 and 1.1 CMs; applies to large packets
- Short grant (IUC 9) — used for transmission of data using smaller FEC codeword size (unicast) for DOCSIS 2.0 CMs
- Long grant (IUC 10) — used for transmission of data using larger FEC codeword size (unicast) for DOCSIS 2.0 CMs
- Long grant (IUC 11) — used for transmission of VoIP packets in 2.0 networks using burst parameters that are optimized for reliable voice transport

For each of the preceding burst profiles (IUCs), the following signaling parameters can be specified:

- Modulation type (QPSK, 8 QAM, 16 QAM, 32 QAM, 64 QAM, and higher)

- Differential encoding (whether on or off)
- Preamble type, length, and value
- Reed–Solomon forward error correction properties (whether on/off, data bytes, parity bytes, last code-word fixed or shortened)
- Scrambler properties (on/off, seed)
- Maximum burst size
- Guard time size
- S-CDMA spreading parameters (spreader on/off, number of codes per subframe, TCM on/off, interleaving step size)

Because the robustness to impairments such as impulse noise can be a function of the packet length, the ability to optimize burst profiles for different packet lengths and applications provides a flexible method of improving robustness while optimizing network efficiency and/or latency. For example, longer packets can be made robust to impulse noise via interleaving, which adds latency but does not reduce network efficiency. Because interleaving in TDMA mode has limited effectiveness on short packets, a better way to improve robustness to impulse noise would be to specify a lower order of QAM, which effectively stretches the packets out in time while keeping latency still fairly low, as would be important for short voice packets. Thus, the DOCSIS MAC permits a system designer to optimize burst profiles against specific impairments separately for legacy modems, new modems, VoIP applications, and best-effort data applications. The last two are notable because an efficient burst profile for best-effort data may result in a packet error rate of up to 1%, which may be acceptable for best-effort data service. On the other hand, VoIP packets may need a burst profile that provides a PER of 0.1%, at most.

Finally, additional IUCs represent other required functions in the protocol:

- Null information element (IUC 7) — defines the end of the time-describing elements in a MAP; the mini-slot offset associated with this element lists the ending offset for the previous grant in the MAP

- Data ACK (IUC 8) — sent to a unicast SID and the minislot offset of this field and is a "don't care" from the CM's perspective
- Reserved (IUC 12–14)
- Expansion (IUC 15)

4.5.2.2 Upstream Description

Each DOCSIS upstream packet consists of a DOCSIS header and an optional payload data unit (PDU). The DOCSIS header is at least 6 bytes long and contains the following fields:

- A frame control byte that determines the type of frame being sent, which may be variable length PDU, ATM cell PDU, reserved PDU, or DOCSIS MAC specific PDU
- A MAC parameter byte whose purpose is determined by the frame control byte
- A 2-byte length field that usually gives the length of the PDU (although when in a request burst, it represents the SID of the cable modem)
- An extended header field (0 to 240 bytes) that provides packet-specific information such as security information, payload header suppression information, and piggyback requests
- A header check sequence (2 bytes) that covers the entire DOCSIS header

A complete upstream packet, including ramp-up time, PHY overhead, unique word, MAC overhead, packet payload, FEC parity bytes, and ramp-down time is shown in Figure 4.27. As was discussed in the section on physical layer technology, when the order of QAM is increased, the preamble must generally also be increased to provide the burst receiver with better estimates of gain, phase, and timing. In addition, the FEC parity overhead may also need to be increased for higher orders of QAM in order to increase the SNR of the demodulated packet. These and other capacity issues will be discussed in detail later.

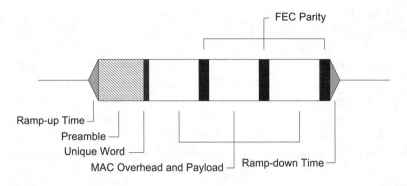

Figure 4.27 Upstream data packet structure.

The first key concept of the MAC protocol for the upstream is the concept of a minislot in DOCSIS. A minislot is a unit of data bandwidth that represents a selectable number of bytes in the upstream for bandwidth allocation purposes. To develop the definition of minislots, the quasisynchronous method of timing used in DOCSIS 1.0, 1.1, and ATDMA 2.0 modems must be described. First, the CMTS sends out periodic sync messages that contain a 32-b time stamp on the downstream, and the CM receives these sync messages and locks the frequency and phase of its local clock so that its local time stamp counter value matches the time stamp value in the sync messages. At this point, the CM has locked to the frequency of the CMTS clock, but not the phase, because of propagation delays in receiving the sync messages; this error is typically a large number of counts.

A minislot is therefore derived from the time stamp message and defined in terms of an integer number of time ticks, where a time tick is a 6.25-μsec period of time. To facilitate assignment of minislots to CMs, each minislot is numbered. Because the time stamp is based on a 10.24-MHz clock, the lower 6 b of the time stamp actually represent a portion of a time tick.

Before initial transmission, the CM loads its ranging offset register with a value to compensate for the known delays (DS interleaver, implementation delays, and so on).

The CM then adjusts its 32-b sync counter by the amount in the ranging offset. Next, the CM bit-shifts the modified count by six plus the number of ticks per minislot to derive an initial minislot count. The CM then selects an initial ranging slot and transmits. Now the CMTS can measure the difference between the received and expected transmission boundaries and then send that information back to the CM as a ranging adjustment. This process is accomplished in the DOCSIS protocol via the ranging request (RNG-REQ) and ranging response (RNG-RSP) messages sent between the CM and CMTS.

At this point, signaling is established and now the modem can establish Internet protocol (IP) connectivity via a temporary modem IP address, initialize configuration information such as time of day (ToD), and go through the process of registration. In this process, the CMTS configures the modem for access to the network and assigns station identifications (SIDs) and associated minislots to the modem for identification during transmissions. The CMTS and CM also establish the capabilities of the modem, e.g., DOCSIS 1.0, 1.1, or 2.0.

Once the modem has ranged and registered on the appropriate downstream and upstream channels, the CMTS may move it to a new upstream channel for the purpose of traffic load balancing via a dynamic channel change command (DCC). Once the modem is logged onto the network, it can begin the process of requesting data bandwidth on the upstream, receiving notification of granted upstream data bandwidth on the downstream, and, finally, transmitting in the minislots granted for use by the modem when the time comes.

Modems may request data bandwidth on the upstream in two common ways. The first and most common in lightly loaded networks is transmission of a request packet in a contention region for which all modems have equal access (and thus request packets may collide if two modems use the same minislots to make the request). The second is to piggyback a request onto the header of a previously granted packet; it is most commonly used in heavily loaded networks in which the modem has more packets already queued up for transport by the time the modem receives the grant for the previous packet.

For contention-based requests, the CMTS scheduler usually groups contention regions in clusters, and it may increase the number of minislots allocated to contention regions if the network traffic load is sufficiently high. For maximizing network capacity, it is desirable to minimize the number of minislots allocated to contention regions; to minimize latency, the number of contention minislots should be high to reduce the probability of collisions.

When collisions do occur, the DOCSIS MAC protocol uses a binary exponential back-off algorithm to increase the probability that subsequent attempts to request bandwidth do not collide. The CM starts with the initial back-off window specified in the downstream MAP and picks a random number within this window. The CM then counts off that many request opportunities (regardless of clustering) and transmits a new request, remembering the transmit time. Next, the CM waits for a grant, grant pending, or the ACK time in a MAP to exceed the transmit time. If the transmit time was exceeded, CM increments the back-off window by a power of two and picks a random number within this new window. (Note that during this process, if the CM hits the maximum back-off window in the MAP, the CM continues the process but does not keep incrementing the back-off window.) The CM then counts off that many request opportunities (again, regardless of clustering) and transmits, remembering the new transmit time and incrementing a retry counter. This process repeats up to 16 times, and if it is still not successful, the packet is dropped and the process begins anew for the next packet in the queue.

For piggybacked requests, there are no collisions because the request is included in a granted region to which only one modem has access. DOCSIS also provides a third way to send small data packets on the network using the REQ-DAT IUC to directly transmit data packets in a shared time slot that may collide with REQ-DAT packets from other modems. As of this writing, REQ-DAT is typically not used on the network.

Version 1.0 of the DOCSIS specification was intended to support best-effort data only, so 1.0 had no QoS considerations. Because VoIP has become desirable to cable operators, version 1.1 of the DOCSIS MAC protocol added the following features:

- Quality of service
- Service flows
- Classifiers
- Scheduling types:
  - Best effort service
  - Unsolicited grant service
  - Real-time polling service
  - Unsolicited grant service with activity detection
  - Non-real-time polling service
- Dynamic service establishment
- Fragmentation (allows segmentation of large packets simplifying bandwidth allocation for CBR-type services)
- Concatenation (allows bundling of multiple small packets to increase throughput)
- Security enhancements (authentication; Baseline Privacy Plus provides authentication as well as in-line DES encryption/decryption)
- Encryption support for multicast signaling (IGMP-Internet group management protocol)
- Payload header suppression (allows suppression of repetitive Ethernet/IP header information for improved bandwidth utilization)

The service flows and classifiers provide mechanisms for DOCSIS to set up and control the different types of services that may flow over a cable modem network. The scheduling types, on the other hand, have a direct impact on the overall performance and capacity of the network. Best-effort service uses the aforementioned contention-based and piggyback grants to request and receive access to the upstream. Unsolicited grant service (UGS) flows are specifically intended for VoIP service in which a call is set up; grants of upstream data bandwidth are automatically provided at some interval, such as every 10 msec, so that voice packets may be transmitted with minimal latency and jitter.

The remaining services are provided for voice and interactive gaming services in which activity may be temporarily paused (such as for silent intervals in speech) and needs to be reinstated quickly when activity restarts. In the case of UGS flows, note that the usual request and grants are not required. Access to the network is automatically provided — much as in TDM telephony systems; therefore, the efficiency of the network for VoIP traffic can be much higher than for best-effort data traffic. The other services, real-time polling, UGS with activity detection, and non-real-time polling, are various trade-offs between the overall network capacity optimization of best-effort data and the network latency minimization of UGS flows.

A more significant network performance enhancement results from the requirement of concatenation and fragmentation in DOCSIS 1.1. Concatenation enhances network capacity via elimination of the repeated PHY overhead in packets: multiple packets queued up for transport are combined into a single upstream burst with one preamble and guard time. As is shown in the next section on performance characterization, concatenation and fragmentation can increase the network performance as much as increasing the order of QAM by two orders.

Another important feature that relates to network capacity is payload header suppression (PHS). This feature was developed specifically for VoIP packets, which typically have IP headers with fields that are fixed throughout a voice call. During call setup, the DOCSIS MAC layer determines which packet fields will be fixed during the call and can be suppressed. The most typical example would be suppression of the 14-byte Ethernet header, the 20-byte IP header, and the 8-byte UDP header; these can be replaced in the DOCSIS MAC header by a 2-byte PHS extended header, for a net reduction of 40 bytes. For a G.711 voice packet with 10-msec packetization and not counting the PHY overhead such as guard time and preamble, this reduces a 149-byte packet to a 109-byte packet, thereby improving spectral efficiency by about 25%. In physical layer terms, this improvement is equivalent to the improvement from increasing the order of QAM from 16 to 32, which would require 3 dB greater SNR in the channel to achieve.

4.5.3 The MAC Scheduler: Evaluation of MAC Protocols, Traffic Models, and Implications for Physical Layer Design

Although the MAC protocol and the physical layer often primarily dictate the performance of the overall network, for the DOCSIS and any other centralized MAC protocol, the MAC scheduler has an enormous impact on overall network performance under loaded traffic conditions. The MAC scheduler may adapt the number of contention minislots allocated to the channel, the MAP rate, and when and how often fragmentation is to be performed; it also must guarantee quality of service for VoIP packets while optimally scheduling best-effort data packets. A high-quality MAC scheduler can easily improve the maximum network load by up 25% merely from optimizing the algorithms used in scheduling traffic under random traffic conditions, which are the most difficult to optimize. Clearly, the performance of the MAC can be at least as important as the performance of the PHY technologies used in the cable modem network.

Protocols and/or schedulers are frequently evaluated via plotting the packet transport delay vs. increasing traffic load on the network. Figure 4.28 depicts DOCSIS protocol simulation results of delay vs. load on the network. The benefits of fragmentation and concatenation are seen in that a greater amount of traffic on the network is possible for the same average delay in transporting packets on the network. In this case, if the network does not support fragmentation or concatenation, the maximum traffic load is reduced by about 25% of the maximum if these features were supported. The maximum throughput and associated latency of a protocol vary considerably with the physical layer specification, the type of traffic assumed, and, for centralized protocols such as DOCSIS, the scheduling algorithms used to adapt the network to changing traffic loads, priorities, and patterns.

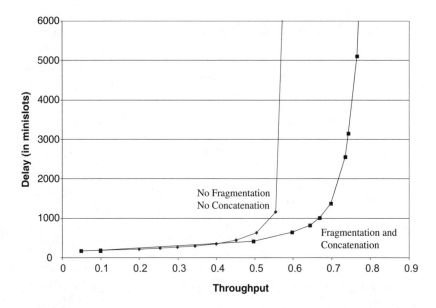

Figure 4.28 Network capacity improvement from fragmentation and concatenation.

Because the initial DOCSIS protocol was designed for best-effort data transmission, the traffic model assumed was similar to the distribution shown in Figure 4.29 for Web traffic. Note that although the most common packets on the upstream are small packets (TCP, ACKs, and SYNs), these only account for 12% of the total capacity on the upstream, as shown in Figure 4.30. The result is that if the capacity of a best-effort data network is to be increased, the medium and large packets will dominate the network performance; thus, effort should be focused on increasing the order of QAM used in the signaling. On the other hand, for a network that transports primarily small packets such as VoIP traffic, techniques such as payload header suppression, which increases the efficiency of small voice packets, and synchronous operation as used in S-CDMA, which improves small packet efficiency via reduction in the length of preamble, are most viable for improving network efficiency.

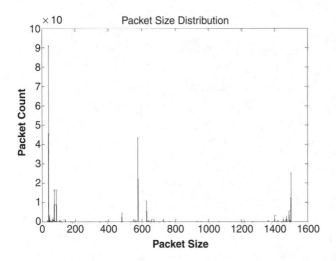

Figure 4.29 Upstream traffic distribution from cable network measurements.

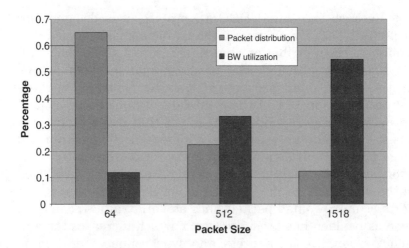

Figure 4.30 Relative frequency of packets vs. packet size on cable upstream channels.

4.6. SYSTEM CONSIDERATIONS FOR CABLE MODEM NETWORKS

In addition to increasing the network capacity via improvements in PHY and in MAC technology, other technologies reside at higher open systems interconnection (OSI) layers that can be used to increase capacity over cable modem networks. Several higher layer techniques are discussed in this section, including system adaptation to dynamic upstream impairments and suppression of TCP acknowledgments.

4.6.1 Dynamic Adaptation to Upstream Impairments

The upstream RF channel in HFC networks is highly dynamic and contains a variety of impairments that can reduce the capacity of the network by requiring use of lower orders of QAM, lower symbol rates, higher FEC overhead, and/or higher latency from interleaving. On the other hand, an intelligent CMTS can be used that has the capability to detect and classify RF impairments and dynamically adapt the channel signaling parameters in accordance with the dynamic impairments. This intelligence permits the total daily capacity of the network to be increased by increasing capacity at the cost of robustness during optimal plant condition periods and restoring the network to a more robust yet lower capacity configuration during impaired plant periods. One method of accomplishing this intelligence and dynamic adaptation would be for the CMTS to contain a spectrum-monitoring subsystem that, in conjunction with a lookup table of recommended burst profile vs. RF impairments, would adapt the upstream burst profiles for the various traffic types transported.

The key benefit of an adaptive system strategy is that the average network capacity can be increased, and if the peak utilization times correspond to times when the channel impairments are diminished, the network planning process can be more cost effective. For example, if RF impairments dictate the robustness of QPSK modulation to achieve a given packet error rate, but this level of impairments only exists for a fraction of the time the network operates, it is likely possible to operate at higher QAM during the majority of the day. If 64 QAM could be supported most of the day, even if QPSK was required during the lunch hour, then the overall capacity of the network could be triple what it would be if worst-case channel conditions were used to determine the system operating parameters. The challenge for such an adaptive CMTS strategy is to detect and characterize the channel impairments properly. This must be done quickly enough and in enough detail to adjust the signaling parameters so that the maximum spectral efficiency (bits per hertz) can be transmitted under the current channel conditions without increasing the average packet error rate or the quality of service on the network.

Dynamic adaptation only makes sense if the channel is changing dynamically. That this is true of HFC upstream channels is evidenced by previous measurements of cable upstream interference.[10] These studies show that impulse and burst noise are often higher at certain times of the day. Ingress is often higher at night, and common path distortion (CPD) varies with temperature, humidity, and wind due to a major variable source of CPD, the cable connectors. These same studies show that many impairments exist on the plant usually well below 10% of the time. Thus, without adaptation strategies in the CMTS, the network capacity could be limited to 33% or less of its potential capacity in order to handle impairments that only occur 10% of the time.

One of the first techniques possible in system adaptation is to adapt the modulation order of QAM and/or the amount of FEC used. This technique is particularly useful in DOCSIS cable modem networks because the order of QAM used in upstream burst profiles can be changed on the fly without requiring the modems to range or equalize again, which could cause VoIP calls to be dropped. First, the level of FEC used on packet transmissions can be increased as impairments increase. Over 6-dB improvement in robustness is possible with this technique, albeit with a 20 to 30% drop in spectral efficiency. Adapting the order of QAM from 64 QAM to QPSK in DOCSIS 2.0 modems provides an increase of 12 dB in robustness; combined with FEC adaptation, a total range of 18 dB of improvement in robustness can be obtained, although at the expense of 75% of network capacity.

The next level of adaptation involves changing the channel center frequency to avoid significant levels of impairments such as ingress. Although the availability of ingress cancellation technologies in the CMTS will reduce the necessity of changing channels much of the time, this adaptation technique remains viable for MSOs with spare RF upstream bandwidth. However, now that higher order QAM is available and likely to be used on cable upstreams, this may reduce the desirability of frequency changes. The reason is that the equalization during the initial ranging process makes higher order QAM possible, so if a frequency change is imposed, it may require the modems on the network to range again, which could cause VoIP calls to be dropped. A possible system tactic against this scenario is to hop to the new channel using a lower order of QAM such as QPSK which does not require the level of equalization; then, as calls are completed, those modems can be ranged again and switched to higher orders of QAM for the burst profiles.

It may also be possible to avoid interference in the time domain. As was briefly mentioned in the physical layer section, periodic impulse noise often arises in cable modem upstreams, and the most common waveform has a repetition frequency that corresponds to the AC power-line frequency (60 Hz in North America). Therefore, one often sees periodic impulses at repetition frequencies of 60 or 120 Hz, or other harmonics. Recall that in Figure 4.12, periodic power-line impulse noise at approximately 120-Hz repetition rate was observed. Although the burst duration is typically 1 to 10 μsec (and thus easily countered by FEC), the duration can be much higher — up to several milliseconds — which exceeds the ability of correction capability of FEC in TDMA or in S-CDMA mode.

However, the time between such bursts is long (8 to 16 msec), so periodic impulse noise could be tracked and avoided by intelligent scheduling in an advanced CMTS. Note also that most periodic impulse noise has a nonuniform spectral signature, as shown in Figure 4.13. This characteristic may be exploited and significant amounts of the noise energy reduced via ingress cancellation processing or via choosing a center frequency that has minimal spectral energy of the impulse events. If either of the latter capabilities is not supported in the advanced PHY CMTS, the benefits of advanced PHY can still be reaped via deployment of data-only service in channels that have high impulse/burst noise. The additional packet loss due to periodic impulse noise may be low enough that the resulting degradation is one that most users would seldom notice.

Another tactic is to reduce the symbol rate for increased robustness against all types of impairments, again at the cost of reduced capacity. Assuming the modem transmit power is maintained at the original level, a reduction in modulation rate by a factor of two will add 3 dB more robustness against AWGN, ingress, and burst noise. Furthermore, the length of an impulse event that can be corrected is doubled by the fact that, in the time domain, the symbols are twice as long as before and therefore fewer TDMA symbols or S-CDMA chips are corrupted by the same impulse event.

Note that special considerations exist for mixed DOCSIS 1.x/2.0 channels. Table 4.6 lists techniques in advanced CMTS receivers (some of which are part of the DOCSIS specification) and also indicates whether or not they apply to legacy modems on the network. Thus, when legacy modems are mixed with new modems on a cable modem network, the CMTS must not select an adaptation technique that only works for 2.0 modems. On the other hand, techniques specific to 2.0 can be applied to the 2.0 modems as long as an alternative for the 1.x modems is applied as well. For example, if moderate impulse noise is detected, the CMTS could increase the interleaving on 2.0 modems while maintaining the order of modulation at 64 QAM and reduce the order of modulation on 1.x modems in order to stretch the 1.x packet's bytes out in time.

Alternatively, the 2.0 modems could switch to S-CDMA mode, if impulse conditions warrant. If the impulse noise is too long for simple constellation changes, the symbol rate of all modems on the network may need to be reduced so that

Table 4.6 Advanced Features in 2.0 CMTS

Feature	Improves 1.x ?
Improved AWGN mitigation	
Lower implementation loss	Yes
Better receive equalization	Yes
Improved burst acquisition	Yes
More FEC (T = 16)	No
Ingress/CPD cancellation	
Cancellation of ingress	Yes
Cancellation of CPD	Yes
Improved mitigation of impulse and burst noise	
Cancellation of spectral peaks in periodic impulse noise	Yes
More FEC (T = 16)	No
RS byte interleaving	No
S-CDMA mode	No

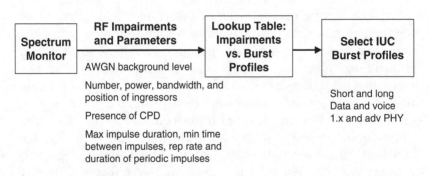

Figure 4.31 Example CMTS adaptation approach.

the 1.x modems stay active. The equalization capabilities of 1.x and 2.0 modems are different, and this may also lead to a different adaptation strategy when mixed networks are deployed.

An example adaptation approach that could be built into a CMTS is shown in Figure 4.31, with key components the spectrum monitor and a lookup table of burst profiles. The spectrum monitor can be internal or external to CMTS, but it is important that the RF impairment detection and classification processes use rules based on plant measurements and impairment models, such as those presented in Howard.[5] This is because different adaptation strategies exist for different impairments.

For example, if the total interference power is used to characterize the channel, then ingress cancellation and FEC/interleaving will not be leveraged to their fullest extent. Consider the case with an AWGN background noise floor that is 22 dB down from the signal power level, but an ingress signal is present that is 10 dB above the signal power. With modern ingress cancellation technology, a 2.0 modem could easily operate at 64 QAM and a 1.x modem could operate at 16 QAM in this level of noise. However, if the total interference power were used to characterize the channel, the system would erroneously assume the channel was unusable due to SNR being too low for even QPSK operation.

Once the RF impairments have been detected and classified, the results may be used to determine the burst profiles for the channel that optimize capacity while maintaining sufficient robustness against the impairments. One approach to this requirement is a lookup table in which the system performance is characterized in a lab against a variety of impairments and levels and optimum burst profiles determined for each impairment and level of impairment. As the FEC overhead is increased and the modulation type reduced, the spectral efficiency will drop, but for the benefit of greater robustness. The actual FEC used in the burst profile will depend on the packet size, quality of service required, and so on.

For example, one set of tables could apply to a packet error rate of less than 0.1%, and another set of tables could allow error rates of up to 1%. The former could then be applied to voice packets and the latter to best-effort data packets. Thus, there could be several lookup tables for each type of service and packet size that optimizes the burst profile subject to the main constraint of tolerating the given level of AWGN with a selected packet error rate.

Similar lookup tables can be developed for each impairment and even combinations of impairments. In this manner, when any previously seen (or postulated) combination of impairments is detected on the cable upstream, the CMTS can use the optimum burst profiles for those particular impairments.

4.6.2 Network Management and Adaptation

As the amount of data and types of services passed through the HFC network increase over time, more intelligent methods of adapting the network dynamically will be needed to maximize its performance. Conditions such as increased bandwidth needs, advent of new services requiring higher levels of QoS, and plant degradations such as impairments or entire outages on portions of the network will require network monitoring and management to ensure smooth operation of the network, as well as transitioning to new plant architectures. These topics, including cable-specific aspects, are discussed in Chapter 8 of this text.

4.7 FUTURE HFC TECHNOLOGIES AND ARCHITECTURES

Several technologies are currently under discussion for improving HFC network performance and capacity. Ultimately, the addition of a major new technology for the DOCSIS standard must be weighed against the alternative of pushing the fiber deeper towards the home, which also increases capacity and reliability of the network. In all cases, the potential of a new technology must be carefully weighed against the cost, application to legacy systems, and performance improvement compared to alternatives.

4.7.1 DOCSIS Downstream

Improvements in downstream throughput are likely in the future as cable multiple system operators (MSOs) compete with digital subscriber loop (DSL) providers for higher bandwidth provision to customers. As of this writing, MSOs are considering higher order modulation schemes for the downstream, up to 1024 QAM, as well as wider downstream channels: 12 MHz instead of the current 6 MHz. At least one silicon vendor already provides a 1024 QAM downstream capability and 12-MHz wide downstream channels.[19]

Other techniques discussed for increasing downstream capacity include multiple simultaneous channels and new downstream modulation schemes based on multitone modulation[20] or ultra-wideband modulation.[21] A key benefit of multiple simultaneous channels is the ability to continue to support legacy modems and set top boxes while new modems and set top boxes are deployed that support much higher burst rates.[22] Some of the proposed benefits of multitone for the downstream include avoidance of CTB and CSO products via elimination of certain tones, simplification of equalization processing, and increased RF efficiency provided by the smaller effective value of α, the Nyquist pulse-shaping factor (also called excess bandwidth), which is currently around 0.12 for the downstream. On the other hand, ultra-wideband modulation has been proposed as an overlay scheme that operates in the presence of existing downstream modulation.

4.7.2 DOCSIS Upstream

Multitone modulation has been proposed in the past for the cable upstream[12] and will likely be proposed again in the future, especially if technology permits the entire upstream spectrum to be considered a single band for use by a single modem. Alternately, technology improvements could lead to a multiple, simultaneous single-carrier approach in which a given CM can transmit on multiple frequencies simultaneously. The latter approach has the advantage of legacy modem support.

4.7.3 Cable Plant Architecture Alternatives

Several approaches for modifying the cable plant have been proposed and/or developed as a means of dramatically increasing capacity on the HFC network. The minifiber node architecture[23] was proposed as a way to push the fiber deeper while still employing the coax plant. In effect, all of the amplifiers in the plant are replaced with minifiber nodes and the remaining coax becomes passive coax plant. In contrast, several cable operators have already deployed fiber-deep architectures that use passive coax runs after the fiber nodes using conventional components.

Another approach is to overlay onto the coax plant a transmission system that uses spectrum between 908 MHz and 1.08 GHz[24] (current DOCSIS coax networks do not use frequencies above 860 MHz). In this approach, the fiber is not pushed more deeply towards the home, but rather the capacity of the existing coax plant is increased via the insertion of large numbers of additional plant components. In this example, the higher frequency signaling is via a new signaling scheme. In contrast, another overlay scheme[25] uses the "bent pipe" approach to modulate any conventional cable RF signal up to the higher frequencies and demodulate them prior to or at arrival at the customer premises, which means that any legacy HFC network component can use the additional bandwidth.

Both of the overlay approaches just described are costly compared to conventional approaches, but do provide significant bandwidth enhancement for the network. Whether either one is eventually deployed by most MSOs or standardized into DOCSIS depends on how rapidly the cost drops over time of fiber to the home (FTTH) technologies.

4.8. SUMMARY

HFC networks are versatile broadband networks that permit a wide range of trade-offs in cost and performance to be made, in terms of current technologies and also considering future improvements. Compared to satellite networks, HFC networks can support scalable, high-speed data service on the upstream and the downstream and can support telephony in addition to video services. Compared to DSL, HFC networks already provide broadband video services to all customers regardless of distance from the network facilities. However, as with all broadband networks, the increasing bandwidth needs of customers and their applications will continue to encourage the development of new technologies and architectures so that HFC networks can remain competitive with other last mile technologies.

ACKNOWLEDGMENTS

The lead author wishes to acknowledge the following Broadcom engineers for contributions to the material in this chapter and to his understanding of HFC networks: Tom Quigley, a leading authority and contributing architect for the MAC layer of the DOCSIS protocol; Lisa Denney and Niki Pantelias for their lunchtime learning sessions on the DOCSIS MAC, certain material of which has been used in the section on MAC protocols; Rich Prodan, for his work in S-CDMA and in characterization of upstream and downstream plant impairments; Gottfried Ungerboeck, whose expertise in the field of broadband communications systems and modulation theory is unmatched; Hans Hsu for contributions to the adaptive CMTS algorithms and for generation of several performance data results used here. Of course, the lead author acknowledges his wife for manuscript review; without her support the writing of this chapter would not have been possible.

GLOSSARY

ACK	acknowledgment
AM-SSB	amplitude-modulated, single side band
A-TDMA	advanced time domain multiple access
ATM	asynchronous transport mode
AWGN	additive white Gaussian noise
BER	bit error rate
CATV	community antenna television
CM	cable modem
CMTS	cable modem termination system
CPD	common path distortion
CPR	centralized priority reservation
CSO	composite second order distortion

CTB	composite triple beat
CW	continuous wave
DC	direct current
DCC	dynamic channel change
DES	data encryption standard
DHCP	dynamic host control protocol
DOCSIS	data-over-cable service interface specifications
DSL	digital subscriber loop
FCC	Federal Communications Commission
FDM	frequency division multiplex
FEC	forward error correction
FFT	fast Fourier transform
FTTH	fiber to the home
HFC	hybrid fiber-coax
HRC	harmonically related carriers
IEEE	Institute of Electrical and Electronic Engineers
IMD	intermodulation distortion
IP	Internet protocol
IPSEC	Internet protocol security
IRC	incrementally related carriers
ISI	intersymbol interference
IUC	interval usage code
MAC	media access control
MAP	media access partitioning
MCNS	multimedia cable network system
MPEG	Motion Pictures Experts Group
MSO	multiple system operator (HFC network operators)
OOK-CW	on–off keyed continuous wave
OSI	open systems interconnection
PCS	personal communications service
PDU	payload data unit
PER	packet error rate
PHS	payload header suppression
PHY	physical layer of the OSI model
PN	pseudonoise
QAM	quadrature amplitude modulation
QoS	quality of service
QPSK	quaternary phase shift keyed

RAM   random access memory
REQ-DAT   request or data packet (seldom used)
RF   radio frequency
RNG-REQ   ranging request
RNG-RSP   ranging response
RS   Reed–Solomon
S-CDM   asynchronous code division multiple access
SID   station identification
SIR   signal to interference ratio
SNR   signal to noise ratio
SSTV   slow scan television
STD   standard frequency plan for HFC networks
SYN   synchronization packet for TCP
TCM   trellis-coded modulation
TCP   transmission control protocol
TDM   time division multiplex
TDMA   time division multiple access
TFTP   trivial file transfer protocol
ToD   time of day
TV   television
UCD   upstream channel descriptor
UGS   unsolicited grant service
VCMT   variable constellation multitone
VoIP   voice over Internet protocol

REFERENCES

1. W. Ciciora, J. Farmer, and D. Large, *Modern Cable Television Technology*, Morgan Kaufmann, San Francisco, 1999.

2. D. Paris and K. Hurd, *Basic Electromagnetic Theory*, McGraw-Hill, New York, 1969.

3. T. Kolze, An approach to upstream HFC channel modeling and physical-layer design, in *Cable Modems: Current Technologies and Applications*, International Engineering Consortium, Chicago, 1999; and see also T. Kolze, HFC upstream channel characteristics, GI IEEE 802.14 contribution, IEEE 802.14-

95/075, Hawaii, July 7, 1995; and T. Kolze, Upstream HFC channel modeling and physical layer design, Broadband Access Systems: Voice, Video, and Data Communications symposium of SPIE, Photonics East Conference, Boston, November 18–22, 1996.

4. DOCSIS 2.0 Radio Frequency Interface Specification: SP-RFIv2.0-I05-040407, available on the Web at www.cablemodem.com/specifications

5. D. Howard, Detection and classification of RF impairments for higher capacity upstreams using advanced TDMA, NCTA Technical Papers, 2001.

6. http://www.fcc.gov/csb/facts/csgen.html

7. http://www.sciatl.com

8. R. Katznelson, Statistical properties of composite distortions in HFC systems and their effects on digital channels, NCTA Technical Papers, 2002.

9. CableLabs' test and evaluation plan on advanced physical layer (Adv PHY) for DOCSIS upstream transmission, May 25, 2001.

10. R. Prodan, M. Chelehmal, and T. Williams, Analysis of two-way cable system transient impairments, NCTA Conference Record 1996.

11. B. Currivan, Cable modem physical layer specification and design, in *Cable Modems: Current Technologies and Applications*, International Engineering Consortium, Chicago, 1999.

12. Cable operator, Knology, has archived the 802.14 committee documents on its Web site at http://home.knology.net/ieee80214/

13. F. Buda, E. Lemois, and H. Sari, An analysis of the TDMA and S-CDMA technologies of DOCSIS 2.0, 2002 NCTA Technical Papers.

14. B. Currivan, T. Kolze, J. Min, and G. Ungerboeck, Physical layer considerations for advanced TDMA CATV return path, *Int. Eng. Consortium Annu. Rev. Commun.*, 55, 2002.

15. M. Lops, G. Ricci, and A.M. Tulino, Narrow-band-interference suppression in multiuser CDMA systems, *IEEE Trans. Commun.*, 46(9), September 1998, pp. 1163–1175.

16. J.A. Young and J.S. Lehnert, Analysis of DFT-based frequency excision algorithms for direct-sequence spread-spectrum communications, *IEEE Trans. Commun.*, 46(8), August 1998, pp. 1076–1087.

17. Y. de Jong, R. Wolters, and H. van den Boom, A CDMA-based bidirectional communication system for hybrid fiber-coax CATV networks, *IEEE Trans. Broadcasting*, 43(2), June 1997.

18. J. Limb and D. Sala, A protocol for efficient transfer of data over hybrid fiber/coax systems, *IEEE/ACM Trans. Networking*, 5(6), December 1997.

19. D. Howard, L. Hall, K. Brawner, H. Hsu, N. Hamilton-Piercy, R. Ramroop, and S. Liu, Methods to increase bandwidth utilization in DOCSIS 2.0 systems, NCTA Technical Papers, 2003.

20. http://www.broadbandphysics.com

21. http://www.pulselink.net

22. S. Woodward, Fast channel: a higher speed cable data service, *Proc. SCTE Conf. Emerging Technol.*, January 2002.

23. O. Sniezko and X. Lu, How much "F" and "C" in HFC?: deep fiber reduces costs, SCTE, *Commun. Tech.*, June 2000, www.ct-magazine.com/ct/archives/0600/0600fe10.htm.

24. http://www.naradnetworks.com

25. http://www.xtendnetworks.com

5

Optical Access:
Networks and Technology

BRIAN FORD AND STEPHEN E. RALPH

5.1 INTRODUCTION

Composed of many technologies and spanning many data rates, the access network is the part of the network that connects the end user to the network, the so-called "last mile." It is also called the first mile to emphasize the importance of the user in any network system. End users include residential as well as businesses with their own internal networks. Thus, the access network necessarily includes access points with a great variety of data types and capacity. Importantly, the capacity of these access network connections has not advanced commensurately with the core of the network or with the end user's capacity. The access network is therefore the limiting portion of the network. The access network is currently the data path bottleneck for many reasons, including economic, technological, and regulatory. However, it is clear that once the access bottle neck is relieved, demands for capacity will likely increase throughout the network. In some sense, then, the access network hampers growth of the entire network.

In this chapter, we review optical access architectures and technologies. In particular we focus on fiber to the home (FTTH)

333

networks and, to some extent, LAN technologies. Large businesses access the network through high-capacity switches and do not use traditional "access" technologies even though the LAN is the access point. On the other hand, the requirements of small business offices and homes are similar.

Ultimately, the goal of FTTH is to provide cost-efficient broadband services for residential users. With respect to the use of fiber, broadband typically implies substantially larger aggregate and end user bandwidth than wired technologies. For example, DSL services typically provide 1.5Mb/s; however, even modest FTTH scenarios envision in excess of 20 Mb/s and up to 1 Gb/s is also possible. The advantages of fiber are well documented and include large bandwidth (tens of gigabits per second), easy upgradeability, ability to support fully symmetric services, long transport distances, and inherent immunity to electromagnetic interference. The primary drawback has been the cost associated with the electrical to optical conversion.

The deployment of optical fiber in the network has advanced rapidly after its initial use in interoffice networks in the 1970s. By the early 1980s, fiber had become the transport technology of choice for interoffice networks and was extending into the feeder network to the remote terminal (RT). As fiber extended closer to the network edge, fiber to the home (FTTH) was envisioned as the ultimate goal. British Telecom,[1] NTT,[2] France Telecom,[3] and BellSouth[4] conducted FTTH field trials in the 1980s to gain practical experience and to identify the obstacles that needed to be overcome to make FTTH a cost-effective access technology.

These early trials demonstrated that FTTH had high potential and that further improvements and cost reductions would allow an economically viable FTTH solution. Since the early trial, the installed cost of a 96-fiber cable has now declined to about $11,000 per kilometer (aerial) and $24,000 per kilometer (buried); the optical loss has dropped to less than 0.5 dB/km; optical transceiver technologies now allow for low-cost mass production; and architectural improvements have decreased the cost of optical fiber access networks. Indeed, a recent assessment of communities in North America revealed that 70 communities in 20 states currently provide FTTH service.[5]

Similarly, in the local area network (LAN) environment, the first Ethernet standard to include a fiber interface was in 1987 for a single 2-km optical repeater that was expected to be used to link different buildings using a large LAN. In 1995 the IEEE fast Ethernet standard (802.3u) 100 base–FX specified 100 Mb/s over 2 km of fiber was adopted. Later, the IEEE 1Gb/s Ethernet (GbE) standard (802.3z, 1998) provided for multimode fiber (MMF) to deliver broadband services and was primarily implemented for connections between routers and switches. More recently the 10 GbE standard (802.3ae) was ratified. This specified MMF and single-mode fiber (SMF) interfaces operating at 10 Gb/s. Importantly, a LAN data rate of 10 Gb/s and a WAN interface compatible with the telecom standard SONET data rate OC-192 are both specified. The success of these standards and the high fraction of Ethernet packets entering the network directly influence the topologies, protocols, and, to some degree, the technologies deployed in FTTH environments.

We first review architectures for FTTH and discuss a variety of passive optical networks. We next consider current as well as future technologies likely to have an impact on access networks. We note that the success of Ethernet has significantly affected FTTH in two ways. First, Ethernet packets are the dominant data format and any access technology must efficiently transport them. Second, the success and simplicity of the architecture and components is likely to lead to similar component characteristics in FTTH networks. Indeed, the Ethernet in the first mile efforts of IEEE is likely to standardize this. Lastly, we note that electronic and optical signal processing will play an increasingly important role in optical networks, including access networks.

5.2 FTTH ARCHITECTURES

Many optical access architectures including star, ring, and bus, have been considered for FTTH[6]; the star architecture is the architecture of choice due to a variety of considerations, including

- Low initial cost
- Scalability
- Flexibility in service provisioning
- Maintenance
- Security

FTTH architectures can be further characterized by (1) location of electronics; (2) location of bandwidth aggregation (3) end-user bandwidth (burst and continuous); and (4) protocols. The challenges of cost-effective fiber-to-the-home deployments include minimizing the cost of the installed fiber and the optical–electronic transition. Thus, the transceiver cost and the number of transceivers per premises are primary measures driving FTTH economics. Today, the majority of suburban communities are served by twisted-pair copper or hybrid fiber coax (HFC) networks, both of which restrict bandwidth and create "bottlenecks" in the last mile from the central office (CO) to the premises. Relieving this bottleneck by a home run fiber (Figure 5.1a) requires a substantial number of hub facilities. For example, in existing HFC deployments, the fiber counts are typically 200 to 300 fibers per headend, and each of these fibers services 500 to 1000 homes. Changing to a home run fiber for each end user would require either unmanageable fiber counts at the head end, or a dramatic increase in the number of active nodes in the field.

The FTTH topologies shown in Figure 5.1 depict four basic architectures of the optical distribution network (ODN). The home run fiber architecture of Figure 5.1a uses the most dedicated fiber per home, extending potentially tens of kilometers. The electronics are only at the CO and the home and are known as the optical line termination (OLT) and the optical network unit/termination (ONU/ONT), respectively. There is no bandwidth sharing after the CO and most upgrade paths remain viable without additional fiber installation. However, providing a separate fiber run, originating in an existing CO, to each customer would involve unmanageable fiber counts.

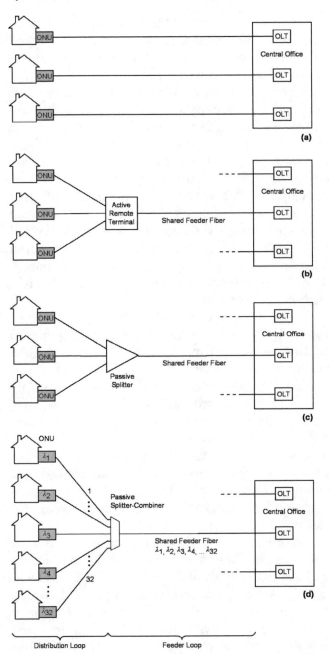

Figure 5.1 Point-to-point topologies for optical access.

The other architectures reduce the amount of dedicated fiber and, ultimately, the maximum bandwidth per user. Figure 5.1b depicts an "active star," which requires remote electronics that need to be conditioned for harsh environments as well as provided with protected power. Active stars enable multiplexing and regeneration at that point and therefore extend the reach of the architecture when needed. Figure 5.1c depicts a passive optical network (PON) created by a simple optical power splitter. The maximum number of connections, typically 32, per PON infrastructure is limited by optical power available at the ONU. Figure 5.1d depicts a PON variant that uses wavelength division multiplexing (WDM) techniques to increase the bandwidth per user. Hybrid combinations of Figure 5.1b through Figure 5.1d are often advocated. These architectures are all optical fiber-based, point-to-multipoint infrastructures based on a bandwidth-sharing model. All architectures require a dedicated network transceiver per premise as well as dedicated optical fiber.

In addition to the cost associated with bringing optical connectivity to the residential market, there is the need to offer voice, data, and video services over a single, high-speed connection. Indeed, "multiservice operators" offer multiple connections from different providers. The FTTH architecture must accommodate the variety of services using a common link layer, the so-called "converged network," or by using different wavelengths to distribute the different services. For example, point-to-point data and voice may be provisioned over one wavelength; however, broadcast data may be provisioned on a separate wavelength. The broadcast wavelength is also called an overlay.

In addition to the simple overlay model used to provision point-to-point services together with broadcast, the possibility of adding true WDM to the PON architecture exists. The most likely candidate for WDM in access is coarse WDM, which incorporates wide channel spacing to accommodate the lack of frequency precision and drift associated with low-cost lasers. Typical wavelength spacing is 20 nm, which is sufficient to accommodate the frequency uncertainty of uncooled lasers.

The advantages of WDM PON are primarily that it has the capacity of a pure home run architecture while still sharing the feeder fiber and maintaining all of the electronics at the CO or end users. Each end user may exploit the full capacity of the dedicated wavelength. Also, WDM PON may be seen as the obvious upgrade path for single-wavelength PON.

5.2.1 Passive Optical Networks

The need to reduce the amount of dedicated optical fiber without incurring additional costs for remote electronics and powering led to the concept of a passive optical network (PON), which minimizes or eliminates active elements between the end user and the CO. Today, the star network is embodied by the various PON architectures that have been standardized.

British Telecom deployed a telephony over PON (TPON) system in the early 1990s.[7] TPON supported a maximum of 128 fiber ends and 294×64 kb/s bidirectional traffic channels using time division multiplexing (TDM) downstream and time division multiple access (TDMA) upstream on a single wavelength. In the mid-1990s, a group of network operators extended the TPON concept to incorporate asynchronous transfer mode (ATM) protocols by launching an initiative for a full services access network (FSAN).[8] The FSAN organization produced specifications for ATM PON or APON, which became ITU-T Recommendation G.983.1. NTT, BellSouth, France Telecom, and British Telecom further extended the FSAN work in 1999 to develop common technical specifications for FTTH.[9–11]

Three of the nation's largest telecommunications service providers — BellSouth, SBC Communications Inc., and Verizon — have adopted a set of common technical requirements based on established industry standards and specifications for a technology known as fiber to the premises (FTTP). FSAN has also identified a number of the FTTx concepts depicted in Figure 5.2.[12] Distinction is made according to the depth that the fiber has penetrated to the end user; however, a common architecture is envisioned for the various networks. More recently Verizon

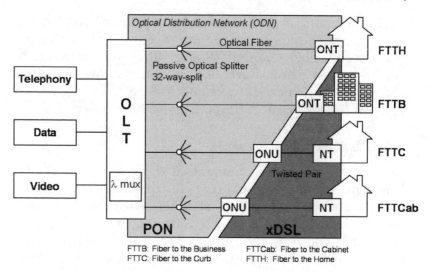

Figure 5.2 FTTx concepts. (Adapted from K. Okada et al., *IEEE Commun. Mag.*, 39, 134–141, December 2001.)

announced deployment of FTTP services with downstream bandwidths as high as 30 Mbps. A total of 12 million connections by 2008 are planned.

5.2.2 PON Power-Splitting Optimization

Figure 5.3 illustrates three variations of the PON access topologies of Figure 5.1c. The top configuration places a passive optical splitter at the traditional RT site to divide the optical power over 32 fibers in the downstream direction and combine the optical signals in the upstream direction. The feeder fiber and the transceiver at the CO are shared over up to 32 customers. Fiber beyond the splitter is dedicated, resulting in higher initial cost than with the other two configurations, but enabling the sharing ratio to be changed in the future by reinforcing the feeder.

The second configuration minimizes fiber requirements, but results in drop fibers that are difficult to manage. Drops usually extend from poles (aerial plant) or pedestals (buried plant) that serve four homes; they extend directly to the customer's home without traversing other properties. Serving

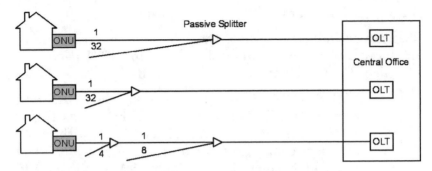

Figure 5.3 Passive optical network topologies for optical access.

32 homes from the splitter in the second configuration results in the difficulty of running drops from pole to pole or from pedestal to pedestal whenever a customer requests service. Furthermore, it provides no upgrade path to increase bandwidth per customer other than increasing the bit rate of the PON.

The third configuration employs two levels of splitting with a four-way split at the pole or pedestal for ease of drop placement and an eight-way splitter at the traditional RT location. This enables the sharing of each feeder fiber over 32 customers and each distribution fiber over 4 customers, while retaining the opportunity to remove the RT splitter in the future to decrease the feeder split ratio as bandwidth needs increase. BellSouth deployed an FTTH network in Dunwoody, Georgia, in 1999, based on the FTTH common technical specifications and using the third configuration in Figure 5.3.[13]

The split ratio, and thus the sharing of the fibers and the sharing of bandwidth, is a primary measure of the FTTH deployment. The PON architecture reduces the fiber count leaving the CO, allows for significant sharing of fiber, reduces the need to service active hubs in the field, and is easily reconfigurable. The reconfiguration allows expanding the number of premises or the bandwidth allocated to particular premises. In contrast, active FTTH architectures, which include electronic switching at remote hubs, offer other advantages such as ease of data aggregation, upgrade, and extended reach due to the regeneration within the RT.

5.2.3 PON Standards

Two major industry groups, the ITU and the IEEE, have produced a number of standards related to FTTH networks, of which three are primary PON standards:

- ATM PON-APON was developed by the FSAN and adopted by ITU-T. The standard is known as ITU-T G.983 and specifies an ATM-based PON, with maximum aggregate downstream rates of 622 Mb/s. The term APON suggested that only ATM services could be provided to end users, so the FSAN broadened the name to broadband PON (BPON).

- Ethernet PON-EPON has been developed by the IEEE Ethernet in the first mile (EFM) group. EPON describes extensions to the IEEE 802.3 media access control (MAC) and MAC control sublayers together with a set of physical (PHY) layers, which enable Ethernet protocols on a point-to-multipoint network topology implemented with passive optical splitters, and optical fiber. In addition, a mechanism for network operations, administration and maintenance (OAM) is included to facilitate network operation and troubleshooting. The EPON standard 802.3ah was ratified in June 2004.

- Gigabit PON-GPON was developed by the FSAN organization. The standard is known as ITU-T G.984. GPON may be viewed as a hybrid of EPON and APON systems and specifies aggregate transmission speeds of 2.488 Gb/s for voice and data applications. The goal of GPON is to support multiple services over PONs with gigabit and higher data rates. Furthermore, enhancements to operation, administration, maintenance, and provisioning (OAM&P) functionality and scalability are included. In January 2003, the GPON standards were ratified by ITU-T and are known as ITU-T G.984.1, G.984.2, and G.984.3.

The EFM effort has addressed three access network topologies and physical layers: point-to-point copper over the existing copper plant at speeds of at least 10 Mb/s up to at least 750 m; point-to-point optical fiber over a single fiber at a speed of 1000 Mb/s up to at least 10 km; and point-to-multipoint fiber at a speed of 1000 Mb/s up to at least 10 km, the Ethernet PON (EPON).

5.3 ATM PASSIVE OPTICAL NETWORKS

The basic characteristics of the system include a tree and branch PON that supports up to 32 optical network terminations with a logical reach up to 20 km. Downstream transmission is a continuous ATM stream at a bit rate of 155.52 or 622.08 Mb/s. Upstream transmission is in the form of bursts of ATM cells. In addition to ATM transport protocols, physical media-dependent (PMD) layer and transmission convergence (TC) layer protocols are specified.

5.3.1 ATM PON System Architecture

The G.983.1 defines a general network architecture shown in Figure 5.4, including the following elements:

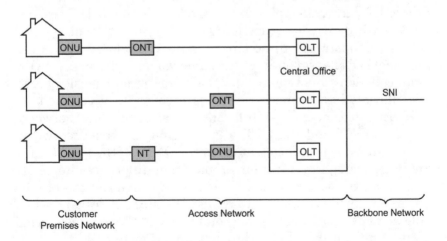

Figure 5.4 ITU G.983.1: general network architecture.

- Optical line termination (OLT)
- Optical network unit (ONU) or optical network termination (ONT), elements considered to be interchangeable in this discussion
- Optical access network (OAN)
- Service node interface (SNI)
- User network interface (UNI)

5.3.2 Upstream and Downstream Transport

A PON system acts as a broadcast network in the downstream direction and a point-to-point network in the upstream direction. The OLT in the central office transmits data through the tree and branch physical network architecture, so all downstream cells are delivered to all ONTs. The ONTs receive all cells but only accept cells specifically addressed to them. In the upstream direction, an ONT transmits on an assigned wavelength during its transmission window. The upstream cells pass through one or more couplers and are delivered to the OLT, but not to other ONTs.

Because the upstream and downstream directions operate differently, OLT and ONT hardware requirements are different. In the downstream direction, the OLT is the only device transmitting, so cell collisions are not of concern. Thus, the OLT can transmit in block mode and the ONT receives data with relatively little change in the receive level.

The upstream direction is more challenging. Because multiple ONTs transmit on the same wavelength on one PON and collisions must be avoided, each upstream timeslot must be allocated to a specific ONT. Moreover, to accommodate differing distances (thus different propagation delay) between various ONT and the OLT, a guard time is needed between adjacent cells originating at different ONTs. If the guard time is large enough to allow for propagation differences up to 20 km, the overhead is very wasteful, so a ranging protocol has been specified in G.983.1 to adjust the start of transmission of each ONT. This adds some complexity to the PON protocol, but significantly reduces the guard band overhead.

Another challenge is related to the OLT receive signal level. The signal from each ONT is attenuated because of optical fiber, connector, and splitter losses. Because the attenuation varies from ONT to ONT, the OLT receiver must be able to adjust quickly to widely different power levels and sync up with the upstream data frames. This requires a burst mode receiver, which is more challenging than the block mode receiver in the ONT.

The different characteristics of the upstream and downstream data paths also result in different protocol requirements. The downstream data path carries a continuous stream of 53-byte timeslots. Each timeslot carries an ATM cell or a physical layer operations/administration/maintenance (PLOAM) cell.

5.3.3 Broadcast Downstream Overlay

One attractive method for delivering video content to end users of PON systems is through the use of a broadcast downstream overlay. Originally ITU-T Recommendation G.983.1 specified the downstream wavelength window to extend from 1480 to 1580 nm. Recommendation G.983.3 updated the G.983.1 wavelength allocation plan to restrict the ATM PON downstream signal to the 1480- to 1500-nm range, to allow for the addition of an enhancement band. This enhancement band can be used for one of several applications. The applications in mind at the time of the specification included unidirectional (e.g., video) and bidirectional (e.g., DWDM) services.

The ATM PON downstream specification was built around a 1490- ± 10-nm center wavelength. The selection of this wavelength was driven by the perceived need for efficient amplification of analog video signal — specifically, for U.S. markets in which loop lengths tended to be longer. Because optical amplifications were most readily achieved in the 1550 to 1560 nm, this band was reserved as the enhancement band for analog video.

An alternative enhancement band was based on the delivery of digital services. This band plan specification has a wavelength allocation extending from 1539 to 1565 nm. The

ITU DWDM grid drove this wavelength allocation. Clearly, WDM technologies are important to future PON expansion and upgrade.[15]

5.4 ETHERNET PASSIVE OPTICAL NETWORKS

In 2001, the IEEE approved a standards project for EFM.[16] The initial goals included defining a standard that supports three subscriber access network topologies and physical layers:

- Point-to-point copper over the existing copper plant at speeds of at least 10 Mb/s up to at least 750 m
- Point-to-point optical fiber over a single fiber at a speed of 1000 Mb/s up to at least 10 km
- Point-to-multipoint fiber at a speed of 1000 Mb/s up to at least 10 km

The simplicity, interoperability, and low cost of Ethernet have allowed this IEEE standard to virtually monopolize LAN technologies. Indeed, the vast majority of data over the Internet begins and ends as Ethernet packets. Thus the potential importance of EPON. We briefly describe the gigabit Ethernet and 10-GbE standards.

5.4.1   Gigabit Ethernet and 10 GbE

This is the reason why EPON is potentially very important. In the next subsection we briefly describe the gigabit Ethernet and 10- GbE standards.

The original Ethernet was created by Bob Metcalfe in 1972 as an extension to Alohanet.[17] It used carrier sense multiple access with collision detection (CSMA/CD) for data communications over a shared coaxial cable. Ethernet was standardized in 1983 by the IEEE 802.3 Working Group; the first IEEE 802.3 Ethernet standard is known as 10BASE5, operating at 10 Mb/s using baseband transmission over a maximum of 500 m of coaxial cable.

Subsequent Ethernet standards have evolved to include speeds as high as 10 Gb/s (10 GbE) (Table 5.1 and Figure 5.5). However, it is interesting to note that some of important features of the original Ethernet are not contained in the later

Table 5.1   Ethernet Physical Interface Standards

Designation	Bit rate (Mb/s)	Media type	Topology	Range (m)	IEEE date
10BASE5	10	Coax	Bus	500	1983
10BASE2	10	Coax	Bus	185	1985
10BROAD36	10	Coax	CATV cable	3600	1985
10BASE-T	10	Cat 3	Pt to Pt	100	1990
10BASE-FB	10	MMF	Bus	2000	1993
10BASE-FP	10	MMF	Passive star	2000	1993
10BASE-FL	10	MMF	Pt to Pt	2000	1993
100BASE-TX	100	Cat 5	Pt to Pt	100	1995
100BASE-FX	100	MMF	Pt to Pt	2000	1995
1000BASE-TX	1000	Cat 5	Pt to Pt	100	1998
1000BASE-SX 62.6 μm	1000	MMF	Pt to Pt	275	1998
1000BASE-SX 50 μm	1000	MMF	Pt to Pt	550	1998
1000BASE-LX 62.6 μm	1000	MMF	Pt to Pt	550	1998
1000BASE-LX 50 μm	1000	MMF	Pt to Pt	550	1998
1000BASE-LX 10 μm	1000	SMF	Pt to Pt	5000	1998
10GBASE-S 850 nm	10,000	MMF	Pt-Pt	26–300	2002
10GBASE-L 1310 nm	10,000	SMF	Pt to Pt	10,000	2002
10GBASE-E 1550 nm	10,000	SMF	Pt to Pt	30,000–40,000	2002
EFM	1000	SMF	Pt to Pt	10,000	2004
EFM	1000	SMF	Pt to M-Pt	10,000	2004

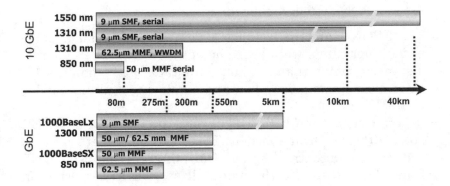

Figure 5.5   GbE and 10 GbE reach standards.

Ethernet standards. For example, Ethernet interfaces at 100 Mb/s or higher use a point-to-point topology rather than a shared medium, obviating carrier sensing and collision detection. These interfaces are intended to connect to an Ethernet switch, which buffers received Ethernet packets and forwards them to the appropriate output ports.

Several of the newer Ethernet interfaces are designed for multimode or single-mode optical fiber. This extends the reach in order potentially to use the interface for optical access. The 100BASE-FX Ethernet has a reach of 2 km over multimode fiber; a call for interest in March 2002 addressed the interest in creating a new single-mode fiber version with a reach of perhaps 15 km (1000BASE-FX Ethernet has a reach of 5 km over single-mode fiber). Although the Ethernet reach is long compared to distances within an enterprise, it is still shorter than the distance from the CO to many customers.

The 10-Gb/s Ethernet standard IEEE 802.3ae (10 Gb Ethernet) was ratified in 2002. LAN and WAN topologies are compatible with 10 GbE, which is likely to be a unifying technology that joins traditional telecom service providers with packet-based end-user needs. The integration of residential and campus networks with the optical backbone will allow for efficient transport of packet-based networking and will support the deployment of wideband access to these end users.

5.4.2 EPON System Architecture

The EPON standard includes a point-to-multipoint (P2MP) passive optical network based on optical Ethernet standards: distance greater than 10 km, data rates at standard GbE rates over single-mode fiber, and employing a minimum 16-to-1 split ratio.

In contrast to APON, EPON data are based on the IEEE 802.3 Ethernet frame format. Thus, variable-length packets are utilized, in contrast to the TDMA protocol that used fixed-frame formatting. Furthermore, all services transported on EPON are carried within a single protocol. Thus EPON provides connectivity for all packetized communication.

Solutions to the inherent incompatibility between the fixed time slots of TDM and the variable packets (64-to-1518 8-bit words) of Ethernet necessarily involve a number of compromises:

- The first approach is to use a time slot that is large enough to fit even the largest possible packet. However, this is an inefficient use of bandwidth because some time slots would contain only 64 octets even though each time slot is capable of handling 1518 octets.
- A second approach is to use aggregate multiple Ethernet frames into a fixed time slot consistent with the TDM protocol. The widely variable Ethernet frame size does not always allow complete filling of the available TDM time slot, but bandwidth efficiency will be significantly improved. Aggregating and unaggregating the frames may lead to an overall increase in complexity.
- The third approach involves segmenting Ethernet frames into fixed-size packets. This would require that a SAR layer be added onto the EPON protocol stack.

One other challenge with EPON packet transport lies in the Ethernet standard. The traditional Ethernet peer-to-peer relationship between network nodes does not exist among ONTs. The MAC and PHYs of the existing IEEE 802.3 CSMA/CD Ethernet standard must be enhanced to adapt to the PON infrastructure and its associated traffic-flow paradigm of downstream broadcasting and upstream TDMA.[18]

Figure 5.6 depicts the downstream traffic flow of an EPON network. Packets are broadcast to all users using variable-length packets. The splitter sends the entire signal (all packets) to all users ONU. Information contained within the header of each packet identifies the intended user. Packets may be intended for single users, groups of users, or all users. Only users that recognize the proper header address accept the packet, disregarding all other packets. As with the APON, upstream data are handled differently to avoid collisions at the splitter/combiner.

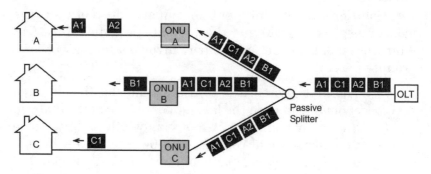

Figure 5.6 EPON downstream traffic.

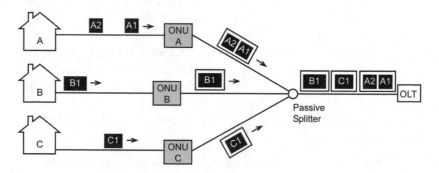

Figure 5.7 EPON upstream traffic.

Figure 5.7 depicts the time division multiplexing strategy. The upstream data comprise synchronized time slots and each user transmits variable-length packets only in a time slot designated for that user. The upstream cells pass through one or more couplers and are delivered to the OLT, but not to other ONUs. Note that this requires a self-synchronizing system because the distance and, thus, time to each user are not predetermined.

EPONs can be configured using two or three wavelengths. Two wavelengths, one for downstream and another for upstream, permit full-duplex transmission and sufficient downstream bandwidth to support voice, video, and data for each user. A downstream broadcast overlay can be implemented with a third wavelength.

5.5 GIGABIT-CAPABLE PASSIVE OPTICAL NETWORK (GPON) SYSTEMS

The GPON standard follows the A/B PON standard and is therefore similar. The chief goals of this standard are to enhance the A/B PON architecture. The enhancement allows support of higher data rates, especially for the transport of data services. The data rates and multiprotocol support of GPON enable a converged network that delivers voice, video, and data services over a PON-based infrastructure.

In addition to the issues described in the APON, GPON includes the following:

- Full service support, including voice (SONET and SDH TDM), Ethernet (10/100 base T), and ATM
- Physical reach of at least 20 km
- Support for multiple bit rates using a single protocol; Nominal line rates of:
  - Downstream: 1.25 or 2.5 Gb/s
  - Upstream: 155 Mb/s, 622 Mb/s, 1.25 Gb/s, and 2.5 Gb/s
- Enhanced operation administration and maintenance and provisioning (OAM&P) capabilities
- Protocol level security for downstream traffic due to the multicast nature of PON

GPON uses class A/B/C optics (described by G.982). The differences are primarily in the transmit and receive power levels, which translate into different link budgets: class A = 5 to 20 dB; class B = 10 to 25 dB; and class C = 15 to 30 dB. Forward error correction can be used as an option. Because GPON is needed to accommodate all services efficiently, a new transmission method called GEM (GPON encapsulation method) was adopted to encapsulate data services and TDM services. GEM provides a variable-length frame with control header.

A goal of the GPON effort is to maintain a common physical layer specification because EPON and, thus, the components for EPON and GPON have similar, although not identical, specifications.

5.6 ADVANCED TECHNOLOGIES

Access technologies necessarily have different cost/performance trade-offs compared to long haul or even metro. Low cost is clearly a primary requirement. This arises because, at the edge of the network, fewer end users share the component, installation, and maintenance costs. This is exemplified by the strategy employed by PON networks, which attempt to maximize the number of users on each fiber segment. With respect to components, passive as well as active, efficient automated packaging methods remain key. This includes enhancing functionality while simultaneously reducing the number of connections, fiber or electronic, of a given packaged device. The optical fiber is an often overlooked element of access technologies. However, some attributes of fiber preferred for access networks may differ from those of long haul or metro.

Additionally, the high-speed electronic signal processing common in wireless links will likely be exploited to compensate for lower cost optical components and thereby enhance system performance. Indeed, some types of electronic signal processing may appear in access links first because the data rates are typically lower than the backbone and therefore more readily implemented in silicon CMOS circuits. Here, the specific focus is on the physical layer, i.e., the so-called layer 1. Equally important advances are needed at other levels.

Attempting to quantify the appropriate bandwidth to provide FTTH subscribers or desktop users is not necessarily useful. A more meaningful question relates to the cost of delivering a particular bandwidth and the market willingness to pay for such bandwidth. Here, however, the only attempt is to address new and future technologies that will enable the deployment of advanced access networks with very high bandwidth at low cost. Note that, in assessing the required bandwidth and network topology, the quality of service (QoS) must be considered. To this end, two metrics can be identified. First is the average continuous bandwidth achievable per user, and second is the maximum burst rate available to the user. The average bandwidth provides some measure

Table 5.2 Approximate Download
Time for a 7-Gb File

DVD movie download times	
Link	Delivery time
Modem (56 kbps)	13 days
Cable modem (1.5 Mb/s)	11 h, 36 min
T-1 (1.54 Mb/s)	11 h, 12 min
DSL (8 Mb/s)	2 h, 12 min
PON OC12/32 (20 Mb/s)	54 min
PON OC48/32 (80 Mb/s)	18 min
Ethernet 100	10 min
GbE	1 min
10 GbE	6 s

Note: Measured bandwidth-distant product (megahertz per kilometer) of identical 1.1-km fiber samples.

Source: Tim Holloway, World Wide Packets presentation, FTTH conference, November, 2002.

of the data aggregation and the maximum rate determines the component capacity needed. In an effort to quantify the impact of various maximum burst rates available, the estimated download times of a full-length DVD movie using various network connection speeds are listed in Table 5.2.

Certainly, one should envision the capability to download a multigigabyte file within a few minutes into portable equipment and be free of the wired network. Furthermore, high QoS with real-time presentation of high-definition television (HDTV) that requires a continuous, average rate of ~20 Mb/s is obviously needed. Indeed current deployments of EPON and GPON systems deliver HDTV bandwidths in addition to voice and data services. Therefore, advanced systems of the future should support burst rates of 1 Gb/s. Certainly, current deployments should be scalable to these rates. Indeed, CATV already provides this bandwidth and EFM proposes to support this minimum rate.

This section examines the component technologies needed to support these rates and greater. Although transmission far exceeding 1-Gb/s rates is routinely used in the long haul, the challenge for access is to enable comparable data rates at a very low cost in components, subsystems, and topologies for point-to-point and point-to-multipoint links. Converged networks supporting voice, video, and data over a single data link layer will require support at >1 Gb/s data rates in addition to the broadcast bandwidth needed. Aggregation within the network requires that even higher single channel rates exist within FTTH networks. Therefore, 10-Gb/s burst rates available in FTTH applications are envisioned for the future. Clearly, these rates are beyond the scope of most plans envisioned today; however, it is useful to consider the technologies that would allow such capacity as a natural growth of the systems installed or planned today.

Note that, although the focus is on the physical layer of local access technologies, this is only one part of a complete infrastructure needed. Delivering high-capacity communications channels to a residence is of limited value if the household infrastructure cannot accept, manipulate, store, and display the incoming information. This additional infrastructure, including the applications that exploit this capability, is another necessary element of a ubiquitous broadband access infrastructure.

5.6.1 Component Requirements

When the focus is on the physical layer of access technologies, the needed features can be identified:

- Enhanced functionality in a single package
- Reduced component size
- Reduced packaging requirements
- Fewer fiber interconnects
- Reduced power consumption
- Amenability to automated manufacturing
- Legacy compatibility
- Standards compatibility

In the near term, optical access will likely expand by continued advances and deployment of PONs and the deployment of systems compatible with GPON and EPON standards. Ethernet-based protocols are likely to achieve increased deployments.

Figure 5.8 CAT 5e cable.

Furthermore, optical access will likely be dominated by wired connections. Note that systems based on free-space optics, which fill the needs of remote users requiring fast setup times, continue to receive attention; however, it is believed that these systems alone cannot support the required QoS for widespread deployment. Here, the focus is on fiber-based optical links.

First, the current status and expected progress of alternate technologies, i.e., the competition, will be briefly examined. As discussed, the optical implementation of fast Ethernet is dominated by all-electrical implementations, except for long-reach applications. Although other technologies are capable of providing broadband access, it is useful to consider the capabilities of a simple copper solution: namely, CAT 5e cable and the RJ45 connector (Figure 5.8).

The analog 3-dB bandwidth-distance product of typical CAT5e cable is less than a few megahertz per kilometer. However, advanced modulation formats, together with signal processing, may allow this technology to support 1-GHz/km performance. These techniques represent advances similar to those used in 1000BASE-T, where analog bandwidths of 100 MHz/km are achieved by a combination of multilevel amplitude coding and using all four twisted pairs in the cable. Indeed, some efforts seek to exploit the advances in high-speed signal processing to allow CAT 5 cable to provide 10-Gb/s links over 100 m.[19] Similarly, ADSL2plus enables twisted pair telephone with a reach of approximately 5,000 feet. Using ADSL2plus, technology service providers can deliver HDTV service over installed twisted pair infrastructure.

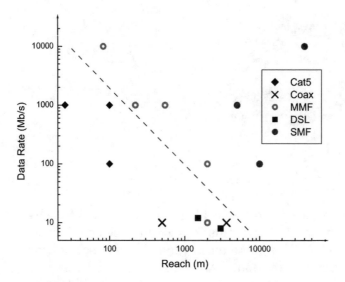

Figure 5.9 Reach standards for various wired link technologies. The dashed line demarks the transition from copper solutions to optical solutions near 0.1 Gb/s/km. Historically, advances in component technologies together with silicon signal processing techniques have moved performance higher.

The message to the optical device and optical network designers is clear: the transition from electrons to photons and back again must provide a solution that cannot be readily replicated by lower cost (copper) solutions. The competition (in some cases) is low-cost electrical connectors, together with silicon CMOS-implemented signal processing. The challenge of optics is then to exploit the large bandwidth distance performance available and to do so only for installations that do and are not likely to have meaningful copper solutions. Fiber to the home is one such installation, although these advanced copper solutions may have an impact on the in-home infrastructure.

Figure 5.9 depicts reach standards for various wired links. The transition from copper to optical is currently near 0.1 Gb/s/km and the transition from MMF to SMF is currently near 1 Gb/s/km. We reiterate that, historically, these performance metrics evolve and, once a cheaper technology can provide similar performance, it will rapidly dominate new deployments. Indeed, 100-Mb optical Ethernet has been superseded by copper.

5.6.2 Fiber

The design and performance of fiber intended for long-haul applications have advanced significantly; however, only recently has fiber intended for metro or access networks experienced notable improvement. Advances have been made in single-mode and multimode fibers. Future advances in fiber relevant to access networks will likely be in MMF and fiber intended for CWDM.

A significant new single-mode fiber type applicable to access is the low water peak fiber. Fiber such as AllWave™ fiber from OFS[20] essentially removes the absorption peak near 1385 nm typically found in optical fiber and thereby opens the entire spectral window from 1200 to 1600 nm. This type of fiber is useful for coarse WDM systems that exploit this entire spectral range. Figure 5.10 depicts the absorption and dispersion for a few fiber types. Note that as link lengths

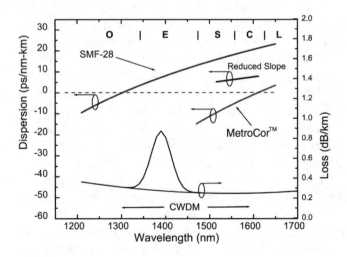

Figure 5.10 Dispersion and loss for various fiber types. MetroCor is a trademark of Corning Inc.

become shorter, the total link loss may be dominated by component and connector loss, including the splitters inherent in a PON system.

Separate from loss, dispersion may become an issue for long wavelength links. This arises in part from the larger line width of VCSELs and FP lasers. For example, some VCSEL line widths are ~1 nm, which corresponds to dispersion as high as 20 ps/km over the 1300- to 1600-nm range. For the 20-km reach desired for PONs, dispersion can severely degrade transmission at 1-Gb/s data rate. Increasing to 10 Gb/s over such lengths may limit the wavelength range to near the dispersion minimum at 1310 nm. Of course, a reduced slope fiber, representing the ability of fiber manufacturers to exercise some control over the dispersion and dispersion slope by proper design of the waveguide dispersion to balance the material dispersion, has been available for some time. A desirable new fiber would exhibit a reduced slope near the dispersion minimum, thereby allowing multiple wavelengths to be implemented at 10 Gb/s without consideration of dispersion compensation.

Multimode fiber has also seen significant advances recently. The ease of use inherent to multimode fiber (MMF) together with the need for low-cost solutions for access technologies in fiber networks has resulted in large deployments of MMF links. Moreover, a resurgence has occurred in research in MMF technologies to expand the useful reach and bandwidth for use in short-haul access links like local area networks, intra-cabinet, and, possibly, fiber-to-the-home. The large optical core and plurality of modes of MMF allow for dramatically simplified packaging and improved packaging yield of optoelectronic components, resulting in economies not possible with SMF.

However, the multiple modes typically exhibit differential mode delay (DMD) — the dispersion in group-delay among the numerous guided modes of the MMF — that can lead to severe intersymbol interference of the transmitted signal, dramatically limiting the bandwidth of the fiber. Standard FDDI-grade MMF has a modal bandwidth of 160 MHz/km, limiting reach to 220 and 26 m for 1- and 10-GbE links respectively. Fortunately, improved design and manufacturing techniques are allowing fiber producers more precise control of the graded index profile used to minimize modal dispersion. Indeed, the new class of MMF has bandwidths of 4000 MHz/km and 500-m reach for 10 GbE. Note that exploitation of the large installed base of fiber for LANs, which use standard FDDI grade MMF, is a key challenge for delivering broadband services.

Another issue related to multimode fiber is also susceptible to modal noise,[21-23] which arises from varying interference effects between the optical modes, i.e., the speckle pattern. A combination of mode-dependent loss and fluctuations in relative power in the different modes leads to modal noise. A variation in the relative phase between the guided modes together with mode selective loss (MSL) also contributes to modal noise. Fortunately, this noise is not fundamental in nature and can be managed with the proper control of connector loss and laser spectrum.

However, the use of SMF in access network is not free of obstacles. Interestingly, nonlinearities may have an impact on access networks. Although the distances are relatively short and the effect of nonlinearities may be expected to be small, two aspects of access links require some attention to nonlinearities. First, in PON systems, the splitting loss is significant and the interest is in allowing launch powers as high as +5 dBm or higher. These powers are sufficient to experience nonlinear effects.

A similar situation exists in HFC systems, which use techniques explicitly to mitigate the effects of high launch power. For example, most HFC systems employ frequency dithering to reduce stimulated Brillouin scattering.[24] SBS is a nonlinear effect caused by high optical power that results in optical power reflected backward. This induced loss limits the maximum launch power. The effect is enhanced with narrow line width sources. Furthermore, different modulation formats may find their way into FFTH links. Indeed, links with broadcast and point-to-point connections are likely to employ different modulation formats on the same fiber. HFC commonly uses QAM and VSB-AM. For these reasons nonlinearities are a significant concern for access fiber. Corning has recently introduced a new fiber specifically for FTTx applications that enables larger launch power compared to conventional single-mode fiber and still avoids SBS.

More aggressive deployments may employ remotely pumped erbium doped fiber (EDF), which allows high splitting ratios while maintaining the passive feature of field deployments. These fibers would allow efficient propagation of the pump wavelengths by reducing loss at 980 nm. Efforts to "design" fibers[25] with a spectrally broad and large Raman gain cross sections are of interest. Although envisioned for metro and long-haul applications, these fibers may be useful in access areas.

Another fiber type that has already found application in metro is the negative dispersion fiber. The negative dispersion is used to mitigate the effects of low-cost, direct-modulated laser sources, which are typically positive chirped. The availability of this negative dispersion fiber might be of interest for CWDM systems.

5.6.3 VCSELs

It is likely that low-cost, long-wavelength VCSELs will play a significant role in delivering broadband optical services at the network edge. The shorter wavelength (~850 nm) VCSELs are somewhat mature; however, they are limited in application due to the higher loss and higher dispersion in conventional fiber at this wavelength. Access technologies strive to avoid amplification, nonlinearities (and the required mitigation methods), and dispersion. Although the distances considered for access technologies are relatively modest by long-haul standards, dispersion issues often arise due to large spectral content of typical low-cost laser sources; for this reason 1310-nm VCSELs are seen as a solution. More generally, high-speed VCSEL technology that covers the entire 1300- to 1600-nm window, together with CWDM architectures and dispersion management, will provide access to the promised bandwidth of optical fiber. The key attributes of low-cost lasers include:

- High efficiency
- Wide operating temperature range

- High-speed direct modulation
- Sufficient coupled output power
- Low relative intensity noise
- Stable spectral and spatial profiles

Wavelength drift must also be considered for lasers useful in access networks. Indeed, this drift may be the dominant consideration for CWDM systems. Wavelength drift for a DFB laser is typically 0.08 nm/°C. These lasers usually include a wavelength monitor and a temperature control to maintain a locked wavelength over a large variation in external temperature. Without temperature control, over a typical operating temperature of −10 to +70°C, the wavelength varies ~6 nm. In addition, laser manufacturers might allow a wafer-to-wafer tolerance of ±3 nm. The 20-nm window of CWDM systems can be met by current technology; however, more stable lasers with primarily passive means of maintaining wavelength stability are needed.

Short-wavelength, 850-nm VCSELS are relatively mature and commonplace due to the optical index of refraction control afforded by the AlGaAs material system. The index variation allowed by this alloy system permits the gain region and the Bragg mirrors to be fabricated using one epitaxial growth. Long-wavelength lasers, on the other hand, cannot be fabricated with a single material system due to the difficulty in producing highly reflective Bragg mirrors with the low index of refraction range available in the InP material system. Various hybrid fabrication schemes have been employed, including wafer bonding. This wafer fusing of dielectric or metamorphic GaAs/AlAs mirrors to InP has met with some success.[26]

The development of VCSELs that can operate over the entire CWDM range of 1300 to 1600 nm is essential. Using lattice-matched AlGaAsSb mirrors and AlGaInAs quantum well-active regions,[27] 1310 and 1500 nm have been demonstrated. These devices require additional development to reach commercial success.

Another materials alternative is the indium gallium arsenide nitride (InGaAsN) system, which allows the same growth process currently used in the mass manufacture of 850-nm VCSELs. Here, the mirrors are formed from AlGaAs similar to the 850-nm VCSELs and the active region is formed from the InGaAsN alloy, which has a bandgap corresponding to the longer wavelengths.

An optical pump approach has been demonstrated in a vertically integrated structure. Using three epitaxial growths, a conventional 850-nm VCSEL laser is bonded to an InP active layer which was also bonded to a lower AlGaAs DBR grown on GaAs. The InP-based 1310-nm active device re-emits back through the 850-nm VCSEL. This double wafer bonding technique yields good high-temperature performance and single transverse mode operation. VCSELs at 1310 nm have been demonstrated with 0.5-mW CW operation up to 85°C and lasing remains CW up to 115°C.[28,29] Moreover, error-free operation at 2.5 Gb/s was demonstrated for a distance of 50 km. Future systems will require 1300-nm VCSELs capable of direction modulation to 10 Gb/s.

VCSEL development must also minimize mode partition noise, which originates from the time-varying spectral content of the laser source (mode partition) and the dispersion in the fiber. This results in a variable pulse shape at the receiver that appears as amplitude and timing noise. Unfortunately, this noise is proportional to the launched power and cannot be overcome by increasing received power. Furthermore, it is difficult to model mode partition noise — particularly at high modulation rates. Single-mode operation of VCSELs may limit mode partition noise; however, the need still exists for higher power, lower noise, high spectral quality laser sources that can be economically fabricated and coupled into optical fiber.

The longer term goal must include the integration of the electronic drivers directly with the VCSEL. Ideally, all electronics other than the CMOS circuitry should be integrated within a single element, preferably comprising a single epitaxial structure. The various methods employed to find the proper gain region and DBR mirrors should be exploited to incorporate the electronic driver function as well. New resonator structures that permit control of the longitudinal and

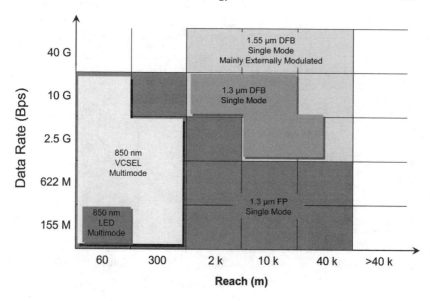

Figure 5.11a Current source and fiber technologies for optical links.

transverse modes may permit higher power vertical cavity lasers to be used in CWDM systems.[30,31]

Figure 5.11a depicts source and fiber technologies appropriate for various reaches and bandwidths. Interestingly, VCSEL and MMF are used for short-reach 10-Gb operation. The short reach is dominated by 850-nm VCSEL and MMF. Typical regimes for FTTx are served primarily by single-mode fiber using 1.3-µm Fabry–Perot lasers for low rates or 1.3- and 1.5-µm DFB lasers for data rates greater than 1 Gb/s. Future systems are likely to rely on advances in VCSEL and MMF technologies. Figure 5.11b shows the possible deployment of VCSEL and MMF technology, which requires advances in 1310-nm VCSELs, continued improvements in MMF bandwidth distance products, and the use of electronic techniques for dispersion compensation.

Although the requisite performance of envisioned FTTx infrastructures can be achieved with MMF and VCSEL, MMF is not perceived to be future proof — that is, upgradeable to

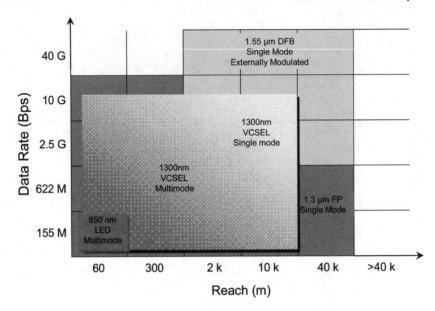

Figure 5.11b Influence of long wavelength VCSELs on future optical links.

data rates by more than an order of magnitude. However, it is likely that MMF links in LAN environments will obtain performance levels previously the domain of SMF and long-haul technologies.

5.6.4   Coarse WDM

Coarse wavelength division multiplexing (CWDM) uses multiple wavelengths to aggregate more signals onto a single optical fiber. Like DWDM, CWDM exploits the same multi-wavelength techniques however; CWDM uses a coarse wavelength spacing. For example DWDM typically uses 100-GHz ($\Delta\lambda$ = 0.8 nm @ λ = 1550 nm) or 50-GHz spacing, whereas CWDM uses 20-nm spacing. CWDM technology is viewed as critical to the success of Ethernet LAN and MAN networks because of the efficiency with which CWDM can exploit the installed fiber infrastructure. Note that these technologies are also central to FTTx deployment in general because they enable overlay wavelengths for transport of broadcast signals as well as provide added flexibility and capacity even if a conventional CWDM system is not installed.

The challenges in deploying a CWDM system include the development and production of long wavelength sources spanning the 1300- to 1600-nm range. Furthermore, dispersion over this range of wavelength varies dramatically; longer reaches may require dispersion engineering or a variable chirp that scales with wavelength. The dramatic variation in attenuation over this band may be obviated by fiber advances such as the low water peak fiber.

The primary assumption in CWDM systems is that access links do not require EDFAs, which are limited to operation around 1550 nm. This allows wavelengths to be distributed over a wide spectral range and spaced sufficiently far apart so that the wavelength drift of the low-cost sources does not result in wavelengths drifting out of their assigned channel. The wide channel spacing reduces cost in allowing the use of uncooled lasers and filters without active thermal management in contrast to the components required for the 100- and 50-GHz channel spacing of DWDM systems. Thus, CWDM technology is compatible with the wavelength-distributed PON.

5.6.5 Packaging, Interconnections, and Manufacturing

Low-cost packaging technologies for active and passive optical components are a cornerstone of optics in the access loop. A number of advanced and emerging packaging technologies are available for array, discrete, and monolithic integrated optical devices. This section only identifies the areas of interest for optics in the access loop and identifies the transition from transponders to transceivers.

Packaging needs include hermetic and nonhermetic varieties, including VCSEL packaging and plastic OE packaging. In addition, many aspects of manufacturing, including optical coupling, materials handling, design, assembly process, and package reliability, are key to low-cost components that can withstand the sometimes harsh environments of FTTH deployments. Advanced materials and novel fabrication/processing technologies (selective growth, dry etching, wafer bonding, etc.) for mass scale production are also areas of need.

Integrated circuits, which provide the optical-to-electronic transition for these network systems, are the most difficult components to design in a fiber-optic transceiver because of the dissimilar semiconductor material requirements for electronics and optics. These optical electronic ICs (OEICs) include the laser driver generating the electrical current to modulate the laser, and the transimpedance amplifier, which receives a very small current from a photodiode and must amplify it with minimal noise effects.

This section does not elaborate on these specific packaging needs but rather illustrates a significant evolution of transceivers. Figure 5.12 shows the transition of a full transponder to a

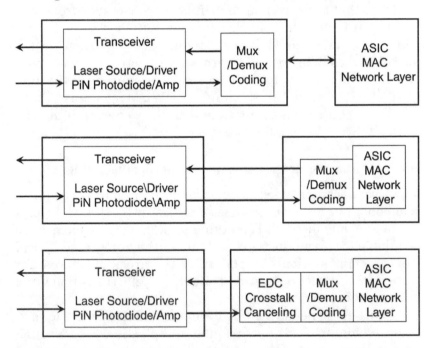

Figure 5.12 Evolution of packaging from transponder to transceiver to integrated electronic dispersion compensation and crosstalk cancellation.

transceiver. Advances in packing have allowed the high-speed serial link to be located between the transceiver and the silicon-based serializer/deserializer. This simplification allows all the silicon-based signal handling to be contained with a single board or package. It also allows for simplification of the variety of optical interfaces that can be selected for a particular need, while maintaining the identical silicon components. The inclusion of additional functionality in the form of electronic dispersion compensation (EDC) and/or crosstalk cancellation, which may originate in any part of the communications link, will enable improved performance of lower cost.

5.6.6   Passive Optical Devices

Passive optical components are clearly central to successful PON topologies. However they are also critical to the success of networks with active switching. Typical passive optics include optical couplers or splitters, tap couplers, and wavelength division multiplexers (WDMs). Gratings, such as fiber Bragg gratings, are also useful elements of optical networks. In addition to being easy to install and maintain, these components must permit an upgrade path and must meet the environmental requirements of outside plant.[32]

Passive components typically fall into three categories: (1) couplers; (2) splitters/combiners; or (3) filters. Note that fiber connections are sufficiently developed that installation and maintenance is not the most significant issue in PONs; however, single-mode fiber connections will always require precision components. Splitters and combiners refer to star couplers and other splitting and combining components that are not wavelength sensitive. On the other hand, filters attempt to remove or add a specific wavelength to a fiber. Although a number of types and sources are available for most passive devices, advances in thermal stability and reduced packaging cost are still needed. Furthermore, network architectures and design should be such that wide performance tolerance windows are acceptable.

Thin film filters provide an extremely cost-effective method of adding and dropping one wavelength, although insertion loss performance is sometimes an issue. The more complex arrayed waveguide devices may be useful in CWDM systems. In the future, more functionality may be accomplished by greater use of planar waveguide circuits. Nano-imprinted devices may offer a means of fabricating such complex structures.[33] Consider, as an example, optical isolators. Return signals are generated from the coupling interfaces of all components in the network. These reflections can have a destabilizing effect on the laser source and add crosstalk noise to the receiver. This requirement for isolation is an often over-looked aspect of optical systems. Particularly for PON systems, which are point-to-multipoint configuration, the aggregate reflected signals may be a significant source of noise.

An optical isolator is a nonreciprocal "one-way" device for optics. It is typically composed of a magnetic crystal having a Faraday effect, a permanent magnet for applying a magnetic field, and polarizing elements. Optical isolators with 0.5 dB insertion loss, approximately, 40 dB in isolation, 60 dB in return loss, and 0.03 p/s in polarization-mode dispersion are available.[34]

When a magnetic field is applied transversely to the direction of light propagation in an optical material with a Faraday effect, a nonreciprocal phase shift occurs and can be used in a polarizing or interferometric configuration to result in unidirectional propagation. Hybrid integration of Faraday rotators and polarizing elements has been demonstrated and thin-film magnets have been integrated on waveguide Faraday rotators; a major advance has been achieved by inserting thin-film half-wave plates and extinction ratios of about 30 dB have been obtained for a wide range of wavelengths around 1.55 μm, comparable to commercially available bulk isolators. However, birefringence control is still needed and the total insertion loss needs improvement (lowest value to date is 2.6 dB). Avoiding phase matching by using a Mach–Zehnder interferometer-based isolator has been demonstrated.[35] Polarization-independent isolators are also possible.[36,37]

5.6.7 Components for Bidirectional Transport

One architectural approach to fiber in the last mile is to separate data and broadcast to the home on two separate fibers or separate wavelengths. The upstream data can be carried on a different wavelength. Simplifying the components required for bidirectional transport allows efficient use of one fiber for incoming, as well as outgoing, data. As described previously, most network architectures for FTTx, such as GPON and Ethernet PONs, support single-fiber bidirectional communication because it reduces the amount of fiber, therefore reducing the cost for each link.[38–41]

One approach for bidirectional transport is to use two different devices, one for the light emitter and the other for the detector, and to use a splitter to couple light into and out of the detector and emitter, respectively. The first approach used biconical couplers and an emitter–detector pair at both ends of the optical fiber to communicate.[42] Alternatively, a number of BiDi transceiver units based on micro-optics with integral beam splitter are becoming available. These devices typically have Fabry–Perot laser diodes for digital transmission at 1310 nm; an InGaAs/InP-PIN-diode for digital receiving at 1490 nm; and a high linearity InGaAs/InP-PIN-diode for analog detection at 1550 nm. These devices enable an analog overlay for broadcast video. However, adding a splitter to the system increases the cost and may have an impact on performance. Fully integrated solutions are likely to yield improved performance and economics.

A simpler approach, which eliminates the need for a coupler or splitter, used a bifunctional device such as an LED, edge-emitting laser (EEL), or vertical cavity surface emitting laser (VCSEL) at both ends of the link.[43–45] These devices are capable of operating as light emitters or photodetectors, depending on how they are biased. Because these devices are inherently light emitters, the optimum structural designs must be altered to improve their sensitivities in detection mode, which compromises the performances of the devices in light emission mode.

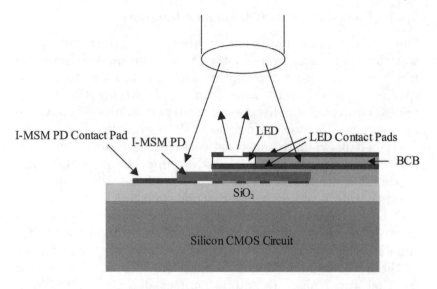

Figure 5.13a   Heterogeneous integration of emitter and detector.

Another approach uses a new structure containing two p–n junctions, which operates as an LED and a heterojunction phototransistor, alternately.[46] Two similar approaches used a photonic integrated circuit (PIC) at each end of the optical link.[47,48] The two p–n junction devices and the PICs were grown monolithically. Even though the emitter and the detector were separated using a monolithic approach, the growth material and structure were limited due to the lattice-matching conditions. To avoid this limitation, a hybrid integration method was used to integrate a thin film GaAs-based LED onto a large silicon CMOS bipolar junction detector.[49] In order to increase the responsivity and speed of the detector, a material other than silicon may be used. Recent results use a heterogeneous integration method to stack two independently grown and optimized thin film devices onto SiO_2-coated silicon and a silicon CMOS transceiver circuit. It should be noted, however, that LEDs are inherently slower devices than lasers.

A fabrication method for stacking thin film emitters and detectors directly on top of one another to realize a bidirectional, co-located emitter/detector pair for bidirectional has

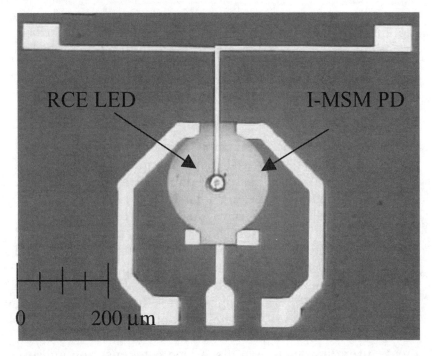

RCE LED   I-MSM PD

0   200 μm

Figure 5.13b Photomicrograph of stacked integration of emitter and detector.

been demonstrated (Figure 5.13). The structure is fabricated by stacking a thin-film GaAs-based inverted metal–semiconductor–metal (I-MSM) photodetector (PD) and a thin film GaAs-based light-emitting diode (LED) that have been independently grown and fabricated into devices. The stacked emitter/detector is bonded onto a SiO_2-coated host substrate.[50,51]

The performance of the co-located emitter and detector retained the performance of the individual devices. The measured responsivity of the I-MSM PD changed from 0.36 A/W before integration to 0.31 A/W after integration. Importantly, the dark current was unchanged. The co-located PD responsivity is smaller due to the shadow effect of the LED. The LED performance was similar before and after co-location. This technology is also compatible with using VCSELS in place of the LED.

Co-located sources and receivers, whether created by single epitaxial methods or the stacking technology described, are viable structures for simplifying bidirectional optical links. Increasing functionality while minimizing fiber connections is a key to lower cost access technologies.

5.6.8 Optical and Electronic Signal Processing

Advanced modulation formats and signal processing are widespread in wireless and wired copper communication links. In contrast, optical links currently deployed employ primitive coding and signal-processing algorithms: binary signaling, threshold detection, and hard decision decoding. Yet, fiber channels suffer from similar constraints of wireless and wired copper links, including intersymbol interference (ISI) and crosstalk. However, it is not possible to transfer the technology of those systems directly to a fiber optic system. The high speeds generally prohibit sophisticated error-control coding and signal processing. Nonetheless, a properly optimized link should exploit optical and silicon-based methods of impairment mitigation, including dispersion compensation and use of different encoding formats.

Traditionally, silicon-based signal processing strategies have been applied exclusively in long-haul links due to the costs associated with the circuits. Interestingly, however, forward error correction (FEC) has recently been included in the GbE standard. It is easy to see the advantage of incorporating FEC in a PON system. "Standard" ITU FEC permits 6 dB of electrical gain for bit error rates of $\sim10^{-15}$. This enables an additional split in a PON system. That is, the 3-dB optical gain of FEC allows nearly twice as many end users, other system parameters remaining equal. This suggests that network-wide FEC may significantly reduce the number of sources needed in a PON network. Although today it is difficult to envision network-wide FEC in FTTH networks, it is likely that the CMOS implementation of FEC will continue the historical gains in performance/price ratio of CMOS circuits. It is therefore inevitable that sophisticated silicon-based processing will eventually affect FTTH networks.

A likely first use of electronic signal processing in access networks may be in channel equalization. INCITS,[52] the International Committee for Information Technology Standards, recently began initial discussions for implementation of an electronic dispersion compensation of MMF for use within the fiber channel standard. The IEEE is also considering EDC for inclusion in various Ethernet standards. As previously mentioned, one of the challenges in implementing CWDM systems is the large variation in dispersion over the 1300- to 1600-nm window. EDC may allow low-cost optics to operate over sufficient distance without significant penalty due to chromatic dispersion.

Optical methods of mitigating channel impairments are also of significance for long-haul and access networks. Indeed, hybrid approaches that optimally exploit optical and electronics technologies are likely to achieve superior performance.[53]

5.6.9 Equalizing Multimode Fiber

To illustrate the impact of optical and electronic signal processing, recent results for multimode fiber will be described. MMF is the dominant media for LANs and has become the media of choice for short-reach high-bandwidth links. Indeed, vertical-cavity, surface-emitting lasers (VCSELs) together with MMF are part of the solution for short reach 1- and 10-Gb/s links. Unfortunately, the large and variable (from fiber to fiber) differential modal delay (DMD) dramatically limits the bandwidth-distance product of MMF links. Figure 5.14 depicts the simple ray picture of this effect.

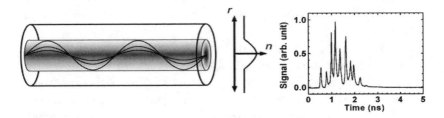

Figure 5.14 Mode propagation in multimode fiber and the measured impulse response.

Table 5.3 Measured Bandwidth-Distant Product (Megahertz per Kilometer) of Identical 1.1-km Fiber Samples

Fiber #	1	2	3	4	5	6	7	8	9
$\lambda = 850$ nm	524	648	948	521	839	612	624	835	649
$\lambda = 1300$ nm	630	622	451	464	504	448	445	500	514

Early in the development of MMF, it was realized that a parabolic graded index (GI) profile would compensate the different paths of the modes. Telecom-grade multimode fiber is available in 62.5- and 50-μm core sizes. These fibers support hundreds of transverse optical modes, which are highly degenerate with respect to group delay. Figure 5.14 also shows the impulse response of a 1.1-km fiber measured using high temporal resolution by the authors' group. The distinct arrival of the modes can easily be determined.

In practice, the benefit of a graded index is limited for three reasons linked to the sensitivity of DMD to index profile. First a wavelength dependence arises from the different dispersion of the undoped cladding and the doped core. Second, variations in the coupling from the source to fiber results in a difference in the net effect of DMD, as is observed with restricted mode launch. Finally, but most importantly, the effect of minor nonuniformities in the refractive index profile, including effective core size, due to manufacturing limitations, accumulates over the length of fiber and results in an unpredictable residual DMD. Recently, fiber manufacturers have improved manufacturing capabilities and new generation MMF is available with increased bandwidth-distance product.

These effects make it difficult to provide a useful model of an arbitrary sample of GI-MMF fiber. Table 5.3 shows the measured bandwidth-distance product of nine different fibers with the same manufacturer specification. These realities point to the need for a postproduction, postinstallation, DMD compensation technique.

A notable method of DMD compensation in GI-MMF is the restricted mode launch (RML), which reduces the effect of DMD by purposeful excitation of a fewer number of modes and by specifying the DMD allowed by this reduced set. RML is enabled by minimal intermodal coupling — the exchange of optical power among modes — along the fiber length. With RML, the optical signal remains in a small group of neighboring modes with similar group velocity, so significantly less pulse broadening occurs, despite the inherent DMD of the fiber.[54] RML is typically achieved by coupling the light into the GI-MMF via a small optical spot, often with an SMF offset from the fiber center.[55] Although improvement is not assured with RML, it has been shown to provide, on average, twofold improvement in link bandwidth. The demonstrated improvement is sufficient to warrant adoption in IEEE 802.3z (gigabit Ethernet) standard (IEEE98); however, the most notable drawback is the significantly increased alignment precision of RML.

Initial efforts using electronic postdetection equalization[56] have attempted to create dynamically equalized links that can adapt to variations in GI-MMF.[57,58] Simpler, hybrid schemes of equalization that are compliant with the motivation of using MMF, cost-effective, optical solution are required.

5.7 SPATIALLY RESOLVED EQUALIZATION

Recently a robust and scalable DMD compensation technique was demonstrated. The method is based upon a multisegment photodetector that does not add significant complexity and can be used with low-cost multimode optical sources, thereby maintaining all the benefits of using multimode fiber. This technique is called spatially resolved equalization (SRE)[59–61] and exploits the idea of obtaining additional information from the optical signal. The technique is somewhat analogous to those applied in multipath wireless links.[62]

The absence of significant intermodal coupling among the modes allows diversity in the temporal response, i.e., the different modes retain the distinct temporal response. In conjunction with the significant difference in mode-field distribution and, in particular mode-filed size, this modal diversity translates into a spatial diversity — a variation in temporal

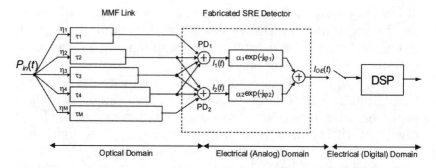

Figure 5.15 Model of MMF and two-segment SRE showing distinct modal channels and subsequent electrical recombination followed by electronic signal processing.

response within the emitted optical spot. This spatial/temporal diversity is exploited using multiple, spatially distributed photodetectors (PDs) and subsequent electrical "processing" to remove the ISI caused by the DMD. More importantly, the DMD compensation is made without *a priori* knowledge of the fiber performance or launch condition.

Figure 5.15 depicts a simple view of the link. The MMF can be viewed as comprising many different channels, each with a distinct temporal delay. Conventional receivers simply sum the temporal response. In contrast, SRE attempts to detect the different modes separately. Although complete separation is not possible without an intervening element, a strong relationship between radius and modal delay is present.

In the most general case, a large number of segments are combined with a dynamically adjustable complex scaling factor (amplitude and phase). The benefits and complexity of such a scheme are in contrast to the demonstrated scalar-SRE approach. A concentric, two-segment, spatially resolved equalization photodetector (SRE-PD) has been fabricated and demonstrated (Figure 5.16).

In Figure 5.17a, the measured signals from the inner and outer detector regions are shown. It is clear that although a great deal of correlation exists between the inner and outer signals, significant diversity is also present. Figure 5.17b shows the SRE result in comparison with the standard detection

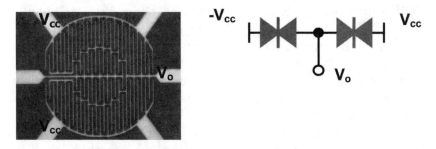

Figure 5.16 (a) Fabricated two-segment photoreceiver used for scalar, spatially resolved equalization; (b) schematic of electrical configuration depicting simple photocurrent subtraction.

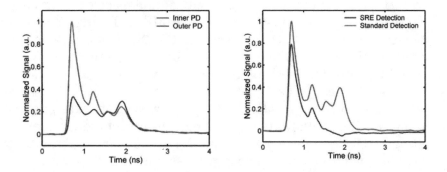

Figure 5.17 Measured impulse response of link (MMF and SRE photoreceiver). Excitation: 1-ps pulse at 1550-nm, 1.1-km fiber length.

result. Clearly, SRE dramatically enhances the impulse response and thus improves the bandwidth. Extensive simulation and experimental demonstrations of eye diagrams and measured bit error rates have shown that the bandwidth distance product can be quadruple for MMF. The photocurrent subtraction yields a 6-dB optical penalty for this bandwidth improvement. Furthermore, SRE is robust to fiber variations and does not need to be optimized for each fiber, although such optimization does improve performance. The strength of SRE includes its simplicity and the fact that it retains the alignment tolerance of conventional MMF receivers. Although

SRE improves MMF performance and hence may primarily impact LANs, it is believed that this class of component advance will impact FTTX deployments of the future.

Purely electronic equalization techniques are also effective against DMD in MMF. Previously, Kasper[63] identified electrical equalization techniques such as a linear equalizer or a decision feedback equalizer (DFE) for MMF systems. Linear equalization techniques have been extremely successful in copper links and recent results suggest that 10-Gb/s analog equalization is technically and commercially viable.[64,65] Optical links are also likely to see similar performance increases due to digital signal processing used to mitigate chromatic and polarization dispersion.[66]

Hybrid techniques that use optical and electronic signal processing have also been applied to MMF, including in combination with SRE to compensate for the ISI. Thus, electrical equalization, optical equalization, or their combinations, together with FEC techniques, are likely to have an impact on access optical links.

Figure 5.18 depicts the BER vs. received electrical signal-to-noise ratio for various MMF equalization techniques operating at 2.5 Gb/s over 1.1 km of 500-MHz/km fiber. The results are obtained from experimentally measured impulse response measurements of the fiber. It should be noted that the signal received by a conventional detector produces a closed eye, so no amount of optical power can improve the BER. The multisegment detector, however, immediately exploits the spatial diversity in MMF to reduce the ISI. Together with DFE or Viterbi codes, SRE can dramatically reduce the ISI penalty.

5.8 CONCLUSIONS

Bringing broadband capability to offices, homes, and the desktop is key to expanding the usefulness of the entire communications network. Indeed, increasing access network speeds to allow real-time transfer of high-definition video, will alter the manner in which users conceptualize and interact with the network.

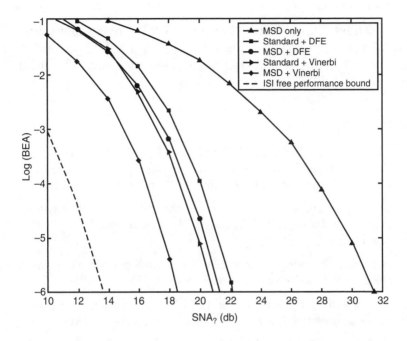

Figure 5.18 MMF link performance with a variety of equalization strategies.

Passive optical networks provide efficient use of fiber and components while retaining significant upgrade capacity and will continue to play a key role in FTTH deployments. A wide range of technologies is needed to implement cost-effective optical access fully, including advances in fibers, sources, and optical and electronic equalization.

Finally, bringing high-capacity links to the home or small office without equal improvements in the home infrastructure will not significantly relieve the access bottleneck. Legacy coax cable and copper wire may not have the capacity to exploit the increased availability of bandwidth and must be addressed as well. Dramatic improvements in organization, distribution, and storage are required if new applications are to exploit large increases in bandwidth fully.

ACKNOWLEDGMENTS

The authors thank Ketan Petal for useful discussions and careful reading of the manuscript. The authors dedicate this chapter to the memory of Dan Spears, a leader and influential proponent of FTTX and a gentleman.

REFERENCES

1. J.R. Fox, D.I. Fordham, R. Wood, and D.J. Ahern, Initial experience with the Milton Keyne's optical fiber cable TV trial, *IEEE Trans. Commun.*, COM-30, 2155–2162, September 1982.

2. K. Sakurai and K. Asatani, A review of broad-band fiber system activity in Japan, *IEEE J. Selected Areas Commun.*, 1, 428–435, April 1983.

3. H. Seguin, Introduction of optical broad-band networks in France, *IEEE J. Selected Areas Commun.*, 4, 573–578, July 1986.

4. R.K. Snelling, J. Chernak, and K.W. Kaplan, Future fiber access needs and systems, *IEEE Commun. Mag.*, 28, 63–65, April 1990.

5. FTTH study conducted by Render Vanderslice & Associates for the FTTH Council, www.ftthcouncil.org.

6. Y.M. Lin, D.R. Spears, and M. Yin, Fiber-based local access network architectures, *IEEE Commun. Mag.*, 27, 64–73, October 1989.

7. T.R. Rowbotham, Local loop developments in the U.K., *IEEE Commun. Mag.*, 29, 50–59, March 1991.

8. D. Faulkner, R. Mistry, T. Rowbotham, K. Okada, W. Warzanskyj, A. Zylbersztejn, and Y. Picault, The full services access networks initiative, *IEEE Commun. Mag.*, 35, 58–68, April 1997 and www.fsanweb/org.

9. ITU (International Telecommunications Union), www.itu.int. ITU-T Rec. G.983.1, broadband optical access systems based on passive optical networks (PON), 1998.

10. D. Spears, B. Ford, J. Stern, A. Quayle, J. Abiven, S. Durel, K. Okada, and H. Ueda, Description of the common technical specifications for ATM PON Systems, NOC'99, 1999.

11. Y. Maeda, K. Okada, and D. Faulkner, FSAN OAN-WG and future issues for broadband optical access networks, *IEEE Commun. Mag.*, 39, 126–133, December 2001.

12. K. Okada, B. Ford, G. Mahony, S. Hornung, D. Faulkner, J. Abiven, S. Durel, R. Ballart, and J. Erickson, Deployment status and common technical specifications for a B-PON system, *IEEE Commun. Mag.*, 39, 134–141, December 2001.

13. D. Kettler, H. Kafka, and D. Spears, Driving fiber to the home, *IEEE Commun. Mag.*, 38, 106–110, November 2000.

14. IEEE Standard 802.3, Carrier sense multiple access with collision detection (CSMA/CD) access method and physical layer specifications, 2000 edition.

15. N.J. Frigo, K.C. Reichmann, and P.P. Iannone, Fiber optic local access architectures, *11th Int. Conf. Integrated Opt. Opt. Fibre Commun., and 23rd Eur. Conf. Opt. Commun.*, V1, 22–25, 135–136, 1997.

16. http://www.ieee802.org/3/efm/.

17. 10 GbE white paper 10 GbE alliance www.10GbEA.com.

18. EPON white paper http://www.iec.org.

19. Solarflare communications, solarflare.com.

20. OFS ALL wave fiber as described in www.ofsinnovations.com.

21. T. Kanada, Evaluation of modal noise in multimode fiber-optic systems, *J. Lightwave Technol.*, 2(1), 11–18, February 1984.

22. R. Dandliker, A. Bertholds, and F. Maystre, How modal noise in multimode fibers depends on source spectrum and fiber dispersion, *IEEE J. Lightwave Technol.*, 3(7), 1985.

23. R.J.S. Bates, D.M. Kuchta, and K.P. Jackson, Improved multimode fiber link BER calculations due to modal noise and non-self-pulsating laser diodes, *Opti. Quantum Electron.*, 27, 203–224, 1995.

24. P.N. Butcher and D. Cotter, *The Elements of Nonlinear Optics*, Cambridge University Press, Cambridge, 1990.

25. R. Stegeman, L. Jankovic, H. Kim, C. Rivero, G. Stegeman, K. Richardson, P. Delfyett, A. Schulte, and T. Cardinal, Tellurite glasses with peak absolute Raman gain coefficients up to 30x fused silica, *Opt. Lett.*, 28, 1126, 2003.

26. K. Streubel, *Opt. Eng.*, 39, 488–497, 2000.

27. M.H.M. Reddy, D.A. Buell, A.S. Huntington, R. Koda, D. Free-
    zell, T. Asano, J.K. Jim, E. Hall, S. Nakagawa, and L.A. Coldren,
    Current status of epitaxial 1.31-1.55 /spl mu/m VCSELs on InP
    all-optical networking, *2002 IEEE/LEOS Summer Topic*, July
    15–17, WG1-58–WG1-59, 2002.

28. V. Jayaraman, T.J. Goodnough, T.L. Beam, F.M. Ahedo, and R.A.
    Maurice, Continuous-wave operation of single-transverse-mode
    1310-nm VCSELs up to 115°C, *Photonics Technol. Lett.*, 12(12),
    1595, 2000.

29. J. Geske, V. Jayaraman, T. Goodwin, M. Culick, M. MacDougal,
    T. Goodnough, D. Welch, and J.E. Bowers, 2.5-Gb/s transmission
    over 50 km with a 1.3-μm vertical-cavity surface-emitting laser,
    Photonics Technol. Lett., 12(12), 1707, 2000.

30. A.J. Fischer, K.D. Choquette, W.W. Chow, A.A. Allerman, D.K.
    Serkland, and K.M. Geib, High single-mode power observed
    from a coupled-resonator vertical-cavity laser diode, *Appl. Phys.
    Lett.*, 79, 4079–4081, 2001.

31. D.M. Grasso and K.D. Choquette, Threshold and modal charac-
    teristics of composite-resonator vertical-cavity lasers, *IEEE J.
    Quantum Electron.*, 39, 1526–1530, 2003.

32. D. Keck, A. Morrow, D. Nolan, and Thompson, Passive components
    in the subscriber loop, *J. Light Technol.*, 7, 1623–1633, 1989.

33. J. Seekamp, S. Zankovych, A.H. Helfer, P. Maury, C.M.S. Torres,
    G. Bottger, C. Liguda, M. Eich, B. Heidari, L. Montelius, and J.
    Ahopelto, Nanoimprinted passive optical devices, *Nanotechnol-
    ogy*, 13(n-5), 581–586, Oct. 2002.

34. FDK Corporation, http://www.fdk.co.jp/index-e.html.

35. J. Fujita et al., *Appl. Phys. Lett.*, 75, 998–1000, 1999.

36. O. Zhuromskyy, M. Lohmeyer, N. Bahlmann, H. Dotsch, P. Her-
    tel, and A.F. Popkov, Analysis of polarization independent
    Mach–Zehnder-type integrated optical isolator, *J. Lightwave
    Technol.*, 17, 1200–1205, 1999.

37. J. Fujita, M. Levy, R.M. Osgood, Jr., L. Wilkens, and H. Dotsch,
    Polarization-independent waveguide optical isolator based on
    nonreciprocal phase shift, *Photon. Technol. Lett.*, 12, 1510–1512,
    2000.

38. G. Kramer and G. Pesavento, Ethernet passive optical network (EPON): building a next-generation optical access network, *IEEE Commun. Mag.*, 40(2), 66–73, 2002.

39. D. Kettler, H. Kafka, and D. Spears, Driving fiber to the home, *IEEE Commun. Mag.*, 38, 106–110, November 2000.

40. T. Shan, J. Yang, and C. Sheng, EPON upstream multiple access scheme, *Proc. ICII 2001 — Beijing, 2001 Int. Conf.*, 2, 273–278, 2001.

41. A. Tan, Super PON-A fiber to the home cable network for CATV and POTS/ISDN/VOD as economical as a coaxial cable network, *J. Lightwave Technol.*, 15(2), 213–218, 1997.

42. B.S. Kawasaki, K.O. Hill, D.C. Johnson, and A.U. Tenne-Sens, Full duplex transmission link over single-strand optical fiber, *Opt. Lett.*, 1(3), 107–108, 1977.

43. T. Ozeki, T. Uematsu, T. Ito, M. Yamamoto, and Y. Unno, Half-duplex optical transmission link using an LED source-detector scheme, *Opt. Lett.*, 2(4), 103–105, 1978.

44. A. Alping and R. Tell, 100Mb/s Semiduplex optical fiber transmission experiment using GaAs/GaAlAs laser transceivers, *J. Lightwave Technol.*, 2(5), 663–667, 1984.

45. M. Dragas, I. White, R. Penty, J. Rorison, P. Heard, and G. Parry, Dual-purpose VCSELs for short-haul bidirectional communication links, *IEEE Photon. Technol. Lett.*, 11(12), 1548–1550, 1999.

46. M. Takeuchi, F. Satoh, and S. Yamashita, A new structure GaAlAs-GaAs device uniting LED and phototransistor, *Jpn. J. Appl. Phys.*, 21(12), 1785, 1982.

47. R. Ben-Michael, U. Koren, B. Miller, M. Young, T. Koch, M. Chien, R. Capik, G. Raybon, and K. Dreyer, A bidirectional transceiver PIC for ping-pong local loop configurations operating at 1.3-μm wavelength.

48. K. Liou, B. Glance, U. Koren, E. Burrows, G. Raybon, C. Burrus, and K. Dreyer, Monolithically integrated semiconductor LED-amplifier for applications as transceivers in fiber access systems, *IEEE Photonics Technol. Lett.*, 8(6), 800–802, 1996.

49. J. Cross, A. Lopez-Lagunas, B. Buchanan, L. Carastro, S. Wang, N. Jokerst, S. Wills, M. Brooke, and M. Ingram, A single-fiber bidirectional optical link using co-located emitters and detectors, *IEEE Photon. Technol. Lett.*, 8(10), 1385–1387, 1996.

50. K.F. Brennan, *The Physics of Semiconductors with Applications to Optoelectronic Devices*, New York: Cambridge University Press, 1999.

51. O. Vendier, N.M. Jokerst, and R.P. Leavitt, High efficiency thin-film GaAs-based MSM photodetectors, *Electron. Lett.*, 32(4), 394–395, 1996.

52. http://www.incits.org.

53. S.E. Ralph, K.M. Patel, C. Argon, A. Polley, and S.W. McLaughlin, Intelligent receivers for multimode fiber: optical and electronic equalization of differential modal delay, *Lasers and Electro-Optics Society, 2002,* LEOS, 1, 295–296, November 10–14, 2002.

54. G. Yabre, Comprehensive theory of dispersion in graded-index optical fibers, *J. Lightwave Technol.*, 18, 166–177, February 2000.

55. L. Raddatz, I.H. White, D.G. Cunningham, and M.C. Nowell, Influence of restricted mode excitation on bandwidth of multimode fiber links, *IEEE Photon. Technol. Lett.*, 10, 534–536, April 1998.

56. J.G. Proakis, *Digital Communications,* 3rd ed., New York: McGraw-Hill, 1995.

57. O. Agazzi, V. Gopinathan, K. Parhi, K. Kota, and A. Phanse, DSP-based equalization for optical channels, presented at IEEE 802.3ae Interim Meeting, New Orleans, September 2000.

58. Fow-Sen Choa, 10 Gb/s Multimode fiber transmissions over any distance using adaptive equalization techniques, *IEEE 802.3ae Interim Meet.*, New Orleans, September 2000.

59. K.M. Patel and S.E. Ralph, Spatially resolved detection for enhancement of multimode-fiber-link performance, *LEOS 2001, 14th Annu. Meet. IEEE*, 2, 483–484, 2001.

60. K.M. Patel and S.E. Ralph, Improved multimode link bandwidth using spatial diversity in signal reception, *CLEO 2001, Tech. Dig.*, 416.

61. K.M. Patel and S.E. Ralph, Enhanced multimode fiber link performance using a spatially resolved receiver, *IEEE Photon. Technol. Lett.*, 14, 393–395, March 2002.

62. R.P. Gooch and B.J. Sublett, Joint spatial and temporal equalization in decision-directed adaptive antenna system, *22nd Asilomar Conf. Signals, Syst. Computers,* 1, 255–259, 1988.

63. B.L. Kasper, Equalization of multimode optical fiber systems, *Bell Syst. Tech. J.,* 61, 1367–1388, September 1982.

64. Gennum Corp., 2001 10 Gb/s backplane equalizer, www.gennum.com.

65. Quellan Inc., www.quellan.com.

66. Big Bear Networks, www.bigbearnetworks.com.

6

Last Mile Wireless Access in Broadband and Home Networks

CARLOS DE M. CORDEIRO AND
DHARMA P. AGRAWAL

6.1 INTRODUCTION

Broadband communications and home networking are becoming household words as more homes utilize many network-enabled devices. The term *broadband* implies high-speed digital communications, requiring wider bandwidth for transmission, and can be employed for the distribution of high-speed data, voice, and video throughout the home. Therefore, the last mile broadband access specifies the connectivity mechanism from the local signal distributor and the home (or the end user). Several companies are providing methods to connect and provide services of voice, data, music, video, and other forms of communication. They include wired solutions such as public switched telephone network (PSTN), digital subscriber line, cable, and fiber optics, as well as wireless options such as fixed wireless and satellite.

Home networking has become a convergence point for the next-generation digital infrastructure. As technology has advanced, household appliances, televisions, stereos, home security systems, and nearly everything that operates on electrical energy have become digitally controlled and potentially connectable to a network. Home networking is becoming a key enabler for a new breed of information appliances to connect to, and communicate over, a rapidly expanding digital network.

Until recently, limitations of the access network have been the major obstacle to the digital networked house. The well-known "last mile problem" has hindered an effortless broadband access at home and therefore has affected the home networking applications. However, promising recent advances in transmission and broadband access technologies are capable of bringing the information superhighway to homes worldwide. More notably, broadband wireless access technologies are warmly accepted by homeowners because of ease of installation, low cost, and high bit rate.

Along with the advances in wireless access to the home, the explosion of the Internet, increasing demand for intelligent home devices, and availability of plenty of in-home computing resources call for home networking solutions. These solutions should provide connectivity between in-home devices, must efficiently exploit the high-speed access to the Internet, and could establish a new home paradigm. More importantly, the truly revolutionary potential for home networking lies in its ability to extend Internet access everywhere directly into the hands of consumers, in what has been called *pervasive computing*. This could also open up a new mass market and could set the stage for a vast range of new home applications. Therefore, last mile broadband access and home networking are tightly coupled, and a thorough discussion of one is incomplete without covering the other.

To make broadband wireless access and home networking widely acceptable, reliability, performance, installation ease, and cost are the most crucial factors to be addressed. Indeed, performance and reliability are known to be key concerns for residential users. Also, complex wiring and expensive networking devices are considered significant barriers to extensive deployment of home-networked products. In particular, installing new wires in an existing home environment is commonly a cumbersome process and usually the source of several problems. For this reason, increased focus has been recently concentrated on the so-called "no-new-wires" innovations. These home-networking solutions include technologies that reuse existing in-home cables or employ wireless connectivity.

This chapter is devoted to the candidate wireless technologies to be employed for the last mile broadband access and for home networking. The purpose is to provide comprehensive information about recent advances and activities in these fields and to serve as a complete and updated reference. In this context, the concepts, challenges, and applications of the technologies that enable broadband wireless access and wireless home networking are addressed in detail.

6.2 CHAPTER ORGANIZATION

In the near future, fourth-generation (4G) wireless technologies will be able to support Internet-like services. This provision will be achieved through a seamless integration of different types of wireless networks with different transmission speeds and ranges interconnected through a high-speed backbone. 4G wireless networks include wireless personal area networks (wireless PANs or simply WPANs); wireless local area networks (wireless LANs or simply WLANs); wireless local loops (WLLs); cellular wide area networks; and satellite networks (see Figure 6.1). Ultimately, the goal is to efficiently reach users directly or through customer premises equipments (CPEs). The widespread use of wireless networks will increase the deployment of new wireless applications, especially multimedia applications such as video on demand, audio on demand, voice over IP, streaming media, interactive gaming, and others.

In this chapter, we investigate the concepts and technologies needed between consumers and the service provider, also known as last mile technologies. In particular, the focus of this chapter is on third- and fourth-generation (3G and 4G) cellular systems, WLL, WLAN, and WPAN systems. We first discuss some general concepts of broadband wireless communication,

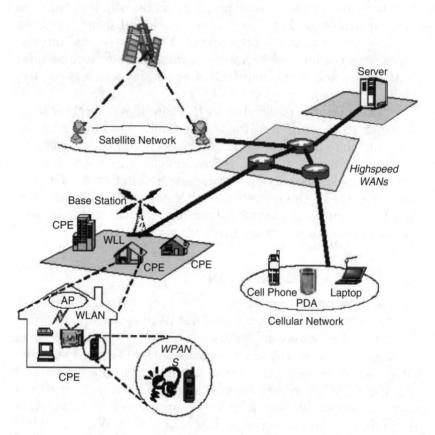

Figure 6.1   The envisioned communication puzzle of 4G.

delve into the enabling technologies of such 3G/4G and WLL
systems, then cover WLANs and, finally, WPAN systems.

6.3 A BROADBAND HOME ACCESS
   ARCHITECTURE

A broadband home access architecture represents an ultimate
technical solution that could bring the vision of pervasive
computing to fruition at home. It serves as an economic con-
duit that connects the next generation of Internet-based vendors with consumers located in the comfort of their own

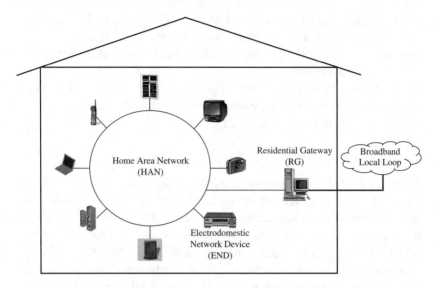

Figure 6.2 Broadband home access architecture.

homes. In general, the broadband home access architecture has four distinct technological components (see Figure 6.2) that must be deployed as an integrated and interoperable system in order to penetrate the mainstream marketplace successfully: broadband local loop; residential gateway (RG); home area network (HAN); and electrodomestic network devices (ENDs). In what follows, the role of each of these components is investigated.

6.3.1 Broadband Local Loop

Originally developed to support telephony traffic, the local loop of a telecommunications network now supports a much larger number of subscriber lines transmitting voice and Internet traffic. Although originally defined in the context of telephony, the term "local loop" (similar to "last mile") is today widely used to describe the connection from the local provider to consumers. Broadly speaking, the five methods of bringing information to the home are: telephone wire; coaxial cable; fiber optics; wireless RF; and satellite communications. The basic requirement for all of these methods is to bring high-speed information to homes as well as offices. Many techniques use a combination of these methods for a complete system to provide complete coverage of stationary and mobile users and applications. The next section covers the most prominent solutions to providing broadband wireless access to home or office environments.

6.3.2 Residential Gateway (RG)

The RG is the interface device that interconnects the broadband local loop to the in-home network. It offers an effective bidirectional communication channel to every networked device in the home. Because it serves as the centralized access point between the home and the outside world, the RG represents an important technological component in the broadband home access architecture. Moreover, this gateway serves as a convergence point bridging the different broadband and LAN technologies, as well as PANs.

6.3.3 Home Area Network (HAN)

The HAN is the high-speed, in-home network that distributes information to the electrodomestic network devices (ENDs). It provides interconnectivity for all ENDs within the home premises. A wide variety of technologies exist for interconnecting devices within the home, but no single technology meets all of the requirements for the diversity of applications that could be envisioned. Although traditional 10Base-T/Cat5 Ethernet offers a robust and proven solution, most consumers do not have the time, interest, or knowledge to rewire their homes. Fortunately, the emergence of "no-new-wires" technologies offers alternatives for solving the mass-market home networking issue, including wireless, phone line, and power line solutions. Section 6.5 discusses wireless home networking solutions said to represent a large fraction of this market.

6.3.4 Electrodomestic Network Devices (ENDs)

ENDs can be described as a set of "intelligent" processing tools used in home environments. They include computers, appliances, and electronics that have embedded intelligence and capability to communicate with other devices. These devices will be able not only to communicate with other in-home devices, but also to connect to the outside world (e.g., Internet) when the in-home network is connected to a broadband local loop. These will enable the development of new applications such as remote administration and Web-based home control and automation.

6.4 LAST MILE BROADBAND WIRELESS ACCESS

Rapid growth in demand for high-speed Internet/Web access and multiline voice for residential and small business customers has created a demand for last mile broadband access. Typical peak data rates for a shared broadband pipe for residential customers and small office/home office (SOHO) are around 5 to 10 Mb/s on the downlink (from the hub to the terminal) and 0.5 to 2 Mb/s on the uplink (from the terminal to the hub). This asymmetry arises from the nature and dominance of Web traffic. Voice- and videoconferencing require symmetric data rates. Although long-term evolution of Internet services and the resulting traffic requirements are hard to predict, demand for data rates and quality of broadband last mile services will certainly increase dramatically in the future. Many wireless systems in several bands compete for dominance of the last mile. Methods considered include point-to-point, point-to-multipoint, and multipoint-to-multipoint for bringing broadband communications information into the home and providing networking capabilities to end users.

Broadband access is currently offered through digital subscriber line (xDSL)[47,48] and cable (both discussed in earlier chapters), and broadband wireless access (BWA), which can also be referred to as fixed broadband wireless access (FBWA) networks. Each of these techniques has its unique cost, performance, and deployment trade-offs. Although cable and DSL are already deployed on a large-scale basis, BWA is emerging as an access technology with several advantages. These include avoiding distance limitations of DSL and high costs of cable; rapid deployment; high scalability; lower maintenance and upgrade costs; and incremental investment to match market growth. Nevertheless, a number of important issues, including spectrum efficiency, network scalability, self-installable CPE antennas, and reliable non-line-of-sight (NLOS) operation, need to be resolved before BWA can penetrate the market successfully. In this section BWA is characterized and its major features outlined, including important issues at physical and medium access control (MAC) layers; the next section discusses the major technologies available for BWA.

6.4.1   Basic Principles

We start our discussions on BWA with a discussion of the basic principles of wireless communications. In wireless technology, data are transmitted over the air and are ideal platforms for extending the concept of home networking into the area of mobile devices around the home. Consequently, wireless technology is portrayed as a new system that complements phone-line and power-line networking solutions. It is not clear whether wireless technology will be used as a home network backbone solution (as suggested by some proponents of the IEEE 802.11 standard); however, it will definitely be used to interconnect the class of devices that could constitute a subnetwork with mobile communications. These mobility subnetworks will interface with other subnetworks and with the Internet by connecting to the home network backbone whether it is wired or wireless.

Wireless networks transmit and receive data over the air, minimizing the need for expensive wiring systems. With a wireless-based home network, users can access and share expensive entertainment devices without installing new cables through walls and ceilings. At the core of wireless communication are the transmitter and the receiver. The user may interact with the transmitter — for example, if someone inputs a URL into his PC, this input is converted by the transmitter to electromagnetic waves and sent to the receiver. For two-way communication, each user requires a transmitter and a receiver. Therefore, many manufacturers build the transmitter and receiver into a single unit called a transceiver.

The two main propagation modes used in wireless networks are infrared and radio frequency and are described next.

6.4.1.1 Infrared (IR)

Most people today are familiar with everyday devices that use IR technology, such as remote controls for TVs, VCRs, and DVD and CD players. IR transmission is categorized as a line-of-sight (LOS) wireless technology. This means that the workstations and digital appliances must be in a direct line to the transmitter in order to establish a communication link successfully. An infrared-based network suits environments in which all the digital appliances that require network connectivity are in one room. However, new diffused IR technologies can work without LOS inside a room, so users should expect to see these products in the near future. IR networks can be implemented reasonably quickly; however, people walking between transmission/reception or moisture in the air can weaken the signals. IR in-home technology is promoted by an international association of companies called IrDA (Infrared Data Association).[1] Further details on the IrDA system are given later in this chapter.

6.4.1.2 Radio Frequency (RF)

Another main category of wireless technology comprises devices that use radio frequency. RF is a more flexible technology, allowing consumers to link appliances that are distributed throughout the house. RF can be categorized as narrow band or spread spectrum. Narrow band technology includes microwave transmissions, which are high-frequency radio waves that can be transmitted to distances up to 50 km. Microwave technology is not suitable for home networks, but could be used to connect networks in separate buildings. Spread spectrum technology is one of the most widely used technologies in wireless networks and was developed during World War II to provide greater security for military applications. Because it entails spreading the signal over a number of frequencies, spread spectrum technology makes the signal harder to intercept.

A couple of techniques are used to deploy spread spectrum technologies. For instance, a system called frequency-hopping spread spectrum (FHSS) is the most popular technology for operating wireless home networks. FHSS systems constantly hop over entire bands of frequencies in a particular sequence. To a remote receiver not synchronized with the hopping sequence, these signals appear as a random noise. A receiver can only process the signal by tuning to the appropriate transmission frequency. The FHSS receiver hops from one frequency to another in tandem with the transmitter. At any given time, a number of transceivers may be hopping along the same band of frequencies. Each transceiver uses a different hopping sequence carefully chosen to minimize interference. Later in this chapter, Bluetooth technology, which employs the FHSS technique, will be covered.

Because wireless technology has roots in military applications, security has been a design criterion for wireless devices. Security provisions are normally integrated with wireless network devices, sometimes making them more secure than most wireline-based networks.

6.4.2 Services, Deployment Scenarios, and Architectures of BWA

Typical BWA services include Internet access, multiline voice, audio, and streaming video. As a consequence, quality of service (QoS) guarantees for some data and voice applications are needed. In addition, carrier requirements include meeting the Federal Communications Commission (FCC) regulations on power emission and radio interoperability. Also, scalability using a cellular architecture wherein throughput per square mile can be increased by cell splitting, low-cost CPE, and infrastructure equipment, high coverage and capacity per cell could reduce infrastructure costs, self-installability of CPE antennas, and, finally, ease/facilitate portability. Three different deployment scenarios can be considered for BWA: supercells, macrocells, and microcells.

6.4.2.1 Supercells

In this scenario, a large service area with a radius of up to 30 mi is covered. The base transceiver station (BTS) antenna height is typically in excess of 1000 ft, and a high-gain rooftop directional CPE antenna is needed with an LOS connection between transmitter and receiver. This is a single-cell config-uration that is not scalable. The same cell frequency reuse in angle and polarization may be feasible with sectorization. Due to LOS propagation, carrier-to-noise (C/N) ratio values of around 30 dB can be sustained, which makes the use of high-order modulation possible. Due to a strict need for LOS links, full coverage cannot generally be guaranteed.

6.4.2.2 Macrocells

Macrocells typically use cellular architecture with spatial fre-quency reuse between cells. The BTS antenna height is sig-nificantly lower than in the supercell case, typically 50 to 100 ft. Low BTS heights induce severe path loss and possibly NLOS propagation. A cell radius of around 5 mi may be possible. Due to NLOS propagation and cochannel interfer-ence (CCI) from other cells, significantly lower C/N and car-rier-to-interference (C/I) ratio values could be supported compared to supercells; lower-order modulation must be used. Directional CPE antennas can still be employed. The archi-tecture is scalable in capacity and coverage, and high coverage is possible because NLOS propagation is supported.

6.4.2.3 Microcells

Microcells are similar to macrocells, except that much smaller cells (typically with a cell radius of 1 mi) are used. BTS towers are lower than in the macrocell case, typically below rooftop, and may be 20 to 40 ft high. To support portability, the CPE architecture has omnidirectional indoor antennas. Small cell size offers sufficient link margin to provide indoor coverage.

6.4.2.4 Challenges in Fixed Wireless Networks

Next, major challenges in fixed wireless networks will be outlined and the main differences between fixed broadband

and current mobile wireless access networks discussed. Essentially, the quality and data rate requirements in the fixed case are *significantly* higher than in the mobile case. Due to the requirements for higher data rates, the link budget shrinks by roughly 15 dB, assuming that fixed transmit power remains the same as in the mobile cellular case. The requirements for higher quality increase needed fade[50] margins for C/N and C/I by roughly 15 dB each. Taking into account that the use of directional antennas provides a gain of roughly 15 dB in link budget against noise, this translates into a 15-dB disadvantage in link budget against noise and a 15-dB disadvantage against CCI. The former means much smaller coverage or cell radius (one fifth of the mobile cell radius); the latter requires much higher reuse factors (20 to 30 instead of 3 in mobile) and thus one sixth cell capacity.

Therefore, new sophisticated physical and radio link layers are needed in order to maintain coverage and retain a reuse factor of three. The use of multiple antennas, discussed in more detail later, provides significant leverage in terms of link budget against noise and CCI and seems to be very promising in meeting these requirements. Coverage influences network economics because good coverage reduces the infrastructure costs during initial setup. Extra capacity improves network cost effectiveness by delaying the need for cell splitting.

6.4.3 BWA Channels

Wireless transmission is limited by available radio spectrum while impaired by path loss, interference, and multipath propagation, which cause fading and delay spread. Because of these limitations, in general much greater challenges are present in wireless than wired systems. BWA channels are discussed in this subsection.

6.4.3.1 Path Loss and Delay Spread

The path loss in BWA channels depends on the type of terrain. It is observed that the COST 231-Hata model gives reasonable estimates of the path loss for a flat terrain and high base station antenna heights for a wide frequency range. In moderate or hilly terrain and for lower base station antenna heights, the Hata model may not be adequate and other models may need to be used. Measurements in the 1.9-GHz band representing the macro/microcell cases have been reported by Erceg et al.[2] Because these measurements show differences in path loss on the order of tens of decibels for different terrain categories, it very important to distinguish these categories.

The amount of delay spread[50] in fixed wireless channels depends strongly on the antenna characteristics. Median root mean square (RMS) delay spreads for directional antennas in suburban environments of approximately 75 ns have been reported; with omnidirectional antennas in the same terrain, a delay spread of 175 ns has been found.[3] The reason for the difference in delay spread for directional and omnidirectional antennas is that echoes at longer delays tend to arrive at angles farther away from the direct path and are thus more attenuated by the side lobes in the case of directional antennas. Measurements conducted mostly in the 900-MHz band[4] with omnidirectional antennas in suburban, urban, and mountainous environments show delay spreads of up to 16 μs. The fading rates[50] encountered in fixed wireless environments are between 0.1 and 2 Hz.

6.4.3.2   K-Factor

The path gain of a fixed BWA channel can be represented as having a fixed component plus a fluctuating (scatter) component. The ratio of the average energy in the fixed component to the average energy in the scatter component is called the K-factor. The value of the K-factor has significant implications on the system design and its performance. Generally, it is found that the K-factor in fixed wireless applications can be very low because of low BTS and CPE antenna heights (under the eave CPE antennas).

Figure 6.3 shows K-factor measurements conducted by the Smart Antennas Research Group at Stanford University. In these measurements performed in the 2.4 GHz band, the transmit antenna was 10 or 20 m high, and the CPE antenna with a 50° 3-dB beamwidth in azimuth was 3 m high. It is

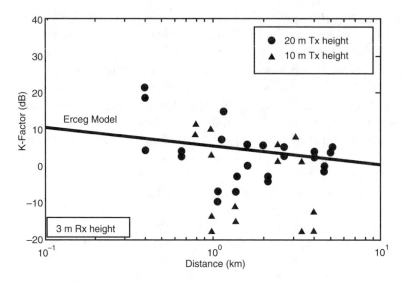

Figure 6.3 K-factor as a function of distance between transmitter and receiver.

found that the K-factor decreases significantly with increasing distance between transmitter and receiver. Note that the K-factor shown in Figure 6.3 has been averaged over time and frequency. In practice, significant fluctuations in K-factor can occur due to wind and car traffic. Figure 6.3 also shows the Greenstein–Erceg model[2] for the median K-factor vs. distance,[5] assuming 20-m transmit and 3-m receive antenna heights. The experimental data and the model are shown to be in an excellent agreement.

To summarize, in a fixed BWA system design, very low K-factors (almost purely Rayleigh fading conditions) must be assumed in order to provide large cell coverage and reliable operation at the edge of the cell.

6.4.4 Physical Layer, MAC Layer, and Radio Link Protocols

Issues regarding the physical layer, MAC layer, and the radio link protocol of a BWA system will be discussed in this subsection.

6.4.4.1 Physical Layer

6.4.4.1.1 Modulation

Consider three alternative modulation formats: single-carrier (SC) modulation with equalization; direct sequence code-division multiple access (DS-CDMA)[50] with a rake receiver; and orthogonal frequency-division multiplexing (OFDM) with interleaving and coding. The new technique of ultrawideband modulation (UWBM) will also be briefly discussed. Assume that the transmitter does not know the channel.

6.4.4.1.2 Single-Carrier Modulation with Equalization

Several equalization options with different performance and implementation trade-offs exist for SC modulation. Maximum-likelihood equalization yields optimum performance, but is computationally very complex. Decision-feedback equalization is generally considered an attractive practical option. Simpler alternatives include linear equalizers such as zero forcing or MMSE. However, linear equalization does not properly exploit the frequency diversity in the channel created by delay spread. In practice, for high delay spread and/or high data rate cases, the computational complexity of SC equalizers and the complexity required for equalizer adaptation can impose limits on the performance of SC systems.

6.4.4.1.3 DS-CDMA

DS-CDMA uses a spreading code sequence multiplied by the digital baseband transmitted symbol. This spreading code sequence has a much higher rate (chip rate) than the information-bearing symbol, which spreads the symbol in frequency. A RAKE receiver can be employed to exploit frequency diversity. (This technique uses several baseband correlators to process multipath signal components individually. The outputs from the different correlators are combined to achieve improved reliability and performance.) With increasing data rate, symbol and chip rates increase, thus allowing the system to resolve finer differences in physical path delays, but leading to an increased number of discrete-time baseband channel impulse response taps; this increases the necessary number of elements in the RAKE combiner, which, in turn, results in increased computational complexity.

6.4.4.1.4   Orthogonal Frequency-Division Multiplexing

OFDM eliminates the need for equalization by inserting a guard interval (cyclic prefix), which is a copy of the first part of the OFDM symbol and must be long enough to accommodate the largest possible delay spread. The transmitter and the receiver employ an inverse fast Fourier transform (inverse FFT, or simply IFFT) and FFT, respectively, and equalization reduces to simple scalar multiplications on a tone-by-tone basis (see DSL chapter). In OFDM, frequency diversity is obtained by coding and interleaving across tones. For increasing delay spread and/or data rate, the cyclic prefix must increase proportionally so that it remains longer than the channel impulse response. Thus, in order to maintain a constant overhead due to the CP, the number of tones, N, must increase proportionally. This results in increased computational complexity due to the increased FFT size. In summary, the ease of equalization seems to favor OFDM over SC and DS-CDMA from the point of view of complexity.

6.4.4.1.5   Ultra-Wideband Modulation (UWBM)

Recently, UWBM has attracted a lot of interest for wireless broadband communications. In UWBM, a train of modulated subnanosecond pulses is used to convey information. Here, a RAKE receiver realizes the path diversity. The result of the pulses transmitted across an ultrawideband spectrum means that UWBM may be able to coexist with other narrowband systems because the interference energy per system may be small and may only increase the noise floor. Recently, UWBM has received some encouragement from the FCC. Several industry efforts are now underway to commercialize the technology.

6.4.4.1.6   Hardware Considerations

From the point of view of complexity, OFDM seems to be more attractive than SC and DS-CDMA. In practice, however, OFDM signals make the system sensitive to power amplifier nonlinearities. Therefore, in the OFDM case, the power amplifier cost is higher. The decision whether SC, DS-CDMA, or OFDM should be used is therefore driven by two factors: the cost of silicon required for transmit and receive signal-processing operations and the cost of power amplifier.

6.4.4.1.7   Channel Coding

Channel coding adds redundancy to the transmitted data to allow the receiver to correct transmission errors. As mentioned earlier, in the OFDM case, channel coding combined with interleaving also provides frequency diversity. Typical BWA channel coding schemes employ concatenated Reed–Solomon (RS)/convolutional coding schemes in which RS codes are used as outer codes and convolutional codes as inner codes. Soft decoding and iterative decoding techniques can provide additional gain.

6.4.4.1.8   Synchronization

The timing and frequency offset sensitivities of SC and DS-CDMA systems are theoretically the same as long as they use the same bandwidth and data throughput. OFDM is more sensitive to synchronization errors than SC and DS-CDMA.

6.4.4.1.9   Link Adaptation

In BWA systems, channel conditions may vary significantly due to fading. It is therefore desirable to adapt the modulation and coding schemes according to the channel conditions. Although voice networks are designed to deliver a fixed bit rate, data services can be delivered at a variable rate. Voice networks are engineered to deliver a certain required worst-case bit rate at the edge of the cell. Most users, however, have better channel conditions. Therefore, data networks can take advantage of adaptive modulation and coding to improve the overall throughput. In a typical adaptive modulation scheme, a dynamic variation in the modulation order (constellation size) and forward error correction (FEC) code rate is possible. In practice, the receiver feeds back information on the channel, which is then used to control the adaptation. Adaptive modulation can be used in up- and downlinks. The adaptation can be performed in various ways and can be specific to the user only, user- and time-specific to user and time, or depend on QoS.

6.4.4.1.10 Multiple Access

Time division multiple access (TDMA)[50] is performed by assigning different disjoint time slots to different users or, equivalently, the transmission time is partitioned into sequentially accessed time slots. Users then take turns transmitting and receiving in a round-robin sequence. For data networks, where channel use may be very bursty, TDMA is modified to reservation-based schemes in which time slots are allocated only if data are to be transmitted.

In CDMA, all the users transmit at the same time with different users employing different quasiorthogonal signature sequences. Little theoretical difference exists in terms of capacity between TDMA and CDMA; however, CDMA offers unique advantages in terms of realizing signal and interference diversity. In BWA, however, fixed spreading CDMA is not attractive because, due to a high spreading factor (typically larger than 32), the operating bandwidth becomes very high. For example, for a data rate of 10 Mb/s, an operating bandwidth of 160 MHz is required for a spreading factor of 32. In third-generation (3G) mobile systems for high-data-rate links, the spreading factor drops to four in order to keep the bandwidth at 4 MHz. Such a low spreading factor makes CDMA look almost like TDMA. Other practical approaches include multicode CDMA modulation.

6.4.4.1.11 TDD vs. FDD

The BWA industry is currently debating the merits of time-division duplexing (TDD) vs. frequency-division duplexing

(FDD) in point-to-multipoint networks. FDD is the legacy used in the fixed wireless industry in point-to-point links originally designed for transporting analog voice traffic, which is largely symmetric and predictable. TDD, on the other hand, is used in the design of point-to-multipoint networks to transport digital data, which are asymmetric and unpredictable. Although TDD requires a single channel for full duplex communications, FDD systems require a paired channel for communication — one for the downlink and one for the uplink.

In TDD, transmit/receive separation occurs in the time domain, as opposed to FDD, where it happens in the frequency domain. Although FDD can handle traffic that has relatively constant bandwidth requirements in both communications directions, TDD effectively handles varying uplink/downlink traffic asymmetry by adjusting time spent on up- and downlinks. Given that Internet traffic is bursty (i.e., time varying), the uplink and downlink bandwidths need to vary with time, which favors TDD. TDD requires a guard time equal to the round-trip propagation delay between the hub and the remote units. This guard time increases with link distance. Timing advance can be employed to reduce the required guard time. In FDD, sufficient isolation in frequency between the up- and downlink channels is required. In brief, FDD seems to be simpler to implement, although it offers less efficient solutions.

6.4.4.2 MAC Layer and Radio Link Protocol

The MAC layer and the radio link protocol work with the physical (PHY) layer to deliver the best possible QoS in terms of throughput, delay, and delay jitter to the users. The major task of the MAC layer is to associate the transport and QoS requirements with different applications and services, and to prioritize and schedule transmission appropriately over up- and downlink. A wireless MAC protocol should therefore provide differentiated grades and quality of service, dynamic bandwidth allocation, and scheduling for bursty data. An important feature of the MAC layer is the ability to perform retransmission, which allows operation at higher error rates and thus better frequency reuse, increases robustness, and improves TCP performance. The major MAC functions are:

- Control up- and downlink *transmission schedules*, thus allowing support of multiple service flows (i.e., distinct QoS) on each CPE–BTS link.
- Provide *admission control* to ensure that adequate channel capacity is available to accommodate QoS requirements of the new flow, and to enforce policy constraints like verifying that a CPE is authorized to receive the QoS requested for a service flow.
- Offer *link initialization* and maintenance like channel choice, synchronization, registration, and various security issues.
- *Support integrated voice/data transport*. Typical data requirements are bandwidth-on-demand, very low packet error rates, and type-of-service differentiation. Voice requirements are bandwidth guarantees, and bounded loss, delay, and jitter.
- Support fragmentation, automatic repeat request (ARQ), and adaptive modulation and coding.

Next, some MAC features that are specifically desirable in the wireless scenario will be summarized:

- *Fragmentation* of packet data units (PDUs) into smaller packets, which helps to reduce the packet error rate and limit the latency for voice communication
- *Retransmission* on the level of fragmented PDUs
- *Scheduling support* for multiple modulation/coding schemes
- Wireless-specific *link maintenance and control*, such as uplink power control and adaptive modulation and coding

6.4.5 Multiple Antennas in BWA

As outlined previously, fixed BWA systems face two key challenges: providing high-data-rate and high-quality wireless access over fading channels, with quality as close to wireline quality as possible. The requirement for high quality arises because wireless BWA systems compete with cable modems and DSL, which operate over fixed channels and thus provide very good quality.

This high-quality requirement constitutes a major difference from existing mobile cellular networks in which customers are accustomed to low QoS. Also, in existing mobile cellular networks, the requirements for data rate are much lower than in the fixed BWA case. The use of multiple antennas at transmit and receive sides of a wireless link in combination with signal processing and coding is a promising means to satisfy all these requirements. Note that, in fixed BWA as opposed to mobile cellular communications, the use of multiple antennas at the CPE is possible.

The benefits provided by the use of multiple antennas at the BTS and CPE are as follows:

- *Array gain*: multiple antennas can coherently combine signals to increase the C/N value and thus improve coverage. Coherent combining can be employed at the transmitter and receiver, but it requires channel knowledge. Because channel knowledge is difficult to obtain in the transmitter, array gain is more likely to be available in the receiver.

- *Diversity gain*: spatial diversity through multiple antennas can be used to combat fading and significantly improve link reliability. Diversity gain can be obtained at the transmitter and receiver. Recently developed space–time codes[6] realize transmit diversity gain without knowing the channel in the transmitter.

- *Interference suppression*: multiple antennas can be used to suppress CCI and thus increase the cellular capacity.

- *Multiplexing gain*: the use of multiple antennas at the transmitter and receiver allows opening up parallel spatial data pipes within the same bandwidth, which leads to a linear (in the number of antennas) increase in data rate.[7–9,52]

In summary, the use of multiple antennas at the BTS and CPE can improve cellular capacity and link reliability. More details on the impact of multiple antennas on cellular networks can be found in Sheikh et al.[10]

6.5 LAST MILE BROADBAND WIRELESS ACCESS TECHNOLOGIES

The previous section presented general concepts and issues involved in broadband data delivery to offices and homes. Now attention is turned to the specific technologies proposed to achieve this. In recent years, there has been increasing interest shown in wireless technologies for subscriber access, as an alternative to traditional wired (e.g., twisted-pair, cable, fiber optics, etc.) local loop. These approaches are generally referred to as WLL, or fixed wireless access, or even last mile broadband wireless access. These technologies are used by telecommunication companies to carry IP data from central locations on their networks to small low-cost antennas mounted on subscribers' roofs. Wireless cable Internet access is enabled through the use of a number of distribution technologies. In the following subsections, these broadband wireless technologies and their characteristics will be investigated.

6.5.1 Multichannel Multipoint Distribution System (MMDS)

Analog-based MMDS[11,12] began in the mid-1970s with the allocation of two television channels for sending business data. This service became popular, and applications were made to allocate part of the instructional television fixed service band to wireless cable TV. Once the regulations had been amended, it became possible for a wireless cable system to offer up to 31 6-MHz channels in the 2.5- to 2.7-GHz frequency band. During this timeframe, nonprofit organizations used the system to broadcast educational and religious programs. In 1983, the FCC allocated frequencies in both of these spectra, providing 200-MHz bandwidth for licensed network

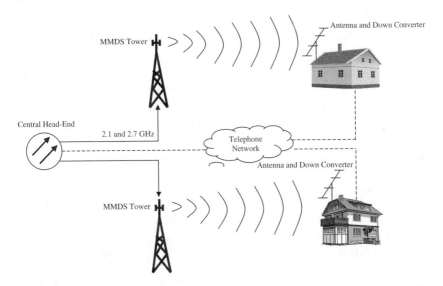

Figure 6.4 MMDS broadband connectivity.

providers and with an output power up to 30 W. The basic components of a digital MMDS system providing broadband connectivity to a home network are shown in Figure 6.4.

An MMDS broadband system consists of a head-end that receives data from a variety of sources, including Internet service providers and TV broadcast stations. At the head-end, data are processed, converted to the 2.1- and 2.7-GHz frequency range, and sent to microwave towers. Quadrature amplitude modulation (QAM) is the most commonly used format employed in sending data over an MMDS network, although some operators use a modulation format called coded orthogonal frequency division multiplexing (COFDM). This format operates extremely well in conditions likely to be found in heavily built-up areas where digital transmissions become distorted by line-of-sight obstacles such as buildings, bridges, and hills. The signals are then rebroadcast from low-powered base stations in a diameter of 35 mi from the subscriber's home. This provides up to 10 Mbps during peak use and can provide speeds up to 37.5 Mbps to a single user. Signals are received with home rooftop antennas.

The receiving antenna has a clear line of site to the transmitting antenna. A down converter, usually part of the antenna, converts the microwave signals into standard cable channel frequencies. From the antenna, the signal travels to a gateway device where it is routed and passed onto the various devices connected to the in-home network. Today, MMDS systems are used throughout the U.S. and in many other countries.

6.5.2   Local Multipoint Distribution Service (LMDS)

LMDS[13,14] is a last mile point-to-multipoint distribution service that propagates communications signals with a relatively short RF range to multiple end users. In this multipoint system, the base station or hub transmits signals in a point-to-multipoint method that resembles a broadcast mode. The return path from the subscriber to the base station or hub is accomplished by a point-to-point link. Overall, the architecture of LMDS is similar to that of the MMDS system.

LMDS combines high-capacity radio-based communications and broadcast systems with interactivity operated at millimeter frequencies. Other systems, however, have been primarily used for analog TV distribution (e.g., MMDS). This started with Cellular Vision and Bernard Bossard proposing a system for TV distribution in central New York City.[19] Digital television opened up a combined mechanism for transport of data representing TV programs, data, and communication. The possibility of implementing a full-service broadband access network by rebuilding a broadcast network as an interactive network by functionally adding a communications channel for the return has been a perfect match with the growth of the Internet and data services. Broadband interactivity has been possible with digitalization.

The transmitter site should be on top of a tall building or on a tall pole, overlooking the service area. The transmitter covers a sector typically 60 to 90° wide. Full coverage of an area thus requires four to six transmitters. The streams transmitted contain 34 to 38 Mb/s of data addressed to everybody (typical TV), subgroups, or individuals (typical communication, Internet). In the coverage zone, the capacities of the point-to-point return channels are determined by the requirements of individual users.

Operation of LMDS in an area normally requires a cluster of cells with separate base stations for co-located transmitter/receiver sites. One of the base station sites will serve as coordination center for the franchise area and connect the LMDS cells to external networks. Intercell networking may be implemented using fiber or short-hop radio relay connections. Co-location with mobile base stations allows for infrastructure sharing.

Operation in the millimeter range imposes some restrictions. Precipitation effects lead to severe attenuation; depending on the climatic zone and the frequency of operation, the reliable range of operation could be limited to 3 to 5 km. Line of sight is also required. Full coverage will not be possible, however, and numbers quoted are normally in the 40 to 70% range, and something in excess of 95% is a minimum for a service offered to the public.

Improved coverage is thus required and may be achieved in different ways. The numbers quoted refer to a single cell. By introducing some overlapping between cells, it may be possible to obtain coverage in shielded areas in one cell from the neighboring cell transmission site. Use of repeaters and reflectors is another possibility, but requires some additional equipment; this could be compensated for by increasing the number of users. Thus, site-dependent modes of operation could solve the coverage problem.

The most severe restriction may be the attenuation caused by transmission through vegetation. Buildings completely shielded by vegetation need an elevated rooftop antenna or some broadband connection to an unshielded site. Propagation issues are by now well understood and are not considered a serious obstacle for reliable operation of millimeter systems. The problems are known and proper precautions can be taken.[20]

6.5.2.1 Operating Frequencies

Even though the capacity in the millimeter part of the spectrum is considerable, many systems compete for frequency allocations, and it has been difficult to obtain a worldwide allocation for LMDS. In the U.S., a 1.3-GHz band in the 28- to 29-GHz band has been allocated, while European countries are allocating frequencies in different bands. The major high-capacity band is presently 40.5 to 42.5 GHz with a possible extension to 43.5 GHz.

Licensing and deployment in Europe indicate that systems will be in different frequency bands from 24 up to 43.5 GHz. The frequency band of 24.5 to 26.6 GHz with subbands of 56 MHz has been opened for point-to-multipoint applications in many European countries. These bands may then be used for LMDS or other related fixed wireless access systems, which can then represent a typical multipoint business system with some capacity for private users. Only systems addressing the business domain are typically based on asynchronous transfer mode (ATM) technology.

The 40-GHz band is normally to be shared among two or three licensees, limiting the available spectrum per operator to 500 to 2000 MHz with two polarizations. The licensing policy may vary from country to country, with stimulation to competition as the main guideline. The LMDS has a potential of becoming the high-capacity access domain for private users.

6.5.2.2   Technologies Employed

Proven technologies required for service startup exist, and different companies have products available addressing the needs of small business customers and, to some extent, demanding private users. In LMDS, a high-capacity broadcast-based downlink is shared among several users in a flexible way. The front-end technology is still expensive at millimeter frequencies, but existing high electron mobility transistor modules offer the required performance. The output power level needed per 36-Mb/s transport beam is about 25 dBm. A technology allowing for final stage amplification of several transport beams reduces the equipment complexity and the cost. The hub transmitters, however, are shared by many users and cost is not that critical.

The front-end technology at 40 GHz is more expensive than at 28 to 29 GHz, and attenuation by precipitation increases with frequency, favoring the lower frequency ranges. Higher capacity offered at 40 GHz may compensate for these effects in the long run. The number of transport streams is determined by demand and limitations set by available spectrum. This gives a scalable architecture, starting with relatively low capacity and adding transmitter modules as demand increases.

The transmission format for digital video broadcasting (DVB) satellite transmission based on quadrature phase shift keying (QPSK) modulation has been adopted by the Digital Audio/Visual Council (DAVIC) and the DVB project, and with the same input frequency interface between the outdoor and indoor units in the range of 950 to 2150 MHz. This allows for application of set-top boxes developed for reception of digital TV by satellite with data included in the transport multiplex. The input frequency is then fed into a set-top box interfacing a TV or a PC or both in parallel, depending on the user's orientation. Both options allow for interactivity because set-top boxes are also equipped with a return channel connection to the PSTN/ISDN.

However, some significant differences exist. In DVB, IP, or ATM data are included in the MPEG transport stream in combination with TV programs. DAVIC has separate high-capacity ATM-based data transmissions. Until now, the PC-oriented user has dominated the interactive world; manipulation of content, inclusion of more advanced text TV, possibilities for e-commerce, different games, and active participation in competition have led to an increased interest in interactive television with a low-capacity return channel.

The uplink is the individual connection, and different technologies may be used depending on the demand. Two of the broadband driving applications — namely, interactive TV and the Internet — will require only low-capacity return links, and technologies like general packet radio service (GPRS)[46] and PSTN/ISDN will be adequate. For more demanding customers, an in-band radio return link with on-demand capacity is required.

Radio-based solutions for small- and medium-sized enterprises do have a radio-link type return link, allowing for symmetric connections or connections that may be asymmetric in either direction. However, it is felt that existing radio return solutions, with their requirement for isolation between transmit and receive implemented through the use of appropriate filtering in allocated bands, impose limitations on operation flexibility and efficient management of resources. A combined use of systems in broadcast and data for private users and business organizations will necessarily result in strong variations in capacity for the two directions. Possible future TDD operation could solve this problem.

The main technological challenge is the large-scale production of a real low-cost, two-way user terminal for the private market as the mass market depends on it. The total capacity of a system is mainly determined by the available frequency resource. In a cellular system employing QPSK modulation, the capacity of a 2-GHz system is around 1.5 Gb/s per cell for down- and uplink channels.

6.5.2.3   Applications

LMDS is the first system with high flexibility, allowing for increased capacity on demand. Reducing the cell size through reduction of cell diameter or illumination angle increases the total capacity. Its flexibility with respect to on-demand high capacity in both directions makes it well suited to home offices and teleteaching in a local domain. The first major applications are oriented to TV, Internet, and business, thus combining professional and entertainment use. In Europe, LMDS has been considered a supplement/alternative to cable TV and has actually been referred to as a *wireless cable*. With digital television, the possibility for converging TV, data, and communications has sparked development of new broadband applications. Hopefully, availability of broadband capacity will stimulate the growth of applications such as telemedicine and teleteaching, which have been recognized for quite some time, although neither has really taken off.

From television to interactive television. The TV business has had strong growth, but the time spent by individuals watching TV has not changed very much. Digital TV introduces new possibilities. The first step is the introduction and development of interactive TV and adding new and interesting functionality. More local TV programs will take advantage of LMDS. Interactive TV will stimulate growth in e-commerce; local activities such as property trading, apartment renting, car buying and selling, and many other transactions may take advantage of the possibilities offered by broadband networking. Telebanking and vacation planning are applications in which interactive TV offers added functionality.

Teleteaching. Education and life-long learning are one of the major challenges in many countries today. Lack of educated and skilled teachers, particularly in current technologies, is a common concern. The local focus of LMDS makes it excellent for high-capacity connections to schools at different levels, connecting a group of local schools as well as providing connections to remote sites. Locally, it would also be possible to connect to homes and have lessons stored for the use of students and parents. Broadband access will offer many possibilities in an educational area in which exploration has barely begun. In this connection, the advantage of LMDS is the flexibility in capacity allocation and the multicast property of the downlink, allowing very efficient delivery for such types of applications.

6.5.3 Satellite Communications

Satellite communications[15,16] allow the most remote places to receive the Internet, telephones, faxes, videos, and telecommunications via satellite signals and is described in detail in a later chapter. For completeness of the text satellite communications are briefly introduced here.

The infrastructure, bandwidth, and the possibility of combining satellite communications with other types of systems make this method an ideal candidate for providing ubiquitous communications to everyone worldwide, thus representing a major player in the BWA arena. Communication satellites are, in effect, orbiting microwave relay stations, located about 35,000 km above the surface of the Earth, used to link two or more Earth-based microwave stations (also called *Earth stations* or *ground stations*). Communications satellite providers typically lease some or all of a satellite's channels to large corporations, which use them for long-distance telephone traffic, private data networks, and distribution of television signals. Leasing these huge communications "pipes" can be very expensive; therefore, they are not suitable for the mass residential marketplace. Consequently, a new suite of services, called the direct broadcast satellite (DBS) system, has been developed to provide consumers with a range of high-speed Internet access services.

A DBS system consists of a minidish that connects in-home networks to satellites with the ability to deliver multimedia data to a home network at speeds in excess of 45 Mbps. However, this speed can only be achieved when downloading content. To upload or send information to the Internet, the system uses a slow telephone line connection. Satellite systems normally use the QPSK modulation scheme to transmit data from the dish in the sky to the minidish located on the top of the roof.

6.5.4   3G and 4G Cellular Systems

The two most important phenomena affecting telecommunications over the past decade have been the explosive parallel growth of the Internet and of mobile telephone services. The Internet brought the benefits of data communications to the masses with e-mail, the Web, and e-commerce; mobile service has enabled "follow-me anywhere/always on" telephony. The Internet helped accelerate the trend from voice-centric to data-centric networking. Data already exceed voice traffic and the data share continues to grow. Now, these two worlds are converging. This convergence offers the benefits of new interactive multimedia services coupled to the flexibility and mobility of wireless. To realize its full potential, however, broadband access connections are needed.

Unlike the older mobile systems used for telecommunication, today's cellular phones have more mobility and are more compact and easier to handle. The technologies governing the mobile devices are improving at lightning speed. Just a few years ago, an expensive, bulky mobile phone was nothing but an analog device mostly used by business people for voice communication. Today, the wireless technology is digital. The so-called second generation (2G) of cellular technology such as GSM (global system for mobile communication), which allows transmitting speech in digital format over a radio path, is in existence. The 2G networks are mainly used for voice transmission and are essentially circuit switched.

GPRS and EDGE (enhanced data rates for GSM evolution) are 2.5G networks, which are an extension of 2G networks in that they use circuit switching for voice and packet switching for data transmission. With GPRS and EDGE, broadband communication can be offered to mobile users with nominal speeds for stationary users up to 171.2 kbps in GPRS and 473.6 kbps in EDGE. Circuit-switched technology requires that the user be billed by airtime rather than the amount of data transmitted because that bandwidth is reserved for the user. Packet-switched technology utilizes bandwidth much more efficiently, allowing each user's packets to compete for available bandwidth and billing users for the amount of data transmitted. Thus, a shift towards using packet-switched, and therefore IP, networks is natural.

To eliminate many problems faced by 2G and 2.5G networks, such as low speeds for many mass-market applications (e.g., multimedia messaging) and incompatible technologies (e.g., TDMA in 2G and CDMA in 2.5G) in different countries, 3G networks (UMTS, IMT-2000) were proposed. Expectations for 3G included increased bandwidth: 128 kbps in a car, and 2 Mbps in fixed applications. In theory, 3G would work over North American as well as European and Asian wireless air interfaces. In reality, the outlook for 3G is neither clear nor certain. Part of the problem is that network providers in Europe and North America currently maintain separate standards' bodies (3GPP for Europe and Asia[53]; 3GPP2 for North America[54]) that mirror differences in air interface technologies.

In addition, financial questions cast a doubt over 3G's desirability. Some countries are concerned that 3G will never be deployed. This concern is grounded, in part, in the growing attraction of 4G wireless technologies.

A 4G, or fourth-generation, network is the name given to an IP-based mobile system that provides access through a collection of radio interfaces. A 4G network promises seamless roaming/handover and best connected service, combining multiple radio access interfaces (such as WLAN, Bluetooth, GPRS) into a single network that subscribers may use. With this feature, users will have access to different services, increased coverage, the convenience of a single device, one bill with reduced total access cost, and more reliable wireless access even with the failure or loss of one or more networks. At the moment, 4G is simply an initiative by research and development labs to move beyond the limitations, and deal with the problems of 3G (which is facing some problems in meeting its promised performance and throughput).

At the most general level, the 4G architecture will include three basic areas of connectivity: PAN (such as Bluetooth); local high-speed access points on the network including WLAN technologies (e.g., IEEE 802.11 and HIPERLAN); and cellular connectivity. Under this umbrella, 4G calls for a wide range of mobile devices that support global roaming. Each device will be able to interact with Internet-based information that will be modified on the fly for the network used by the device at that moment. In short, the roots of 4G networks lie in the idea of *pervasive computing*. In summary, the defining features of 4G networks are:

- High speed — 4G systems should offer a peak speed of more than 100 Mb/s in stationary mode with an average of 20 Mb/s when traveling.
- High network capacity – 4G system capacity should be at least ten times that of 3G systems. This will quicken the download time of a 10-Mbyte file to 1 s on 4G, from 200 s on 3G, enabling high-definition video to stream to phones and create a virtual reality experience on high-resolution handset screens.

- Fast/seamless handover across multiple networks — 4G wireless networks should support global roaming across multiple wireless and mobile networks.
- Next-generation multimedia support — the underlying network for 4G must be able to support fast speed and large volume data transmission at a lower cost than today's cost.

A candidate glue for all this could be software defined radio (SDR).[55] SDR enables devices such as cell phones, PDAs, PCs, and a whole range of other devices to scan the airwaves for the best possible method of connectivity, at the best price. In an SDR environment, functions that were formerly carried out solely in hardware — such as the generation of the transmitted radio signal and tuning of the received radio signal — are performed by software. Thus, the radio is programmable and able to transmit and receive over a wide range of frequencies while emulating virtually any desired transmission format.

6.5.5 IEEE Standard 802.16

To provide a standardized approach to WLL, the IEEE 802 committee set up the 802.16 working group[43] in 1999 to develop broadband wireless standards. IEEE 802.16[17] standardizes the WirelessMAN air interface and related functions for wireless metropolitan area networks (MANs). This standard serves as a major driving force in linking businesses and homes to local telecommunication networks.

A WirelessMAN provides network access to buildings through exterior antennas, communicating with central radio base stations (BSs). The WirelessMAN offers an alternative to cabled-access networks, such as fiber-optic links, coaxial systems using cable modems, and DSL links. This technology may prove less expensive to deploy and may lead to more ubiquitous broadband access because wireless systems have the capacity to address broad geographic areas without the costly infrastructure development required in deploying cable links to individual sites. Such systems have been in use for several years, but the development of the new standard marks the maturation of the industry and forms the basis of new industry success using second-generation equipment.

In this scenario, with WirelessMAN technology bringing the network to a building, users inside the building can be connected to it with conventional in-building networks such as Ethernet or wireless LANs. However, the fundamental design of the standard may eventually allow for an efficient extension of the WirelessMAN networking protocols directly to the individual user. For instance, a central BS may someday exchange MAC protocol data with an individual laptop computer in a home. The links from the BS to the home receiver and from the home receiver to the laptop would likely use quite different physical layers, but design of the WirelessMAN-MAC could accommodate such a connection with full QoS. With the technology expanding in this direction, it is likely that the standard will evolve to support nomadic and increasingly mobile users such as a stationary or slow-moving vehicle.

IEEE Standard 802.16 was designed to evolve as a set of air interfaces based on a common MAC protocol but with physical layer specifications dependent on the spectrum of use and the associated regulations. The standard, as approved in 2001, addresses frequencies from 10 to 66 GHz, where a large spectrum is currently available worldwide but at which the short wavelengths introduce significant deployment challenges. A recent project has completed an amendment denoted IEEE 802.16a.[18] This document extends the air interface support to lower frequencies in the 2- to 11-GHz band, including licensed and license-exempt spectra. Compared to the higher frequencies, such spectra offer a less expensive opportunity to reach many more customers, although at generally lower data rates. This suggests that such services will be oriented toward individual homes or small- to medium-sized enterprises.

6.5.5.1 MAC Layer

The IEEE 802.16 MAC protocol was designed to support point-to-multipoint broadband wireless access applications. It addresses the need for very high bit rates, both uplink and downlink. Access and bandwidth allocation algorithms must accommodate hundreds of terminals per channel, with terminals that may be shared by multiple end users. The services required by these end users are varied and include legacy TDM voice and data; IP connectivity; and packetized voice over IP (VoIP). To support this variety of services, the 802.16 MAC must accommodate continuous and bursty traffic. Additionally, these services are expected to be assigned QoS in keeping with the traffic types. The 802.16 MAC provides a wide range of service types analogous to the classic ATM service categories as well as newer categories such as guaranteed frame rate (GFR).

The 802.16 MAC protocol must also support a variety of backhaul requirements, including ATM and packet-based protocols. Convergence sublayers are used to map the transport layer-specific traffic to a MAC that is flexible enough to carry any traffic type efficiently. Through such features as payload header suppression, packing, and fragmentation, the convergence sublayers and MAC work together to carry traffic in a form that is often more efficient than the original transport mechanism.

Issues of transport efficiency are also addressed at the interface between the MAC and the PHY layer. For example, modulation and coding schemes are specified in a burst profile that may be adjusted to each subscriber station adaptively for each burst. The MAC can make use of bandwidth-efficient burst profiles under favorable link conditions but shift to more reliable, though less efficient, alternatives as required to support the planned 99.999% link availability.

The request-grant mechanism is designed to be scalable, efficient, and self-correcting. The 802.16 access system does not lose efficiency when presented with multiple connections per terminal, multiple QoS levels per terminal, and a large number of statistically multiplexed users. It takes advantage of a wide variety of request mechanisms, balancing the stability of connectionless access with the efficiency of contention-oriented access.

Along with the fundamental task of allocating bandwidth and transporting data, the MAC includes a privacy sublayer that provides authentication of network access and connection establishment to avoid theft of service, and it provides key exchange and encryption for data privacy. To accommodate more demanding physical environment and different service requirements of the frequencies between 2 and 11 GHz, the 802.16a project is upgrading the MAC to provide ARQ and support for mesh, rather than only point-to-multipoint, network architectures.

6.5.5.1.1 MAC Layer Details

The MAC includes service-specific convergence sublayers that interface to higher layers, above the core MAC common part sublayer that carries out the key MAC functions. Below the common part sublayer the privacy sublayer is located.

6.5.5.1.1.1 Service-Specific Convergence Sublayers

IEEE Standard 802.16 defines two general service-specific convergence sublayers for mapping services to and from 802.16 MAC connections. The ATM convergence sublayer is defined for ATM services, and the packet convergence sublayer is defined for mapping packet services such as IPv4, IPv6, Ethernet, and virtual local area network (VLAN). The primary task of the sublayer is to classify service data units (SDUs) to the proper MAC connection, preserve or enable QoS, and enable bandwidth allocation. The mapping takes various forms depending on the type of service. In addition to these basic functions, the convergence sublayers can also perform more sophisticated functions such as payload header suppression and reconstruction to enhance airlink efficiency.

6.5.5.1.1.2 Common Part Sublayer

Introduction and General Architecture. In general, the 802.16 MAC is designed to support a point-to-multipoint architecture with a central BS handling multiple independent sectors simultaneously. On the downlink, data to the subscriber stations (SSs) are multiplexed in TDM fashion. The uplink is shared between SSs in TDMA fashion.

The 802.16 MAC is connection oriented. All services, including inherently connectionless services, are mapped to a connection. This provides a mechanism for requesting bandwidth, associating QoS and traffic parameters, transporting and routing data to the appropriate convergence sublayer, and all other actions associated with the contractual terms of the service. Connections are referenced with 16-b connection identifiers and may require continuous availability of bandwidth or bandwidth on demand.

Each SS has a standard 48-b MAC address, which serves mainly as an equipment identifier because the primary addresses used during operation are the connection identifiers. Upon entering the network, the SS is assigned three management connections in each direction that reflect the three different QoS requirements used by different management levels. The first of these is the basic connection, which is used for the transfer of short, time-critical MAC and radio link control (RLC) messages. The primary management connection is used to transfer longer, more delay-tolerant messages such as those used for authentication and connection setup. The secondary management connection is used for the transfer of standard-based management messages such as dynamic host configuration protocol (DHCP), trivial file transfer protocol (TFTP), and simple network management protocol (SNMP).

The MAC reserves additional connections for other purposes. One connection is reserved for contention-based initial access. Another is reserved for broadcast transmissions in the downlink as well as for signaling broadcast contention-based polling of SS bandwidth needs. Additional connections are reserved for multicast, rather than broadcast, contention-based polling. SSs may be instructed to join multicast polling groups associated with these multicast polling connections.

MAC PDU Formats. The MAC PDU is the data unit exchanged between the MAC layers of the BS and its SSs. A MAC PDU consists of a fixed-length MAC header, a variable-length payload, and an optional cyclic redundancy check (CRC). Two header formats, distinguished by the HT field, are defined: the generic header (Figure 6.5) and the bandwidth request header. Except for bandwidth containing no payload, MAC PDUs have MAC management messages or convergence sublayer data.

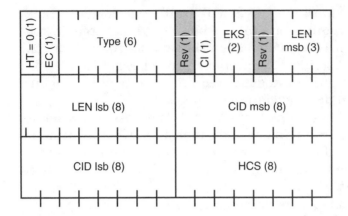

Figure 6.5   Generic header for MAC PDU.

Three types of MAC subheader may be present. A grant management subheader is used by SS to convey bandwidth management needs to its BS. A fragmentation subheader indicates the presence and orientation within the payload of any fragments of the SDUs. The packing subheader is used to indicate packing of multiple SDUs into a single PDU. Immediately following the generic header, a grant management and fragmentation subheaders may be inserted in MAC PDUs if so indicated by the Type field. The packing subheader may be inserted before each MAC SDU if shown by the Type field.

Transmission of MAC PDUs. The IEEE 802.16 MAC supports various higher layer protocols such as ATM or IP. Incoming MAC SDUs from corresponding convergence sublayers are formatted according to the MAC PDU format, possibly with fragmentation and/or packing, before they are conveyed over one or more connections in accordance with the MAC protocol. After traversing the air link, MAC PDUs are reconstructed back into the original MAC SDUs so that the format modifications performed by the MAC layer protocol are transparent to the receiving entity.

IEEE 802.16 takes advantage of packing and fragmentation processes, whose effectiveness, flexibility, and efficiency are maximized by the bandwidth allocation process. Fragmentation is the process in which a MAC SDU is divided into one or more MAC SDU fragments. Packing is the process in which multiple MAC SDUs are packed into a single MAC PDU payload. Both processes may be initiated by a BS for a downlink connection or for an SS for an uplink connection. IEEE 802.16 allows simultaneous fragmentation and packing for efficient use of the bandwidth.

PHY Support and Frame Structure. The IEEE 802.16 MAC supports TDD and FDD. In FDD, continuous as well as burst downlinks are possible. Continuous downlinks allow for certain robustness enhancement techniques, such as interleaving. Burst downlinks (FDD or TDD) allow the use of more advanced robustness and capacity enhancement techniques, such as subscriber-level adaptive burst profiling and advanced antenna systems.

The MAC builds the downlink subframe starting with a frame control section containing the DL-MAP (downlink MAP) and UL-MAP (uplink MAP) messages. These indicate PHY transitions on the downlink as well as bandwidth allocations and burst profiles on the uplink. The DL-MAP is always applicable to the current frame and is always at least two FEC blocks long. To allow adequate processing time, the first PHY transition is expressed in the first FEC block. In TDD and FDD systems, the UL-MAP provides allocations starting no later than the next downlink frame. The UL-MAP can, however, start allocating in the current frame, as long as processing times and round-trip delays are observed.

Radio Link Control. The advanced technology of the 802.16 PHY requires equally advanced RLC, particularly a capability of the PHY to change from one burst profile to another. The RLC must control this capability as well as the traditional RLC functions of power control and ranging. RLC begins with periodic BS broadcast of the burst profiles that have been chosen for the uplink and downlink. Among the several burst profiles used on a channel, one in particular is chosen based on a number of factors, such as rain region and equipment capabilities. Burst profiles for the downlink are each tagged with a downlink interval usage code (DIUC) and those for the uplink are tagged with an uplink interval usage code (UIUC).

During initial access, the SS performs initial power leveling and ranging using ranging request (RNG-REQ) messages transmitted in initial maintenance windows. The adjustments to the SS's transmit time advance, as well as power adjustments, are returned to the SS in ranging response (RNG-RSP) messages. For ongoing ranging and power adjustments, the BS may transmit unsolicited RNG-RSP messages instructing the SS to adjust its power or timing. During initial ranging, the SS can also request service in the downlink via a particular burst profile by transmitting its choice of DIUC to the BS. The selection is based on received downlink signal quality measurements performed by the SS before and during initial ranging. The BS may confirm or reject the choice in the ranging response. Similarly, the BS monitors the quality of the uplink signal it receives from the SS. The BS commands the SS to use a particular uplink burst profile simply by including the appropriate burst profile UIUC with the SS's grants in ULMAP messages.

After initial determination of uplink and downlink burst profiles between the BS and a particular SS, RLC continues to monitor and control the burst profiles. Harsher environmental conditions, such as rain fades, can force the SS to request a more robust burst profile. Alternatively, exceptionally good weather may allow an SS to operate temporarily with a more efficient burst profile. The RLC continues to adapt the SS's current UL and DL burst profiles, always striving to achieve a balance between robustness and efficiency. Because the BS is in control and directly monitors the uplink signal quality, the protocol for changing the uplink burst profile for an SS is simple by BS merely specifying the profile's associated UIUC whenever granting the SS bandwidth in a frame. This eliminates the need for an acknowledgment because the SS will always receive both the UIUC and the grant or neither. Thus, no chance of uplink burst profile mismatch between the BS and the SS exists.

In the downlink, the SS is the entity that monitors the quality of the receive signal and therefore knows when its downlink burst profile should change. The BS, however, is the entity in control of the change. Two methods are available to the SS to request a change in downlink burst profile, depending on whether the SS operates in the grant per connection (GPC) or grant per SS (GPSS) mode. The first method would typically apply (based on the discretion of the BS scheduling algorithm) only to GPC SSs. In this case, the BS may periodically allocate a station maintenance interval to the SS. The SS can use the RNG-REQ message to request a change in downlink burst profile. The preferred method is for the SS to transmit a downlink burst profile change request (DBPC-REQ). In this case, which is always an option for GPSS SSs and can be an option for GPC SSs, the BS responds with a downlink burst profile change response (DBPC-RSP) message confirming or denying the change.

Because messages may be lost due to irrecoverable bit errors, the protocols for changing an SS's downlink burst profile must be carefully structured. The order of the burst profile change actions is different when transitioning to a more robust burst profile than when transitioning to a less robust one. The standard takes advantage of the fact that any SS is always required to listen to more robust portions of the downlink as well as the profile that has been negotiated.

Channel Acquisition. The MAC protocol includes an initialization procedure designed to eliminate the need for manual configuration. Upon installation, SS begins scanning its frequency list to find an operating channel. It may be programmed to register with one specific BS, referring to a programmable BS ID broadcasted by each. This feature is useful in dense deployments in which the SS might hear a secondary BS due to selective fading or when the SS picks up a sidelobe of a nearby BS antenna.

After deciding on which channel or channel pair to start communicating, the SS tries to synchronize to the downlink transmission by detecting the periodic frame preambles. Once the physical layer is synchronized, the SS looks for periodic DCD and UCD broadcast messages that enable the SS to learn the modulation and FEC schemes used on the carrier.

IP Connectivity. After registration, the SS attains an IP address via DHCP and establishes the time of day via the

Internet time protocol. The DHCP server also provides the address of the TFTP server from which the SS can request a configuration file. This file provides a standard interface for providing vendor-specific configuration information.

6.5.5.2 Physical Layer

6.5.5.2.1 10–66 GHz

In the design of the PHY specification for 10 to 66 GHz, line-of-sight propagation has been deemed a practical necessity. With this condition assumed, single-carrier modulation could be easily selected to be employed in designated air interface WirelessMAN-SC. However, many fundamental design challenges remain. Because of a point-to-multipoint architecture, BS basically transmits a TDM signal, with individual subscriber stations allocated time slots sequentially. Access in the uplink direction is by TDMA.

Following extensive discussions on duplexing, a burst design has been selected that allows TDD (in which the uplink and downlink share a channel but do not transmit simultaneously) and FDD (the uplink and downlink operate on separate channels, sometimes simultaneously) to be handled in a similar fashion. Support for half-duplex FDD subscriber stations, which may be less expensive because they do not simultaneously transmit and receive, has been added at the expense of some slight complexity. TDD and FDD alternatives support adaptive burst profiles in which modulation and coding options may be dynamically assigned on a burst-by-burst basis.

6.5.5.2.2 2–11 GHz

Licensed and license-exempt 2- to 11-GHz bands are addressed in the IEEE Project 802.16a. This currently specifies that compliant systems implement one of three air interface specifications, each of which can provide interoperability. Design of the 2- to 11-GHz physical layer is driven by the need for NLOS operation. Because residential applications are expected, rooftops may be too low for a clear sight line to the antenna of a BS, possibly due to obstruction by trees. Therefore, significant multipath propagation must be expected. Furthermore, outdoor-mounted antennas are expensive because of hardware and installation costs.

The three 2- to 11-GHz air interfaces included in 802.16a, draft 3, specifications are:

- *WirelessMAN-SC2* uses a single-carrier modulation format.
- *WirelessMAN-OFDM* uses orthogonal frequency-division multiplexing with a 256-point transform. Access is by TDMA. This air interface is mandatory for license-exempt bands.
- *WirelessMAN-OFDMA* uses orthogonal frequency-division multiple access with a 2048-point transform. In this system, multiple access is provided by addressing a subset of the multiple carriers to individual receivers.

Because of the propagation requirements, the use of advanced antenna systems is supported. It is premature to speculate on further details of the 802.16a amendment prior to its completion. The draft seems to have reached a level of maturity, but the contents could significantly change by ballots. Modes could even be deleted or added.

6.5.5.2.3 Physical Layer Details

The PHY specification defined for 10 to 66 GHz uses burst single-carrier modulation with adaptive burst profiling in which transmission parameters, including the modulation and coding schemes, may be adjusted individually to each SS on a frame-by-frame basis. TDD and burst FDD variants are defined. Channel bandwidths of 20 or 25 MHz (typical U.S. allocation) or 28 MHz (typical European allocation) are specified, along with Nyquist square-root raised-cosine pulse shaping with a roll-off factor of 0.25. Randomization is performed for spectral shaping and ensuring bit transitions for clock recovery.

The FEC uses Reed–Solomon GF (256), with variable block size and appropriate error correction capabilities. This is paired with an inner block convolutional code to transmit critical data robustly, such as frame control and initial accesses. The FEC options are paired with QPSK, 16-state QAM (16-QAM) and 64-state QAM (64-QAM) to form burst profiles of varying robustness and efficiency. If the last FEC block is not filled, that block may be shortened. Shortening in the uplink and downlink is controlled by the BS and is implicitly communicated in the uplink map (UL-MAP) and downlink map (DL-MAP).

The system uses a frame of 0.5, 1, or 2 ms divided into physical slots for the purpose of bandwidth allocation and identification of PHY transitions. A physical slot is defined to be four QAM symbols. In the TDD variant of the PHY, the uplink subframe follows the downlink subframe on the same carrier frequency. In the FDD variant, the uplink and downlink subframes are coincident in time but carried on separate frequencies.

The downlink subframe starts with a frame control section that contains the DL-MAP for the current downlink frame as well as the UL-MAP for a specified time in the future. The downlink map specifies when physical layer transitions (modulation and FEC changes) occur within the downlink subframe. The downlink subframe typically contains a TDM portion immediately following the frame control section. Downlink data are transmitted to each SS using a negotiated burst profile. The data are transmitted in order of decreasing robustness to allow SSs to receive their data before being presented with a burst profile that could cause them to lose synchronization with the downlink.

In FDD systems, the TDM portion may be followed by a TDMA segment that includes an extra preamble at the start of each new burst profile. This feature allows better support of half-duplex SSs. In an efficiently scheduled FDD system with many half-duplex SSs, some may need to transmit earlier in the frame than they are received. Due to their half-duplex nature, these SSs may lose synchronization with the downlink. The TDMA preamble allows them to regain synchronization.

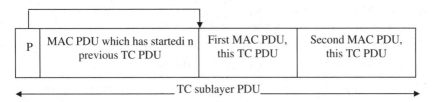

P	MAC PDU which has startedi n previous TC PDU	First MAC PDU, this TC PDU	Second MAC PDU, this TC PDU

←——————————————— TC sublayer PDU ———————————————→

Figure 6.6 TC PDU format.

Due to the dynamics of bandwidth demand for a variety of services that may be active, the mixture and duration of burst profiles and the presence or absence of a TDMA portion vary dynamically from frame to frame. Because the recipient SS is implicitly indicated in the MAC headers rather than in the DL-MAP, SSs listen to all portions of the downlink subframes that they are capable of receiving. For full-duplex SSs, this means receiving all burst profiles of equal or greater robustness than they have negotiated with the BS.

Unlike the downlink, the UL-MAP grants bandwidth to specific SSs. The SSs transmit in their assigned allocation using the burst profile specified by the uplink interval usage code (UIUC) in the UL-MAP entry granting them bandwidth. The uplink subframe may also contain contention-based allocations for initial system access and broadcast or multicast bandwidth requests. The access opportunities for initial system access are sized to allow extra guard time for SSs that have not resolved the transmit time advances necessary to offset the round-trip delay to the BS.

Between the PHY and MAC is a transmission convergence (TC) sublayer. This layer performs the transformation of variable length MAC PDUs into the fixed length FEC blocks (plus possibly a shortened block at the end) of each burst. The TC layer has a PDU sized to fit in the FEC block currently being filled. It starts with a pointer indicating where the next MAC PDU header starts within the FEC block (see Figure 6.6). The TC PDU format allows resynchronization to the next MAC PDU in the event that the previous FEC block had irrecoverable errors. Without the TC layer, a receiving SS or

BS could potentially lose the entire remainder of a burst when an irrecoverable bit error occurs.

6.5.6   IEEE 802.11 as a Last Mile Alternative

Some companies are exploring the possibility of employing IEEE 802.11 WLAN standard (described in detail in the next section) as an alternative delivery platform to last mile broadband communication.[21] Originally designed to be used in office and home environments, 802.11 (in particular, 802.11b) standard is now starting to be considered as a last mile technology because of its extremely low cost compared to existing BWA technologies.

Clearly, if an 802.11b wireless network is used as a last mile distribution technology, it will not solve all BWA problems. However, it does certainly emphasize the importance of its business case. Because the IEEE 802.11b standard operates in the unlicensed 2.4-GHz band, the cost effectiveness of 802.11b and its largely unregulated operating band becomes a double edged-sword. Therefore, good network design approach and friendly cooperation ought to be employed among carriers.

Recent experiments[21] have shown that by employing special components and a readily available antenna, an 802.11b network achieves reliable broadband access. At distances of 1 km, a terminal with a standard 802.11b wireless card and a fixed antenna can receive data at the maximum 7.66-Mb/s data rate (802.11b portends to operate at 11 Mb/s, but with the protocol overheads this effectively means the maximum data transfer rate is around 7.66 Mb/s). As a matter of fact, the range has been shown to be extendable to up to 7 km, although the quality of service suffers over such distances. The fact that the network is operating at high-quality speed over distances of greater than 1 km offers a great potential for the BWA market.

6.5.7   Various Standards

Broadband radio access standards and architectures are currently under serious discussion in Europe, Japan, and the U.S. Different regions and countries use different terms when

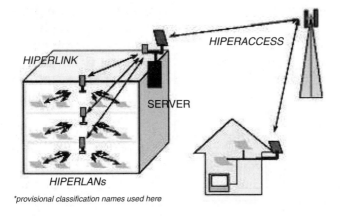

Figure 6.7 Broadband wireless access.

referring to these standards. In Europe, they are referred to as broadband radio access networks (BRAN) and, in the U.S., as LMDS, IEEE 802.16, and BWA systems, among other terms. In Canada and some other countries, they are also referred to as local multipoint communication systems (LMCS). Their applications, however, are varied from fixed to mobile and local to wide area, and include promising applications such as high-speed Internet access, two-way data communications (peer-to-peer or client–server), private or public telephony, two-way multimedia services such as videoconferencing and video commerce, and broadcast video.

For BRAN, broadband access consists of what is termed *high-performance radio access* (HIPERACCESS), HIPERLAN (covered in detail later in this chapter), and HIPERLINK, as shown in Figure 6.7. HIPERACCESS systems connect mainly residential, SOHO, and small to medium enterprise premises and allow access to a variety of telecommunications services, such as voice, data, and multimedia services, with transmission rates varying from about 2 to 25 Mb/s. HIPERACCESS is primarily to be used as a broadband remote access network. The radio spectrum can be in the 2- to 40-GHz range.

HIPERLAN provides local access with controlled QoS for broadband applications (e.g., Internet and videoconferencing) to portable computers for use within buildings and on campus; it uses mainly unlicensed radio spectrum in the 5-GHz band. HIPERLINK is primarily a network–network radio interconnection that will support a variety of protocols and all the aforementioned traffic scenarios. This application would use bit rates of up to 155 Mb/s in parts of the 17-GHz radio spectrum.

IEEE 802.16 covers more issues than HIPERACCESS, including WirelessMAN and wireless high-speed unlicensed metropolitan area networks (HUMANs), which include frequencies from 2 up to 66 GHz. The International Telecommunication Union (ITU) initiated a working group called ITU JRG 8A-9B in charge of broadband wireless access system standardization. This group receives input from BRAN and IEEE 802.16 and tries to deliver a global consensus on this technology from the standpoint and function of the ITU. For details, see the ITU wireless access system Web site.[44]

6.6 WIRELESS LOCAL AREA NETWORKING

Wireless network technologies are expected to become more widespread than the current popular wired solutions. In the context of home networking, wireless communications present an ideal framework, despite a variety of technical and deployment obstacles. Among the technologies for wireless home networking, WLAN and WPAN systems are the key enablers, each of which has its own characteristics and suitability for specific areas. Therefore, this section investigates the most prominent solutions for WLAN systems and later delves into WPAN technologies.

6.6.1 Wireless Home Networking Application Requirements

Today's home networking applications are driving the need for high-performance wireless network protocols with highly usable (effective) speed and isochronous, multimedia-capable services. Factors driving the need for high performance are:

- *Home networks incorporating multimedia.* Existing and emerging digital devices such as televisions, DVD players, digital video recorders, digital audio/MP3 players, DBS systems, flat-panel displays, digital set-top boxes, and PCs create a need to support multimedia content at home. The home network ought to support all types of digital content, including local content (e.g., DVD, MP3) and broadcast content (e.g., video on demand, streaming media). Such multimedia traffic encompasses video, audio, voice, and data. Internet multimedia broadcasting is already prevalent. An ability to support multimedia is expected to be the "killer app" that will encourage a massive acceptance and adoption of home networks. Therefore, a home network must support the coexistence of data (e.g., printing, file transfer) and isochronous content (e.g., voice, video). The consumer needs to choose products today that could provide a solid foundation for multimedia network services in the near future.

- *Consumers adding more nodes to their home networks.* As mentioned earlier, the rapid growth of homes with multiple PCs indicates that the number of nodes in a PC network will continue to rise as new home network appliances are introduced. Wireless appliances may share the same bandwidth. Consequently, higher network throughput and adequate speed are necessary to accommodate additional home network devices. Choosing a wireless network with an access mechanism that supports multiple nodes without significantly degrading performance is essential to supporting a growing home wireless network.

- *The need to preserve high-speed broadband Internet access.* As the preceding sections showed, the desire for faster Internet access is driving a massive deployment of high-speed broadband access. Thus, consumers need to avoid any bottleneck, and a high-performance wireless network is needed to maintain high-speed broadband access to devices.

- *Evolving personal computer applications.* File sizes of typical personal computer applications are growing rapidly with each generation of new software applications. The file size of a word processing document with the same content has doubled over the last few years. Sending graphics and digital photography over e-mail is now commonplace; therefore, a wireless network with high performance is mandatory in order to move files in a timely fashion. This need is further enlarged as more nodes are added and the network volume expands.

6.6.2   IEEE Standard 802.11 for WLANs

The sky appears to be the limit for WLAN technologies. WLANs provide an excellent usage model for high-bandwidth consumers and are quite appealing for their low infrastructure cost and high data rates compared to other wireless data technologies such as cellular or point-to-multipoint distribution systems.

In June 1997, the IEEE approved the 802.11 standard[22,35,42] for WLANs, and in July 1997, IEEE 802.11 was adopted as a worldwide International Standards Organization (ISO) standard. The standard consists of three possible PHY layer implementations and a single common MAC layer supporting data rates of 1 or 2 Mb/s. The alternatives for PHY layer in the original standard include an FHSS system using 2 or 4 Gaussian frequency-shift keying (GFSK) modulation; a direct sequence spread spectrum (DSSS) system using differential binary phase-shift keying (DBPSK) or differential quadrature phase-shift keying (DQPSK) baseband modulation; and an IR physical layer.

Later in 1999, the IEEE 802.11b working group extended the IEEE 802.11 standard with the IEEE 802.11b addition and decided to drop the FHSS to use only DSSS. In addition, other working groups, the IEEE 802.11a and the IEEE 802.11g, significantly modified the PHY to replace the spread spectrum techniques used in the IEEE 802.11 to implement the OFDM, which effectively combines multicarrier, multisymbol,

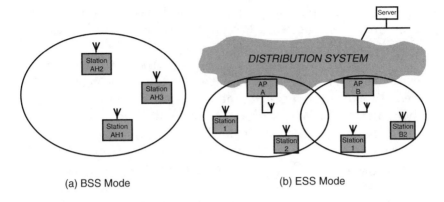

(a) BSS Mode   (b) ESS Mode

Figure 6.8 Possible network topologies.

and multirate techniques. Per the protocol stack, the MAC layer is common across all standards, although they are not always compatible at the PHY layer.

The IEEE 802.11 standard has been widely employed because it can be easily adapted for business or residential use and for low-mobility environments such as airports, coffee shops, hotels, and other locations where a need for broadband Internet access exists. Therefore, this standard, its peculiarities, and its relevance to the field of broadband wireless access will now be discussed. To this end, the IEEE standards 802.11, 802.11a, 802.11b, 802.11g, 802.11e, and 802.11i will be examined. For a complete overview of the current activities within the IEEE 802.11 working group, refer to its Web site.[57]

6.6.2.1 Network Architecture

WLANs can be used to replace wired LANs or as extensions of the wired LAN infrastructure. The basic topology of an 802.11 network is shown in Figure 6.8(a). A basic service set (BSS) consists of two or more wireless nodes, or stations (STAs), which have established communication after recognizing each other. In the most basic form, stations communicate directly with each other in a peer-to-peer mode, sharing a given cell coverage area. This type of network is often formed on a temporary instantaneous basis and is commonly referred to as an ad hoc network,[42] or independent basic service set (IBSS).

The main function of an access point (AP) is to form a bridge between wireless and wired LANs. In most instances, each BSS contains an AP analogous to a BS used in cellular phone networks. When an AP is present, stations do not communicate on a peer-to-peer basis. All communications between stations or between a station and a wired network client go through the AP. APs are not mobile and form a part of the wired network infrastructure. A BSS in this configuration is said to be operating in the *infrastructure mode*. The extended service set (ESS) shown in Figure 6.8(b) consists of a series of overlapping BSSs (each containing an AP) connected together by means of a distribution system, which can be any type of network and is almost invariably an Ethernet LAN. Mobile nodes can roam between APs and seamless coverage is possible.

6.6.2.2   MAC Layer

The IEEE 802 group has adopted the same MAC layer for standards including 802.11a,[23,35,42] 802.11b,[22,35] and 802.11g.[22] The basic access method for 802.11 is the distributed coordination function (DCF), which uses carrier sense multiple access with collision avoidance (CSMA/CA).[15] This requires each station to listen for other potential users. If the channel is idle, the station may transmit. However, if it is busy, each station waits until the current transmission completes and then enters into a random back-off procedure. This prevents multiple stations from seizing the medium immediately after completion of the preceding transmission.

Packet reception in DCF requires acknowledgment as shown in Figure 6.9(a). The period between completion of packet transmission and start of the ACK frame is one short interframe space (SIFS). ACK frames have a higher priority than other traffic. Fast acknowledgment is one of the salient features of the 802.11 standard because it requires ACKs to be handled at the MAC sublayer. Transmissions other than

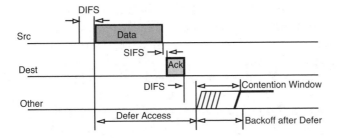

(a) CSMA/CD Backoff Algorithm

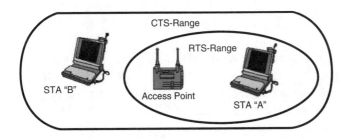

(b) RTS/CTS Procedure Eliminates the "Hidden
Node" Problem

Figure 6.9 CSMA/CD and RTS/CTS exchange in 802.11.

ACKs are delayed at least one DCF interframe space (DIFS). If a transmitter senses a busy medium, it determines a random back-off period by setting an internal timer to an integer number of slot times. Upon expiration of a DIFS, the timer begins to decrement. If the timer reaches zero, the station may begin transmission. However, if the channel is seized by another station before the timer reaches zero, the timer setting is retained at the decremented value for subsequent transmission.

The preceding method relies on the *physical carrier sense*. The underlying assumption is that every station can "hear" all other stations; however, this is not always the case. Referring to Figure 6.9(b), the AP is within range of the STA-A, but STA-B is out of range. STA-B would not be able to detect transmissions from STA-A, so the probability of collision is greatly increased. This is known as the *hidden node.*

To combat this problem, a second carrier sense mechanism is available. *Virtual carrier sense* enables a station to reserve the medium for a specified period of time through the use of request to send (RTS)/clear to send (CTS) frames. In the case described previously, STA-A sends an RTS frame to the AP. The RTS will not be heard by STA-B. The RTS frame contains a duration/ID field that specifies the period of time for which the medium is reserved for a subsequent transmission. The reservation information is stored in the network allocation vector (NAV) of all stations detecting the RTS frame.

Upon receipt of the RTS, the AP responds with a CTS frame, which also contains a duration/ID field specifying the period of time for which the medium is reserved. Although STA-B did not detect the RTS, it will detect the CTS and update its NAV accordingly. Thus, collision is avoided even though some nodes are hidden from other stations. The RTS/CTS procedure is invoked according to a user-specified parameter. It can be used always, never, or for packets exceeding an arbitrarily defined length.

A PHY layer convergence procedure (PLCP) maps a MAC PDU into a frame format. Figure 6.10(c) shows the format of a complete packet (PPDU) in 802.11a, including the preamble, header, and PHY layer service data unit (PSDU or payload):

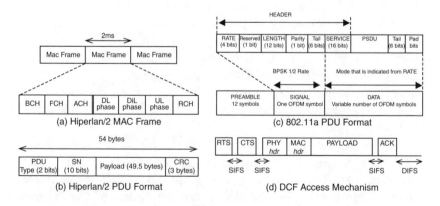

Figure 6.10 MAC structures for HIPERLAN/2 and 802.11a.

- The header contains information about the length of the payload and the transmission rate, a parity bit, and six zero-tail bits. The header is always transmitted using the lowest-rate transmission mode in order to ensure robust reception. Thus, it is mapped onto a single binary phase shift keying (BPSK)-modulated orthogonal frequency-division multiplexed (OFDM) symbol.

- The rate field conveys information about the type of modulation and the coding rate used in the rest of the packet.

- The length field takes a value between 1 and 4095 and specifies the number of bytes in the PSDU.

- The parity bit is a positive parity for the first 17 b of the header.

- The six tail bits are used to reset the convolutional encoder and terminate the code trellis in the decoder.

- The first 7 b of the service field are set to zero and used to initialize the descrambler. The remaining 9 b are reserved for future use.

- The pad bits are used to ensure that the number of bits in the PPDU maps to an integer number of OFDM symbols.

As mentioned earlier, DCF is the basic media access control method for 802.11 and is mandatory for all stations. The point coordination function (PCF) is an optional extension to DCF. PCF provides a time division duplexing capability to accommodate time-bounded, connection-oriented services such as cordless telephony.

6.6.2.3 Physical Layer

Knowing what some of the physical layer terminology means is essential to understanding the intricacies of 802.11:

- GFSK is a modulation scheme in which the data are first filtered by a Gaussian filter in the baseband and then modulated with a simple frequency modulation. The number of frequency offsets used to represent data symbols of 1 and 2 b are represented by "2" and "4," respectively.

- DBPSK is phase modulation using two distinct carrier phases for data signaling providing 1 b per symbol.
- DQPSK is a type of phase modulation using two pairs of distinct carrier phases, in quadrature, to signal 2 b per symbol. The differential characteristic of the modulation schemes indicates the use of the difference in phase from the last change or symbol to determine the current symbol's value, rather than any absolute measurements of the phase change.

The FHSS and DSSS modes are specified for operation in the 2.4-GHz industrial–scientific–medical (ISM) band, which has sometimes been jokingly referred to as the "interference suppression is mandatory" band because it is heavily used by various electronic products. The third physical layer alternative is an infrared system using near-visible light in the 850- to 950-nm range as the transmission medium.

Two supplements to the IEEE 802.11 standard are at the forefront of the new WLAN options that will enable much higher data rates: 802.11b and 802.11a, as well as a European Telecommunications Standards Institute (ETSI) standard called high-performance LAN (HIPERLAN/2). 802.11 and HIPERLAN/2 have similar physical layer characteristics operating in the 5-GHz band and use the modulation scheme OFDM, but the MAC layers are considerably different. The focus here, however, is to discuss and compare the physical layer characteristics of 802.11a and 802.11b. HIPERLAN/2 shares several of the same physical properties as 802.11a and will be covered later in this chapter.

Another standard that warrants mention in this context is the IEEE 802.11g standard. With a ruling from the FCC that allows OFDM digital transmission technology to operate in the ISM band, and the promise of interoperability with a large installed base of the IEEE 802.11b products, the IEEE 802.11g extension to the standard, formally ratified in June 2003, begins to garner the attention of WLAN equipment providers. The IEEE 802.11g provides the same maximum speed of 802.11a, coupled with backwards compatibility with 802.11b devices. This backwards compatibility will make upgrading wireless LANs simple and inexpensive.

IEEE 802.11g specifies operation in the 2.4-GHz ISM band. To achieve the higher data rates found in 802.11a, 802.11g-compliant devices utilize OFDM modulation technology. These devices can automatically switch to QPSK modulation in order to communicate with the slower 802.11b- and 802.11-compatible devices. In theory, 802.11a and 802.11b use almost the same PHY specification, but in practice this may not be completely true because of the backward compatibility requirement with 802.11b. Given their similarities in the use of OFDM at the PHY layer, discussion in this chapter is confined to the 802.11a PHY layer only. Despite all of its apparent advantages, the use of the crowded 2.4-GHz band by 802.11g could prove to be a disadvantage.

6.6.2.3.1   *802.11b Details*

Approved by the IEEE in 1999, 802.11b is an extension of the 802.11 DSSS system mentioned earlier and supports 5.5 and 11 Mb/s of higher payload data rates in addition to the original 1- and 2-Mb/s rates. Many commercial products are now available, and the systems base is growing rapidly. 802.11b also operates in the highly populated 2.4-GHz ISM band (2.40 to 2.4835 GHz), which provides only 83 MHz of spectrum to accommodate a variety of other products, including cordless phones, microwave ovens, other WLANs, and PANs. This makes susceptibility to interference a primary concern. The occupied bandwidth of the spread-spectrum channel is 22 MHz, so the ISM band accommodates only three nonoverlapping channels spaced 25 MHz apart. To help mitigate interference effects, 802.11b designates an optional frequency agile or hopping mode using the three nonoverlapping channels or six overlapping channels spaced at 10 MHz.

To achieve higher data rates, 802.11b uses eight-chip complementary code keying (CCK) as the modulation scheme. Instead of the Barker codes used to encode and spread the data for the lower rates, CCK uses a nearly orthogonal complex code set called complementary sequences. The chip rate remains consistent with the original DSSS system at 11 Mchip/s, but the data rate varies to match channel conditions by changing the spreading factor and/or the modulation scheme.

To achieve data rates of 5.5 and 11 Mb/s, the spreading length is first reduced from 11 to 8 chips. This increases the symbol rate from 1 to 1.375 Msym/s. For the 5.5-Mb/s bit rate with a 1.375-MHz symbol rate, it is necessary to transmit 4 b per symbol (5.5 Mb/s/1.375 Msym/s) and for 11 Mb/s, 8 b per symbol. The CCK approach taken in 802.11b keeps the QPSK spread-spectrum signal and still provides the required number of bits per symbol. It uses all but two of the bits to select from a set of spreading sequences and the remaining 2 b to rotate the sequence. The selection of the sequence, coupled with the rotation, represents the symbol conveying the 4 or 8 b of data. For all 802.11b payload data rates, the preamble and header are sent at the 1-Mb/s rate.

6.6.2.3.2   *802.11a Details*

Although 802.11a was approved in September 1999, new product development has proceeded much more slowly than it did for 802.11b. This is due to the cost and complexity of implementation. This standard employs 300-MHz bandwidth in the 5-GHz unlicensed national information infrastructure (UNII) band. The spectrum is divided into three "domains," each with restrictions imposed on the maximum allowed output power. The first 100 MHz in the lower frequency portion is restricted to a maximum power output of 50 mW. The second 100 MHz has a higher 250-mW maximum, and the third 100 MHz is mainly intended for outdoor applications and has a maximum of 1.0-W power output.

OFDM, employed by 802.11a, operates by dividing the transmitted data into multiple parallel bit streams, each with relatively lower bit rates and modulating separate narrowband carriers, referred to as subcarriers. These subcarriers are orthogonal, so each can be received with almost no interference from another. 802.11a specifies eight nonoverlapping 20-MHz channels in the lower two bands; each of these is divided into 52 subcarriers (4 of which carry pilot data) of 300-kHz bandwidth each. Four nonoverlapping 20-MHz channels are specified in the upper band. The receiver processes the 52 individual bit streams, reconstructing the original high-rate data stream. Four complex modulation methods are employed, depending on the data rate that can be supported by channel conditions between the transmitter and the receiver. These include BPSK, QPSK, 16-QAM, and 64-QAM.

Quadrature amplitude modulation is a complex modulation method in which data are carried in symbols represented by the phase and amplitude of the modulated carrier. For example, 16-QAM has 16 symbols; each represents four data bits. There are 64 symbols in 64-QAM, with each representing six data bits.

BPSK modulation is always used on four pilot subcarriers. Although it adds a degree of complication to the baseband processing, 802.11a includes forward error correction (FEC) as a part of the specification. FEC, which does not exist within 802.11b, enables the receiver to identify and correct errors occurring during transmission by sending additional data along with the primary transmission. This nearly eliminates the need for retransmissions when packet errors are detected. The data rates available in 802.11a are noted in Table 6.1, together with the type of modulation and the coding rate.

Some of the companies developing 802.11a chipset solutions are touting the availability of operational modes that exceed the 54 Mb/s stated in the specification. Of course, because faster data rates are out of the specification's scope, they require the use of equipment from a single source throughout the entire network.

Considering the composite waveform resulting from the combination of 52 subcarriers, the format requires more linearity in the amplifiers because of the higher peak-to-average power ratio of the transmitted OFDM signal. In addition, enhanced phase noise performance is required because of the closely spaced, overlapping carriers. These issues add to the implementation complexity and cost of 802.11a products. Application-specific measurement tools aid in the design and troubleshooting of OFDM signals and systems.

Table 6.1   802.11a Data Rate Description

Data rate (Mb/s)	Modulation type	Coding rate (convolutional encoding & puncturing)	Coded bits per subcarrier symbol	Coded bits per OFDM symbols	Data bits per OFDM symbol
6[a]	BPSK	1/2	1	48	24
9	BPSK	3/4	1	48	36
12[a]	QPSK	1/2	2	96	48
18	QPSK	3/4	2	96	72
24[a]	16-QAM	1/2	4	192	96
36	16-QAM	3/4	4	192	144
48	64-QAM	2/3	6	288	192
54	64-QAM	3/4	6	288	216

[a] Support for these data rates is required by the 802.11a standard.

6.6.2.3.3   Pros and Cons of 802.11a and 802.11b

The 5-GHz band has received considerable attention, but its drawback is the shorter wavelength. Higher frequency signals will have more trouble propagating through physical obstructions encountered in an office (walls, floors, and furniture) than those at 2.4 GHz. An advantage of 802.11a is its intrinsic ability to handle delay spread or multipath reflection effects. The slower symbol rate and placement of significant guard time around each symbol, using a technique called cyclical extension, reduces the intersymbol interference (ISI) caused by multipath interference. (The last quarter of the symbol pulse is copied and attached to the beginning of the burst. Due to the periodic nature of the signal, the junction at the start of the original burst will always be continuous.). In contrast, 802.11b networks are generally range limited by multipath interference rather than the loss of signal strength over distance.

When it comes to deployment of a wireless LAN, operational characteristics have been compared to those of cellular systems in which frequency planning of overlapping cells minimizes mutual interference, supports mobility, and provides seamless channel handoff. The three nonoverlapping frequency channels available for IEEE 802.11b are at a disadvantage compared to the greater number of channels available to 802.11a. The additional channels allow more over-lapping access points within a given area while avoiding additional mutual interference.

Both 802.11a and 802.11b use dynamic rate shifting in which the system will automatically adjust the data rate based on the condition of the radio channel. If the channel is clear, then the modes with the highest data rates are used. However, as interference is introduced into the channel, the transceiver will fall back to a slower, albeit more robust, transmission scheme.

6.6.2.4   IEEE 802.11e

The IEEE 802.11e[56] is an extension of the 802.11 standard for provisioning of QoS. This new standard provides the means of prioritizing the radio channel access within an infrastructure BBS of the IEEE 802.11 WLAN. A BSS that supports the new priority schemes of the 802.11e is referred to as QoS supporting BSS (QBSS).

In order to provide effectively for QoS support, the 802.11e MAC defines the enhanced DCF (EDCF) and the hybrid coordination function (HCF). Stations operating under the 802.11e are called QoS stations; a QoS station, which works as the centralized controller for all other stations within the same QBSS, is called the hybrid coordinator (HC). A QBSS is a BSS that includes an 802.11e-compliant HC and QoS stations. The HC will typically reside within an 802.11e AP. In the following, an 802.11e-compliant QoS station is referred to simply as a station. Similar to DCF, the EDCF is a contention-based channel access mechanism of HCF.

With 802.11e, there may still be the two phases of operation within the superframes, i.e., a CP and a CFP, which alternate over time continuously. The EDCF is used in the CP only, and the HCF is used in both phases, thus making this new coordination function hybrid.

6.6.2.5   IEEE 802.11i

This is a supplement to the MAC layer to improve security that will apply to 802.11 physical standards a, b, and g. It provides an alternative to the existing wired encryption privacy (WEP) with new encryption methods and authentication procedures. Consisting of a framework for regulating access control of client stations to a network via the use of extensible authentication methods, IEEE 802.11x forms a key part of 802.11i.

Security is a major weakness of WLANs. Weakness of WEP encryption is damaging the perception of the 802.11 standard in the market. Vendors have not improved matters by shipping products without setting default security features. In addition, WEP algorithm weaknesses have been exposed. The 802.11i specification is part of a set of security features that should address and overcome these issues. Solutions will start with firmware upgrades using the temporal key integrity protocol, followed by new silicon with AES (an iterated block chipper) and TKIP backwards compatibility.

6.6.3 HIPERLAN/2 Standard for WLANs

Although 802.11 is the standard defined by IEEE, the ETSI BRAN has developed the HIPERLAN/2 standard,[24] which also operates at 5-GHz frequency band similarly to 802.11a. These two standards primarily differ in the MAC layer[25-28]; however, some minor differences also occur in the PHY layers. Here, the HIPERLAN/2 standard is discussed as a means to providing a foundation to broadband access to the home and office environment.

The HIPERLAN/2 radio network is defined in such a way that core-independent PHY and data link control (DLC) layers are present as well as a set of convergence layers (CLs) for interworking. The CLs include Ethernet, ATM, and IEEE 1394 infrastructure,[29] and technical specifications for HIPERLAN/2–third-generation (3G) interworking have also been completed. IEEE 802.11a defines similarly independent PHY and MAC layers (with the MAC common to multiple PHYs within the 802.11 standard). A similar approach to network protocol convergence is expected.

Basically, the network topology of HIPERLAN/2 is the same as in 802.11 (Figure 6.8). Therefore, following the same approach adopted for 802.11, we first discuss the MAC layer characteristics of HIPERLAN/2 will first be discussed and then its physical layer details. Because we now have an understanding of 802.11, the approach is to compare the properties of HIPERLAN/2 continuously with that of 802.11.

6.6.3.1 MAC Layer

As mentioned earlier, the main differences between the IEEE 802.11 and HIPERLAN/2 standards occur at the MAC layer. In HIPERLAN/2, medium access is based on a TDMA/TDD approach using a MAC frame with a period of 2 ms.[30] This frame comprises uplink (to the AP), downlink (from the AP), and direct-link (DiL, directly between two stations) phases. These phases are scheduled centrally by the AP, which informs STAs, at which point in time in the MAC frame they are allowed to transmit their data. Time slots are allocated dynamically depending on the need for transmission resources. The HIPERLAN/2 MAC is designed to provide the QoS support essential to many multimedia and real-time applications.

On the other hand, IEEE 802.11a uses the distributed CSMA/CA MAC protocol that obviates the requirement for any centralized control. The use of a distributed MAC makes IEEE 802.11a more suitable for ad hoc networking and non-real-time applications. Another significant difference between the two standards is the length of the packets employed. HIPERLAN/2 employs fixed length packets, and 802.11a supports variable length packets.

The HIPERLAN/2 MAC frame structure (Figure 6.10a) comprises time slots for broadcast control (BCH); frame control (FCH); access feedback control (ACH); and data transmission in downlink (DL), uplink (UL), and direct-link (DiL) phases, which are allocated dynamically depending on the need for transmission resources. An STA first must request capacity from the AP in order to send data. This is performed in the random access channel (RCH), where contention for the same time slot is allowed.

DL, UL, and DiL phases consist of two types of PDUs: long and short. The long PDUs (illustrated in Figure 6.10b)

Table 6.2 IrDA Data Protocol Stack

IrTran-P	IrOBEX	IrLAN	IrCOMM	IrMC
LM-IAS	Tiny transport protocol (tiny TP)			
Ir link mgmt (IrLMP)				
Ir link access protocol (IrLAP)				
Async serial — IR 9600–115.2 kbps	Sync serial — IR 1.152 Mbps		Sync 4PPM 4 Mbps	

have a size of 54 bytes and contain control or user data. The payload comprises 48 bytes and the remaining bytes are used for the PDU type, a sequence number (SN), and CRC-24. Long PDUs are referred to as the long transport channel (LCH). Short PDUs contain only control data and have a size of 9 bytes. They may contain resource requests, ARQ messages, etc., and are referred to as the short transport channel (SCH).

Traffic from multiple connections to or from one STA can be multiplexed onto one PDU train, which contains long and short PDUs. A physical burst is composed of the PDU train payload preceded by a preamble and is the unit to be transmitted via the PHY layer.[24]

6.6.3.2 Physical Layer

The PHY layers of 802.11a and HIPERLAN/2 are very similar and are based on the use of OFDM. As already discussed, OFDM is used to combat frequency selective fading and to randomize the burst errors caused by a wideband fading channel. The PHY layer modes (similar to Table 6.2) with different coding and modulation schemes are selected by a link adaptation scheme.[29,31] The exact mechanism of this process is not specified in the standards.

Data for transmission are supplied to the PHY layer in the form of an input PDU train or PPDU frame, as explained earlier. This is then input to a scrambler that prevents long runs of ones and zeros in the input data sent to the remainder of the modulation process. Although 802.11a and HIPERLAN/2 scramble the data with a length 127 pseudorandom sequence, the initialization of the scrambler is different.

The scrambled data are input to a convolutional encoder consisting of a 1/2 rate mother code and subsequent puncturing. The puncturing schemes facilitate the use of code rates 1/2, 3/4, 9/16 (HIPERLAN/2 only), and 2/3 (802.11a only). In the case of 16-QAM, HIPERLAN/2 uses rate 9/16 instead of rate 1/2 in order to ensure an integer number of OFDM symbols per PDU train. The rate 2/3 is used only for the case of 64-QAM in 802.11a. Note that there is no equivalent mode for HIPERLAN/2, which also uses additional puncturing in order to keep an integer number of OFDM symbols with 54-byte PDUs.

The coded data are interleaved in order to prevent error bursts from being input to the convolutional decoding process in the receiver. The interleaved data are subsequently mapped to data symbols according to a BPSK, QPSK, 16-QAM, or 64-QAM constellation. OFDM modulation is implemented by means of an inverse fast Fourier transform (FFT); 48 data symbols and four pilots are transmitted in parallel in the form of one OFDM symbol.

In order to prevent ISI and intercarrier interference due to delay spread, a guard interval is implemented by means of a cyclic extension. Thus, each OFDM symbol is preceded by a periodic extension of that symbol. The total OFDM symbol duration is $T_{total} = T_g + T$, where T_g represents the guard interval and T the useful OFDM symbol duration. When the guard interval is longer than the excess delay of the radio channel, ISI is eliminated.

The OFDM receiver basically performs the reverse operations of the transmitter. However, the receiver is also required to undertake automatic gain control, time and frequency synchronization, and channel estimation. Training sequences are provided in the preamble for the specific purpose of supporting these functions. Two OFDM symbols are provided in the preamble in order to support the channel estimation process. Prior knowledge of the transmitted preamble signal facilitates the generation of a vector defining the channel estimate, commonly referred to as the channel state information.

The channel estimation preamble is formed so that the two symbols effectively provide a single guard interval of length 1.6 ms. This format makes it particularly robust to ISI. By averaging over two OFDM symbols, the distorting effects of noise on the channel estimation process can also be reduced. HIPER-LAN/2 and 802.11a use different training sequences in the preamble; the training symbols used for channel estimation are the same, but the sequences provided for time and frequency synchronization are different. Decoding of the convolutional code is typically implemented by means of a Viterbi decoder.

6.7 WIRELESS PERSONAL AREA NETWORKING

This section first explores the consumer applications requirements of wireless home networking. It then presents a detailed description of the technologies, companies, and industry groups seeking to tap into this vast consumer market opportunity, with emphasis on the WPAN systems.

6.7.1 Bluetooth and WPANs

The past quarter century has seen the rollout of three generations of wireless cellular systems attracting end-users by providing efficient mobile communications. On another front, wireless technology became an important component in providing networking infrastructure for localized data delivery. This later revolution was made possible by the induction of new networking technologies and paradigms, such as WLANs and WPANs.

WPANs are short- to very short-range (from a couple of centimeters to a couple of meters) wireless networks that can be used to exchange information between devices in the reach of a person. WPANs can be used to replace cables between computers and their peripherals; to establish communities helping people do their everyday chores making them more productive; or to establish location-aware services. WLANs, on the other hand, provide a larger transmission range.

Although WLAN equipment usually carries the capability to be set up for ad hoc networking, the premier choice of deployment is a cellular-like infrastructure mode to interface wireless users with the Internet. The best example representing WPANs is the recent industry standard, Bluetooth[32,42]; other examples include Spike,[33] IrDA,[1] and in the broad sense HomeRF.[34] As has been seen, for WLANs the most well-known representatives are based on the standards IEEE 802.11 and HIPERLAN with all their variations.

The IEEE 802 committee has also realized the importance of short-range wireless networking and initiated the establishment of the IEEE 802.15 WG for WPANs[36] to standardize protocols and interfaces for wireless personal area networking. The 802.15 WG is formed by four task groups (TGs):

- IEEE 802.15 WPAN/Bluetooth TG 1. The TG 1 was established to support applications that require medium-rate WPANs (such as Bluetooth). These WPANs will handle a variety of tasks ranging from cell phones to PDA communications and have a QoS suitable for voice applications.
- IEEE 802.15 Coexistence TG 2. Several wireless standards (such as Bluetooth and IEEE 802.11b) and appliances (such as microwaves) operate in the unlicensed 2.4-GHz ISM frequency band. The TG 2 is developing specifications on the ISM band due to the unlicensed nature and available bandwidth. Thus, the IEEE 802.15 Coexistence TG 2 (802.15.2) for wireless personal area networks is developing recommended practices to facilitate coexistence of WPANs (e.g., 802.15) and WLANs (e.g., 802.11).
- IEEE 802.15 WPAN/High Rate TG 3. The TG 3 for WPANs is chartered to draft and publish a new standard for high-rate (20Mb/s or greater) WPANs. In addition to a high data rate, the new standard will provide for low-power, low-cost solutions addressing the needs of portable consumer digital imaging and multimedia applications. This developing standard is discussed later in this chapter.

- IEEE 802.15 WPAN/Low Rate TG 4. The goal of the TG 4 is to provide a standard that has ultralow complexity, cost, and power for a low-data-rate (200 kb/s or less) wireless connectivity among inexpensive fixed, portable, and moving devices. Location awareness is being considered as a unique capability of the standard. The scope of the TG 4 is to define PHY and MAC layer specifications. Potential applications are sensors, interactive toys, smart badges, remote controls, and home automation. Further comments on 802.15.4 are provided later in this chapter.

One key issue in the feasibility of WPANs is the interworking of wireless technologies to create heterogeneous wireless networks. For instance, WPANs and WLANs will enable an extension of the 3G cellular networks (i.e., UMTS and cdma2000) into devices without direct cellular access. Moreover, devices interconnected in a WPAN may be able to utilize a combination of 3G access and WLAN access by selecting the best access for a given time. In such networks 3G, WLAN, and WPAN technologies do not compete against each other, but rather enable the user to select the best connectivity for his or her purpose. Figure 6.11[42] clearly shows the operating space of the various 802 wireless standards and activities still in progress.

Given the importance within the WPAN operating space, intensive research activities, and availability of devices, a little time will now be devoted to giving a brief introduction on Bluetooth and then providing an overview of the Bluetooth standard as defined by the Bluetooth SIG (special interest group).

6.7.1.1 Brief History and Applications of Bluetooth

In the context of wireless personal area networks, the Bluetooth technology came to light in May 1998, and since then the Bluetooth SIG has steered the development of an open industry specification, called Bluetooth, including protocols as well as applications scenarios. A Micrologic Research study

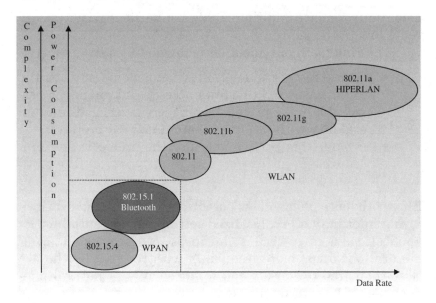

Figure 6.11 The scope of the various WLAN and WPAN standards.

in July 2001 forecasts that, in 2005, 1.2 billion Bluetooth chips will be shipped worldwide.

The Bluetooth SIG is an industry group consisting of leaders in the telecommunications and computing industries that are driving the development of Bluetooth WPAN technology. Bluetooth wireless technology has become a de facto standard, as well as a specification for small-form factor, low-cost, short-range radio links among mobile PCs, mobile phones, and other portable devices.[37] The Bluetooth SIG includes promoter companies such as 3Com, Ericsson, IBM, Intel, Microsoft, Motorola, Nokia, and Toshiba and many more adopter companies. The goal of Bluetooth is to enable users to connect a wide range of computing and telecommunications devices easily, without a need to buy, carry, or connect cables. It enables rapid ad hoc connections and the possibility of automatic, unconscious connections between devices. Because Bluetooth can be used for a variety of purposes, it will also potentially replace multiple cable connections via a single radio link.[37]

The Bluetooth specification is divided into two parts:

- *Core.* This portion specifies components such as the radio, baseband (medium access), link manager, service discovery protocol, transport layer, and interoperability with different communication protocols.
- *Profile.* This portion specifies the protocols and procedures required for different types of Bluetooth applications.

Bluetooth has a tremendous potential in moving and synchronizing information in a localized setting. The potential for its applications is huge because business transactions and communications occur more frequently with the people who are close compared to those who are far away — a natural phenomenon of human interaction.

6.7.1.2 Bluetooth Details

Bluetooth[32] operates in the ISM frequency band starting at 2.402 GHz and ending at 2.483 GHz in the U.S. and most European countries. A total of 79 RF channels of 1 MHz width are defined, where the raw data rate is 1 Mb/s. A TDD technique divides the channel into 625-μs slots; with a 1-Mb/s symbol rate, a slot can carry up to 625 b. Transmission in Bluetooth occurs in packets that occupy one, three, or five slots. Each packet is transmitted on a different hop frequency with a maximum frequency-hopping rate of 1600 hops/s. Therefore, an FHSS technique is employed for communication.

The goals of the specifications are to eliminate the need for wires and to simplify ad hoc networking among devices. Bluetooth utilizes small inexpensive radios for the physical layer and for the master–slave relationship at the MAC layer between devices. The master periodically polls the slave devices for information and only after receiving such a poll is a slave allowed to transmit. Therefore, the master is responsible for controlling access to the network, providing services to slave nodes, and allowing them to conserve power.

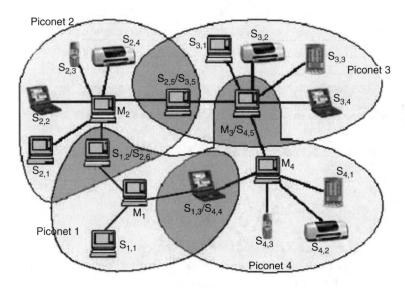

Figure 6.12   Four piconets forming a scatternet.

A master device can directly control seven active slave devices in what is defined as a *piconet*. Piconets are small WPANs formed when Bluetooth-enabled devices are in proximity to each other and share the same hopping sequence and phase. A total of 79 carriers with 1-MHz spacing and a slot of size 625 µs allow Bluetooth to support piconets with up to a 1-Mbps data rate. Transmitting power levels near 100 mW allows devices to be up to 10 m apart; if special transceivers are used, networking up to a 100-m range is also possible.

A node can enter and leave a piconet at any time without disrupting the piconet. More than eight nodes can be allowed to form a network by making a node act as a bridge between two piconets and create a larger network called a *scatternet*. Figure 6.12[42] illustrates a scatternet composed of four piconets, in which each piconet has several slaves (indicated by the letter $S_{i,j}$) and one master (indicated by the letter M_i). Figure 6.13 depicts the Bluetooth protocol stack, which also shows the application "layer" at the top where the profiles would reside. The protocols that belong to the core specifications are:

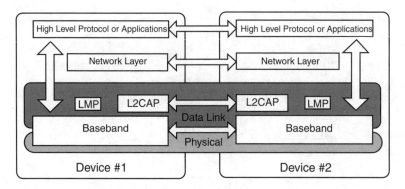

Figure 6.13 Bluetooth protocol architecture.

- The radio. The radio layer, which resides below the baseband layer, defines the technical characteristics of the Bluetooth radios, which come in three power classes. Class 1 radios have transmit power of 20 dBm (100 mW); class 2 radios have transmit power of 4 dBm (2.5 mW); and class 3 radios have transmit power of only 0 dBm (1 mW).
- The baseband. The baseband defines the key medium access procedures that enable devices to communicate with each other using the Bluetooth wireless technology. The baseband defines the Bluetooth piconets and their creation procedure as well as the low-level packet types.
- The link manager protocol (LMP). The LMP is a transactional protocol between two link management entities in communicating Bluetooth devices.
- The logical link control and adaptation protocol (L2CAP). The L2CAP layer shields specifics of the Bluetooth lower layers and provides a packet interface to higher layers. At L2CAP, the concept of master and slave devices does not exist.

The Bluetooth specification defines two distinct types of links for the support of voice and data applications: SCO (*synchronous connection oriented*) and ACL (*asynchronous connectionless*). The first link type supports point-to-point voice switched circuits and the latter supports symmetric as well as asymmetric data transmission. ACL packets are intended to support data applications and do not have prescribed time slot allocations as opposed to SCO packets, which support periodic audio transmission at 64 Kb/s in each direction.

Bluetooth addresses many of the issues, such as power, cost, size, and simplicity, that make it very appealing for use in WPANs. IEEE is looking very carefully at the Bluetooth and how it addresses the functional requirements of 802.15. As a matter of fact, large portions of Bluetooth have been adopted for recommendation to the 802 Group as the 802.15.1 standard.[36]

6.7.2   Infrared Data Association's (IrDA) Serial Infrared (SIR) Data and Advanced Infrared (AIR) Specifications

The IrDA[1] was founded as a nonprofit organization in 1993 and is an industry-focused group of device manufacturers that have developed a standard for transmitting data via infrared light waves. Their goal has been to create a low-cost, low-power, half-duplex serial data interconnection standard that supports a walk-up, point-to-point user model. The latest IrDA SIR core specification is divided into three parts: IrDA data, IrDA control, and IrDA PC99. Each one is used differently, depending on the type of device to be connected. IrDA PC99 is intended for low-speed devices such as keyboards, joysticks, and mouse. IrDA control is recommended in-room cordless peripherals to host PCs such as printers or scanners. IrDA data is recommended for high-speed, short-range, line-of-sight, point-to-point cordless data transfer such as local area networking or file sharing.

The IrDA data architecture has some serious limitations. First, although the architecture can accommodate a point-to-multipoint mode of operation, the IrDA data specification has never been extended to define the protocols to enable this multipoint functionality. Second, within a given field of view, the establishment of an IrDA data connection between a single pair of devices inhibits the establishment of connections

Table 6.3 IrDA Data Optional Protocols

Optional IrDA protocol	Function
Tiny TP	TCP/IP like flow control
IrOBEX	Object exchange services
IrCOMM	Emulates serial and parallel ports to support legacy applications
IrTran-P	Image exchange services for digital photo devices
IrLAN	Provides infrared access to local area networks (LANs)
IrMC	Protocol to provide voice communication and messaging services for mobile devices

between other independent devices whose fields of view intersect with that of an established connection. Thus, the use of the medium becomes dedicated to a single pair of devices. Having realized, among other things, that IrDA is not well suited to WPAN environments, the members of the IrDA community have extended the IrDA data architecture to enable true multipoint connectivity while at the same time limiting the investment in upper layer applications and services. This has resulted in the IrDA AIR specification, which does offer some improvement on mobility freedom to IrDA devices.

6.7.2.1   IrDA Data and IrDA AIR Details

IrDA data is one of three protocols in the Infrared Data Association's (IrDA) serial infrared (SIR) data specification. IrDA data is recommended for high-speed, short-range, line-of-sight, point-to-point cordless data transfer. This protocol has required protocols and optional protocols to meet the connectivity needs of WPAN devices if both are implemented. The three required protocols for IrDA data are PHY (physical signaling layer); IrLAP (link access protocol); and IrLMP (link management protocol and information access service). Table 6.3 illustrates the IrDA data protocol stack[4] with the mandatory and optional protocols.

IrDA data utilizes optical signaling in the 850-nm range for the PHY layer and uses a polling channel access scheme. Data rates for IrDA data range from 2400 bps to 4 Mbps. The typical distance for IrDA data devices is about 1 m and up to 2 m can be reached. IrDA data has built-in tolerances so that devices must be perfectly aligned to communicate. IrDA data is able to support data by utilizing point-to-point protocol (PPP)[38] and voice by encapsulating voice into IP packets. IrDA IrLAP is based on high-level data link control (HDLC)[38] and provides a device-to-device connection and ordered transfer of data.

IrLAP also has procedures that allow devices to discover each other when first turned on. IrLMP supports a method to multiplex multiple connections onto the IrLAP layer. For example, this feature allows multiple file transfers to exist between two or more IrDA data ports. The IrLMP also has a service discovery protocol called information access services (IASs) that enables devices to understand the capabilities of nearby devices. This feature is very useful in an ad hoc networking environment in which devices need services but do not know exactly which device is a printer or an Internet gateway. IrDA data's optional protocols are summarized in Table 6.2.

IrDA SIR is a very mature specification that has been included in over 100 million devices. The protocol has all the features needed for WPAN today and is very cost effective. The only limitations preventing IrDA from taking off are the short-range and blockage issues of infrared optical communication that has been implemented by IrDA. Despite these shortcomings, it is still employed in applications in which no interference occurs from an adjacent devices. Also, high-rate devices of up to 16 Mbps are currently being developed.

To cope up with the limitations of IrDA data such as line of sight requirement, the IrDA association defined the IrDA AIR protocol architecture. IrDA AIR adds a protocol entity called IrLC (link controller) that provides multipoint link layer connectivity alongside an IrLAP protocol entity, which provides legacy connectivity to IrDA devices. To control access to the shared medium, IrDA AIR employs a MAC protocol called IrMAC that uses a burst reservation CSMA/CA MAC protocol.

IrDA AIR has an interesting feature of actively monitoring the symbol error rates at an AIR decode by which it is possible to estimate the SNR (signal-to-noise ratio) for the channel between the source and sink of a packet. This SNR estimate can be fed back to the sending station in order to maintain a good-quality channel. AIR's base rate is 4 Mbps, reducing through successive halving of data rates to 256 kbps to yield a doubling range. AIR prototypes have been demonstrated at 4 Mbps up to 5 m, and at 256 kbps within 10 m.

Therefore, IrDA AIR offers to improve the freedom of movement to IrDA devices and freedom from 1-m restriction, 15 to 30° half-angle coverage profile of IrDA DATA, and operation over a greater range and a wider angle.

6.7.3 HomeRF Working Group's (HRFWG) Shared Wireless Access Protocol — Cordless Access (SWAP-CA)

The HomeRF Working Group (HRFWG)[34] is a consortium of over 100 companies from the computer, electronics, and consumer electronics industries formed in early 1997 to establish an open industry specification for wireless digital communication between PCs and a broad range of interoperable consumer electronic devices anywhere around the house.[39] The HRFWG has developed a specification called the shared wireless access protocol — cordless access (SWAP–CA)[40] that uses RF in the ISM band to support managed and ad hoc networks. The specification has combined the data networking elements of 802.11 and voice elements of the digital European cordless telephone (DECT) standard to allow mobile devices to communicate via voice and data. SWAP-CA has built-in mechanisms for small mobile devices in WPANs; it is focused on allowing users to maximize connectivity to the Internet and the PSTN in as many home devices as possible.[22,39]

6.7.3.1 SWAP-CA Details

The HomeRF Working Group sees SWAP-CA as one of the methods that could provide connectivity to devices at home. This method supports isochronous clients that are slaves to PCs and an asynchronous network of peer devices that is effectively a wireless Ethernet LAN. The HomeRF Working

Group's connectivity vision is to allow computers, cordless phones, and other electronic devices to share resources such as the Internet and the PSTN.

A SWAP-CA system is designed to carry voice and data traffic and to interoperate with PSTN and data networks such as the Internet. Like Bluetooth, the SWAP protocol operates in the 2.4-GHz ISM band. SWAP-CA utilizes frequency hopping spread spectrum for its relaxed implementation of IEEE 802.11 for data networking services. HRFWG has eliminated the complexities of 802.11 to make SWAP-CA cheaper to implement and manufacture. Voice support in SWAP-CA is modeled after the DECT standard, which has been adapted to the 2.4-GHz ISM band. SWAP-CA utilizes TDMA to provide isochronous voice services and other time-critical services, and CSMA/CA to provide asynchronous services for packet data.

SWAP-CA can support as many as 127 devices per network at a distance of up to 50 m. Four types of devices can operate in a SWAP-CA network[1,40]:

- A connection point (CP) similar to an 802.11 AP, which functions as a gateway among a SWAP-CA-compatible device, the PSTN, and a personal computer that could be connected to the Internet
- Isochronous nodes (I-nodes), which are voice focused devices such as cordless phones
- Asynchronous nodes (A-Nodes), which are data-focused devices such as personal digital assistants (PDAs) and smart digital pads
- Combined asynchronous–isochronous nodes (AI-nodes)

A SWAP-CA network can operate as a managed network or as a peer-to-peer ad hoc network. In the managed network implementation, the CP controls the network and is the gateway to other devices like the Internet and the PSTN. The CP provides simultaneous support for voice and data communications by controlling access to the network. In an ad hoc scenario, the SWAP-CA network can only support data and a CP is not required. Figure 6.14 illustrates the two scenarios.

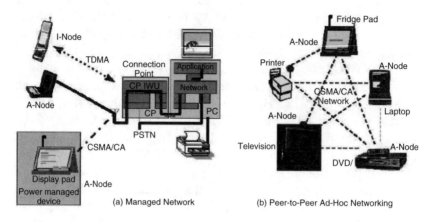

(a) Managed Network   (b) Peer-to-Peer Ad-Hoc Networking

Figure 6.14 SWAP-CA managed and ad hoc networks.

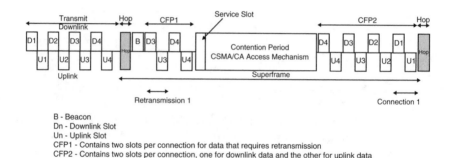

B - Beacon
Dn - Downlink Slot
Un - Uplink Slot
CFP1 - Contains two slots per connection for data that requires retransmission
CFP2 - Contains two slots per connection, one for downlink data and the other for uplink data

Figure 6.15 SWAP-CA hybrid TDMA/CSMA frame.

When a managed network topology is implemented, SWAP-CA utilizes the frame structure of Figure 6.15. The existence of contention-free periods (CFP1 and CFP2) using TDMA and a contention period using CSMA/CA allows the managed SWAP-CA network to support voice and data. The start of the frame is marked by a connection point beacon (CPB) that can be used for network synchronization, controlling the framing format, polling devices to allow access to the TDMA slots, and power management.

The HomeRF Working Group is looking at variations of the SWAP-CA protocol. One is a multimedia protocol SWAP-MM that would focus on video and audio requirements of home networking such as home theater systems. The second is SWAP-lite, which focuses on the simple wireless connectivity needs for devices such as keyboards, mouse, remote controls, and joysticks. SWAP-lite is seen as a direct competitor with IrDA, a very established technology, and therefore might not receive as much attention as SWAP-MM.

6.7.4 IEEE 802.15

As mentioned earlier, the goal for the 802.15 WG is to provide a framework for the development of short-range (less than 10 m), low-power, low-cost devices that wirelessly connect the user within his or her communication and computational environment. WPANs are intended to be small networks in the home or office with no more than 8 to 16 nodes. Because the 802.15.1 standard is a derivate of Bluetooth which has already been explained, and 802.15.2 is a recommended practice rather than a standard, the discussion here will be confined to the 802.15.3 and 802.15.4 developing standards.

6.7.4.1 802.15.3

The 802.15.3 Group[36] has been tasked to develop an ad hoc MAC layer suitable for multimedia WPAN applications and a PHY capable of data rates in excess of 20 Mbps. The current draft of the 802.15.3 standard (dubbed Wi-Media) specifies data rates up to 55 Mbps in the 2.4-GHz unlicensed band. The technology employs an ad hoc PAN topology not entirely dissimilar to Bluetooth, with roles for "master" and "slave" devices. The draft standard calls for drop-off data rates from 55 to 44, 33, 22, and 11 Mbps. Note that 802.15.3 is not compatible with Bluetooth or the 802.11 family of protocols although it reuses elements associated with both.

6.7.4.1.1 802.15.3 MAC and PHY Layer Details

The 802.15.3 MAC layer specification is designed from the ground up to support ad hoc networking, multimedia QoS provisions, and power management. In an ad hoc network, devices can assume master or slave functionality based on

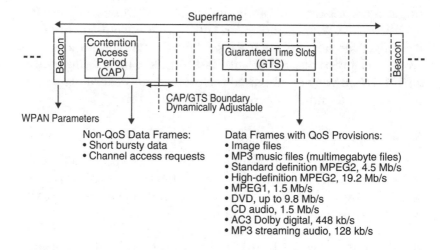

Figure 6.16   IEEE 802.15.3 MAC superframe.

existing network conditions. Devices in an ad hoc network can join or leave an existing network without complicated set-up procedures. The 802.15.3 MAC specification provides provisions for supporting multimedia QoS. Figure 6.16 illustrates the MAC superframe structure, which consists of a network beacon interval and a contention access period (CAP), reserves for guaranteed time slots (GTSs). The boundary between the CAP and GTS periods is dynamically adjustable.

A network beacon is transmitted at the beginning of each superframe, carrying WPAN-specific parameters, including power management, and information for new devices to join the ad hoc network. The CAP period is reserved for transmitting non-QoS data frames such as short bursty data or channel access requests made by the devices in the network. The medium access mechanism during the CAP period is CSMA/CA. The remaining duration of the superframe is reserved for GTS to carry data frames with specific QoS provisions. The type of data transmitted in the GTS can range from bulky image or music files to high-quality audio or high-definition video streams. Finally, power management is one of the key features of the 802.15.3 MAC protocol, which is designed to lower the current drain significantly while connected to a

WPAN. In the power-saving mode, the QoS provisions are also maintained.

The 802.15.3 PHY layer operates in the unlicensed frequency band between 2.4 and 2.4835 GHz and is designed to achieve data rates of 11 to 55 Mb/s, which are commensurate with the distribution of high-definition video and high-fidelity audio. The 802.15.3 systems employ the same symbol rate, 11 Mbaud, as used in the 802.11b systems. Operating at this symbol rate, five distinct modulation formats are specified, namely, uncoded QPSK modulation at 22 Mb/s and trellis coded QPSK, 16/32/64-QAM at 11, 33, 44, 55 Mb/s, respectively (TCM).[41] The base modulation format is QPSK (differentially encoded). Depending on the capabilities of devices at both ends, the higher data rates of 33 to 55 Mb/s are achieved by using 16, 32, 64-QAM schemes with eight-state 2D trellis coding. Finally, the specification includes a more robust 11 Mb/s QPSK TCM transmission as a drop-back mode to alleviate the well-known hidden node problem. The 802.15.3 signals occupy a bandwidth of 15 MHz, which allows for up to four fixed channels in the unlicensed 2.4-GHz band. The transmit power level complies with the FCC rules with a target value of 0 dBm.

The RF and baseband processors used in the 802.15.3 PHY layer implementations are optimized for short-range transmission limited to 10 m, enabling low-cost and small-form-factor MAC and PHY implementations for integration in consumer devices. The total system solution is expected to fit easily in a compact flash card. The PHY layer also requires low current drain (less than 80 mA) while actively transmitting or receiving data at minimal current drain in the power-saving mode.

From an ad hoc networking point of view, it is important that devices have the ability to connect to an existing network with a short connection time. The 802.15.3 MAC protocol targets connection times much less than 1 s. Reviewing the regulatory requirements, it should be noted that operation of WPAN devices in the 2.4-GHz band is highly advantageous because these devices cannot be used outdoors in Japan while operating in the 5-GHz band. The outdoor use of most portable

WPAN devices prohibits the use of 5-GHz band for worldwide WPAN applications.

6.7.4.1.2   802.15.3 and Bluetooth

On the face of it, 802.15.3 could be seen as a source of competition to Bluetooth because it is also a WPAN technology using an ad hoc architecture. In reality, this is not the case. Admittedly, the concept of 802.15.3 is to allow for a chipset solution that would eventually be approximately 50% more expensive than a Bluetooth solution. Furthermore, the power consumption and size would be about 50% greater than a Bluetooth solution. However, on the flip side, 802.15.3 would allow for data rates considerably in excess of current sub-1 Mbps Bluetooth solutions. This is the critical differentiating element. In effect, 802.15.3 is being positioned as a complementary WPAN solution to Bluetooth. This is particularly the case because the Bluetooth SIG is going slowly on its efforts to develop the next-generation Bluetooth Radio 2, which would allow for data rates between 2 and 10 Mbps.

6.7.4.1.3   802.15.3 and WLANs

Some see more potential for 802.15.3 to be seen as overlapping with 802.11-based protocols than with Bluetooth. With 802.11-based wireless LANs pushing 54 Mbps and the work being done by the 802.11e TG, it is clear that wireless LANs are also looking to become a serious contender for multimedia applications. Even though 802.15.3 is being designed from scratch and would theoretically offer superior bandwidth for multimedia applications at favorable cost and power consumption metrics, it will have a challenge distinguishing itself from full-fledged 802.11-based wireless LANs. Even so, one source of differentiation is that 802.15.3 is meant to be optimized for PAN distances (up to 10 m), but WLAN range is clearly larger.

6.7.4.2   802.15.4

IEEE 802.15.4[36,51] defines a specification for low-rate, low-power wireless personal area networks (LR-WPANs). It is extremely well suited to home networking applications in which the key motivations are reduced installation cost and low power consumption. The home network has varying requirements. Some applications require high data rates like shared Internet access, distributed home entertainment, and networked gaming. However, an even bigger market exists for home automation, security, and energy conservation applications, which typically do not require the high bandwidths associated with the former category of applications. Instead, the focus of this standard is to provide a simple solution for networking wireless, low data rate, inexpensive, fixed, portable, and moving devices. Application areas include industrial control, agricultural, vehicular and medical sensors, and actuators that have relaxed data rate requirements.

Inside the home, such technology can be applied effectively in several areas: PC-peripherals including keyboards, wireless mice, low-end PDAs, joysticks; consumer electronics including radios, TVs, DVD players, and remote controls; home automation including heating, ventilation, air conditioning, security, lighting, control of windows, curtains, doors, and locks; and health monitors and diagnostics. These will typically need less than 10 kbps, while the PC-peripherals require a maximum of 115.2 kbps. Maximum acceptable latencies will vary from 10 ms for the PC peripherals to 100 ms for home automation.

Although Bluetooth was originally developed as a cable replacement technology, it has evolved to handle more typical and complex networking scenarios. It has some power-saving modes of operation; however, it is not seen as an effective solution for power-constrained home automation and industrial control applications. On the same note, 802.11 is overkill for applications like temperature or security sensors mounted on a window. Both technologies would require frequent battery changes, which are not suitable for certain industrial applications, like metering systems, that require a battery change once in 2 to 20 years. The trade-off is a smaller, but adequate, feature set in 802.15.4.

As has been seen, 802.15.1 and 802.15.3 are meant for medium and high data rate WPANs, respectively. The

802.15.4 effort is geared towards applications that do not fall into the preceding two categories, which have low bandwidth requirements and very low power consumption and are extremely inexpensive to build and deploy. These are referred to as LR-PANs. In 2000, two standards groups, the Zigbee Alliance (a HomeRF spinoff) and the IEEE 802 Working Group, came together to specify the interfaces and the working of the LR-PAN.

In this coalition, the IEEE group is largely responsible for defining the MAC and the PHY layers; the Zigbee Alliance, which includes Philips, Honeywell, and Invensys Metering Systems among others, is responsible for defining and maintaining higher layers above the MAC. The alliance is also developing application profiles, certification programs, logos, and a marketing strategy. The specification is based on the initial work done mostly by Philips and Motorola for Zigbee (previously known as PURLnet, FireFly, and HomeRF Lite).

Like all other IEEE 802 standards, the 802.15.4 standard specifies layers up to and including portions of the data link layer. The choice of higher level protocols is left to the application, depending on specific requirements. The important criteria would be energy conservation and the network topology. The draft, as such, supports networks in the star and the peer-to-peer topology. Multiple address types —physical (64 b) and network assigned (8 b) — are allowed. Network layers are also expected to be self-organizing and self-maintaining to minimize cost to the customer.

Currently, the PHY and the data link layer have been more or less clearly defined. The focus now is on the upper layers and this effort is largely led by the Zigbee Alliance. In the following sections the MAC and PHY layer issues of 802.15.4 are described.

6.7.4.2.1 802.15.4 Data Link Layer Details

The data link layer is split into two sublayers: the MAC and the logical link control (LLC). The LLC is standardized in the 802 family and the MAC varies depending on the hardware requirements. Figure 6.17 shows the correspondence of the 802.15.4 to the ISO-OSI reference model.

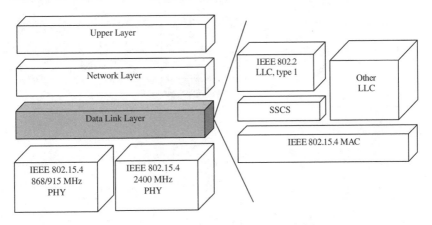

Figure 6.17 802.15.4 in the ISO-OSI layered network model.

The IEEE 802.15.4 MAC provides services to an IEEE 802.2 type I LLC through the service-specific convergence sublayer (SSCS). A proprietary LLC can access the MAC layer directly without going through the SSCS. The SSCS ensures compatibility between different LLC sublayers and allows the MAC to be accessed through a single set of access points. MAC protocol allows association and disassociation; acknowledged frame delivery; channel access mechanism; frame validation; guaranteed time slot management; and beacon management. The MAC sublayer provides the MAC data service through the MAC common part sublayer (MCPS-SAP) and the MAC management services through the MAC layer management entity (MLME-SAP). These provide the interfaces between the SSCS (or another LLC) and the PHY layer. MAC management service has only 26 primitives compared to IEEE 802.15.1, which has 131 primitives and 32 events.

The MAC frame structure has been designed in a flexible manner so that it can adapt to a wide range of applications while maintaining the simplicity of the protocol. The four types of frames are beacon, data, acknowledgment, and command frames. The overview of the frame structure is illustrated in Figure 6.18.

The MAC protocol data unit (MPDU), or the MAC frame, consists of the MAC header (MHR), MAC service data unit

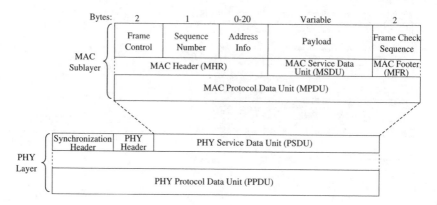

Figure 6.18 The general MAC frame format.

(MSDU), and MAC footer (MFR). The MHR consists of a 2-byte frame control field that specifies the frame type and the address format and controls the acknowledgment; 1-byte sequence number that matches the acknowledgment frame with the previous transmission; and a variable sized address field (0 to 20 bytes). This allows only the source address — possibly in a beacon signal — or source and destination addresses as in normal data frames or no address at all as in an acknowledgment frame. The payload field is variable in length but the maximum possible size of an MPDU is 127 bytes. The beacon and the data frames originate at the higher layers and actually contain some data; the acknowledgment and the command frames originate in the MAC layer and are used simply to control the link at a peer-to-peer level. The MFR completes the MPDU and consists of a frame check sequence field, which is basically a 16-b CRC code.

Under certain conditions, IEEE 802.15.4 provides dedicated bandwidth and low latencies to certain types of applications by operating in a *superframe* mode. One of the devices — usually one less power constrained than the others — acts as the PAN coordinator, transmitting superframe beacons at predetermined intervals that range from 15 to 245 ms. The time between the beacons is divided into 16 equal time slots independent of the superframe duration. The device may transmit at any slot, but must complete its transmission before the end of the superframe. Channel access is usually contention based, although the PAN may assign time slots to a single device. As seen before, this is known as a guaranteed time slot (GTS) and introduces a contention-free period located immediately before the next beacon. In a beacon-enabled superframe network, a slotted CSMA/CA is employed; in nonbeacon networks, the unslotted or standard CSMA/CA is used.

An important function of MAC is to confirm successful reception of frames. Valid data and command frames are acknowledged; otherwise they are (i.e., the frames) simply ignored. The frame control field indicates whether a particular frame must be acknowledged. IEEE 802.15.4 provides three levels of security: no security, access control lists, and symmetric key security using AES-128. To keep the protocol simple and the cost minimum, key distribution is not specified, but may be included in the upper layers.

6.7.4.2.2   802.15.4 PHY Layer Details

IEEE 802.15.4 offers two PHY layer choices based on the DSSS technique that share the same basic packet structure for low duty cycle low-power operation. The difference lies in the frequency band of operation. One specification is for the 2.4-GHz ISM band available worldwide and the other is for the 868/915 MHz for Europe and the U.S., respectively. These offer an alternative to the growing congestion in the ISM band due to large-scale proliferation of devices like microwave ovens, etc. They also differ with respect to the data rates supported. The ISM band PHY layer offers a transmission rate of 250 kbps while the 868/915 MHz offers 20 and 40 kbps. The lower rate can be translated into better sensitivity and larger coverage area, and the higher rate of the 2.4-GHz band can be used to attain lower duty cycle, higher throughput, and lower latencies.

The range of LR-WPAN depends on the sensitivity of the receiver, which is −85 dB for the 2.4-GHz PHY and −92 dB for the 868/915 MHz PHY. Each device should be able to transmit at least 1 mW, but actual transmission power depends on the application. Typical devices (1 mW) are expected to cover a range of 10 to 20 m; however, with good sensitivity and a moderate increase in power, it is possible to cover the home in a star network topology.

The 868/915 MHz PHY supports a single channel between 868.0 and 868.6 MHz and ten channels between 902.0 and 928.0 MHz. Because these are regional in nature, it is unlikely that all 11 channels should be supported on the same network. It uses a simple DSSS in which each bit is represented by a 15-chip maximal length sequence (m-sequence). Encoding is done by multiplying the m-sequence with +1 or −1, and the resulting sequence is modulated by the carrier signal using BPSK.

The 2.4-GHz PHY supports 16 channels between 2.4 and 2.4835 GHz with 5-MHz channel spacing for easy transmit and receive filter requirements. It employs a 16-ary quasi-orthogonal modulation technique based on DSSS. Binary data are grouped into 4-b symbols, each specifying one of 16 nearly orthogonal 32-b chip pseudonoise (PN) sequences for transmission. PN sequences for successive data symbols are concatenated and the aggregate chip is modulated onto the carrier using minimum shift keying (MSK). The use of "nearly orthogonal" symbol sets simplifies the implementation, but incurs a minor performance degradation (<0.5 dB). In terms of energy conservation, orthogonal signaling performs better than differential BPSK. However, in terms of receiver sensitivity, the 868/915 MHz has a 6- to 8-dB advantage.

The two PHY layers, although different, maintain a common interface to the MAC layer, i.e., they share a single packet structure as shown in Figure 6.19. The packet or PHY protocol data unit (PPDU) consists of the synchronization header, a PHY header for the packet length, and the payload, which is also referred to as the PHY service data unit (PSDU). The synchronization header is made up of a 32-b preamble, used for acquisition of symbol and chip timing and possible coarse frequency adjustment, and an 8-b start of packet delimiter signifying the end of the preamble. Out of the 8 b in the PHY

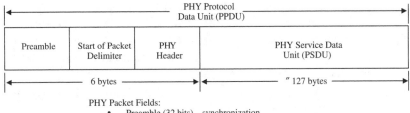

PHY Packet Fields:
- Preamble (32 bits) – synchronization
- Start of packet delimiter (8 bits) – signify end of preamble
- PHY header (8 bits) – specify length of PSDU
- PSDU (d 127 bytes) – PHY layer payload

Figure 6.19 802.15.4 PHY layer packet structure.

header, 7 b are used to specify the length of the PSDU, which can range from 0 to 127 bytes. Channel equalization is not required for either PHY layer because of the small coverage area and the relatively low chip rates. Typical packet sizes for monitoring and control applications are expected to be in the order of 30 to 60 bytes.

Because the IEEE 802.15.4 standard specifies working in the ISM band, it is important to consider the effects of the interference that is bound to occur. The applications envisioned by this protocol have little or no QoS requirements. Consequently, data that do not go through on the first attempt will be retransmitted and higher latencies are tolerable. Too many transmissions also increase the duty cycle and therefore affect the consumption of power. Once again, the application areas are such that transmissions will be infrequent, with the devices in a passive mode of operation for most of the time.

6.7.5 Comparison of WPAN Systems

To understand suitability of the aforementioned systems for WPAN applications, several criteria have been identified for a comparison between different systems while keeping the primary goal of forming ad hoc networks as the use of simple, low-power, small, and cost-effective devices. After a careful analysis of the WPAN requirements, the following criteria have been selected for comparison:

- *Range*: — the communication range of the device
- *Data rate*: — the maximum data rate possible in the network
- *Support for voice*: — support a protocol or method to allow voice communication
- *Power management*: — a method for devices to conserve power
- *LAN integration*: — a method to integrate the WPAN device with a standard LAN such as 802.3 or 802.11

Table 6.4 presents a tabular comparison of the various WPAN technologies discussed so far; however, the following descriptive comparison study is confined to Bluetooth (baseline for 802.15.1), IrDA, and SWAP-CA.

6.7.5.1   Range

WPAN attempts to address networking issues in small areas if 10 m satisfies the needs of users to form an ad hoc network for meetings in small rooms, study sessions in libraries, or home networking for computers or consumer devices. This distance allows devices to have some flexibility in terms of how close they are. Bluetooth can support up to 10 m and, when special transceivers are utilized, a distance of 100-m range can be achieved. SWAP-CA supports distances up to 50 m, which exceeds the requirements of WPAN systems. The IrDA data specification can reach up to 2 m, but is typically used for less than a meter, and IrDA AIR achieves up to 10-m range.

The RF communication used in Bluetooth and SWAP-CA provides a better range and has the ability to pass through the minor obstructions most likely to be present if devices are 10 m apart. IrDA data uses optical signaling and inexpensive transceivers limit the protocol's ability to meet the 10-m diameter desired for WPANs. On the other hand, IrDA AIR aims at achieving the 10-m range, but at the expense of a data rate of only 256 kbps. Another limitation that might have an impact on the range of IrDA data is the use of infrared light that makes IrDA data susceptible to blockage from objects or

Table 6.4 Comparison of WPAN Systems

Technology	IrDa	HomeRF (SWAP-CA)	Bluetooth (802.15.1)	802.15.× 3	802.15.× 4
Operational spectrum	Infrared; 850 nm	2.4 GHz ISM band	2.4 GHz ISM band	2.402–2.480 GHz ISM band	2.4 GHz and 868/915 MHz
Physical layer details	Optical rays	FHSS with FSK	FHSS; 1600 hops per second	Uncoded QPSK trellis-coded QPSK or 16/32/64-QAM scheme	DSSS with BPSK or MSK (O-QPSK)
Channel access	Polling	CSMA-CA and TDMA	Master–slave polling, time division duplex (TDD)	CSMA-CA, and guaranteed time slots (GTSs) in a superframe structure	CSMA-CA, and guaranteed time slots (GTS) in a superframe structure
Maximum data rate	4 Mbps	10 Mbps	Up to 1 Mbps	11–55 Mbps	868 MHz–20 kbps; 915 MHz–40 kbps; 2.4 GHz–250 kbps
Coverage	<10 m	>50 m	<10 m	<10 m	<20 m
Power level issues	Distance based	<300 mA peak current	1–60 mA	<80 mA	Very low current drain (20–50 µA)
Interference	Present	Present	Present	Present	Present
Price	Low (<$10)	Medium	Low (<$10)	Medium	Very low

if the angles between the two communicating devices are off by more than the 15° viewing angle.

6.7.5.2 Data Rate

Data rate is also important to a WPAN system. The typical use for a WPAN system might require a very low data rate to transmit text between two devices, but if the concept of a WPAN is extended for home theater systems or for sharing Internet access, the data rate can become an issue. All techniques support adequate data rates for many WPAN scenarios, ranging from Internet sharing to large file transfers. In terms of absolute numbers, it is difficult to select one data rate that would be an adequate for a WPAN. The concept of a WPAN is relatively new and applications for the technology have not matured enough to push the limits of the data rates available. The developers of SWAP-CA and IrDA have hinted support for higher data rates, and it can be assumed that once all 802.15 standards are approved, higher data rate versions of the standards could be developed.

Bluetooth allows for up to eight devices to operate in a single piconet and transmit data in symmetric (up to 432.6 kbps in both directions) or asymmetric (up to 721 kbps and 57.6 kbps in opposite directions) mode. The piconet devices must share the bandwidth and cannot transmit at that rate simultaneously. SWAP-CA supports 1.6 Mbps using a CSMA/CA scheme, which provides a peak effective throughput of 1 Mbps under lightly loaded conditions. This data rate is acceptable for WPAN systems, but seems a little low for the protocol that allows connectivity between home PCs. The HomeRF Working Group argues that 1-Mbps effective throughput is available even when four voice calls are active simultaneously. The IrDA provides the fastest data rate at 4 Mbps, which is more than enough for large file transfers, print jobs, or Internet sharing. As for IrDA AIR, the rate of 4 Mbps is only achieved for up to a 5-m range. By increasing the range, the data rate is progressively halved up to 256 kbps for a distance of 10 m. This is less than the throughput feasible for the same distance in Bluetooth.

6.7.5.3   Support for Voice

A WPAN technology is most likely to be embedded into existing devices such as mobile phones and pagers; voice communication as well as integration with the PSTN is highly desirable. A possible scenario could be to use two mobile phones as short wave radios using WPAN. Another scenario could be to allow a friend or colleague to utilize one's phone as a PSTN gateway to make calls because one has a surplus of minutes to use before the end of the month.

SWAP-CA's key feature is to support voice services and to allow home users to merge voice and data services into one protocol. SWAP-CA provides toll-quality voice using 32 kbps adaptive differential pulse code modulation (ADPCM) for voice encoding. SWAP-CA has utilized the DECT standard for voice support and allows four simultaneous calls to the PSTN. The only limitation in voice support is that a connection point must be used; thus, it is difficult to set up ad hoc networks.

Bluetooth's voice support is provided by the telephony control protocol specification binary, which is based on ITU-T Recommendation Q.931 for voice. Bluetooth matches standard telephony with a 64-kbps data rate and can support calls for all eight members of a piconet. Bluetooth is able to provide voice support without the need of infrastructure such as a connection point as required in a SWAP-CA network. In a Bluetooth WPAN, a single Bluetooth-enabled voice device (mobile phone) can act as a gateway for all other devices.

IrDA data's optional IrDA mobile communications (MCs) protocol allows IrDA to handle communications from mobile devices such as mobile phones, pagers, and PDAs. The protocol allows three different scenarios to be handled over infrared communication:

- Atomic information exchange — exchanging phone numbers, text message between mobile devices
- Stream-oriented information exchange — ability to use a mobile phone as a modem or an infrared device
- Time-bounded information exchange — support for real-time audio services

Neither IrDA Data nor IrDA AIR specifies any support for voice services, especially in terms of integration with the PSTN and the type of encoding scheme. Therefore, it is difficult to determine how the two would work together. Most likely, an IrDA-enabled mobile or PSTN phone would be able to transmit and receive real-time audio from another IrDA-enabled device using IP running over PPP. The number of calls supported would be based on the data rate between the mobile phone(s) and IrDA device(s).

Of the three protocols, only SWAP-CA and Bluetooth can provide adequate support for voice. SWAP-CA's implementation of DECT and integration with the PSTN are very useful. The only drawback is that a connection point must be used. For the purpose of home networking, SWAP-CA is adequate because the location of a connection point can be planned. For an ad hoc network that needs an additional device for voice services, it is not ideal. Bluetooth provides the same services as SWAP-CA but without the need for a specialized device. In a Bluetooth WPAN, a single node (master node) can act as the aggregator to the PSTN and supports the communication needs of other piconet nodes. In ad hoc networking, Bluetooth is found to be better suited for supporting voice.

6.7.5.4 Support for LAN Integration

The ability to communicate with local area networks allows WPAN devices to take advantage of services such as printing, Internet, and file sharing. All three systems have protocols that enable LAN access. Bluetooth has a profile that allows LAN access using PPP over RFCOMM. The profile does not provide LAN emulation, ad hoc networking, or other methods of LAN access. The standard features in PPP include compression, encryption, authentication, and multiprotocol encapsulation. The profile offers a method for providing the following services using PPP:

- LAN access for single Bluetooth device
- LAN access for multiple Bluetooth devices
- PC to PC using PPP over serial cable emulation

To access LAN services, a LAN access point is needed. A LAN access point device is Bluetooth-enabled and also has access to LAN media such as Ethernet, token ring, 802.11, etc. The device advertises its services using RFCOMM and allows PPP connections to be initiated through it. The SWAP-CA system is adapted from 802.11 and LAN access is readily available. SWAP-CA can use 802.3 or other wired LANs by using a connection point, which acts as a gateway between a wired and SWAP-CA network. SWAP-CA users also have the benefit of using a SWAP-CA PC as a gateway to wired LAN services. The SWAP-CA PC can be the gateway to the Internet, file, or print services on the wired LAN.

IrDA data/AIR has an optional protocol known as infrared LAN (IrLAN) access extensions that allows IrDA data/AIR-enabled devices to access a LAN. IrLAN supports the following scenarios:

- Enables an IrDA device to access a network through an access point device
- Enables two computers to emulate a LAN connection over an IrDA data/AIR connection
- Enables a computer with an IrDA port to connect to a LAN by using a gateway computer that has an IrDA port and network interface card

The three systems provide equal support for LAN integration. All three systems require some type of device that is aware of WPAN and LAN protocols and must be used as a gateway to the LAN. The IEEE 802.15 WPAN Group is looking at implementing protocols in 802.15 to access an 802.11 directly; however, the need for additional hardware may have an impact on the size and power constraints of WPAN devices.

6.7.5.5   Power Management

Because battery power is shared by display, transceiver, and processing electronics, a method to manage power is definitely needed in a WPAN system. All systems have power-saving methods that could possibly aid WPAN devices, but only Bluetooth and SWAP-CA systems offer true power management facilities to prolong battery life.

Bluetooth has a standby and peak power range of <1 to 60 mA and allows devices to enter low-power states without losing connectivity to the WPAN piconet. Bluetooth has three low-power states called PARK, HOLD, and SNIFF, as well as a normal power state when the device is transmitting. During the PARK state, the device is using the lowest amount of power, releasing its MAC address but still maintaining synchronization with the piconet. In the PARK low-power mode, the device can check for broadcast messages, but if it needs to transmit, it must wait until it has acquired a MAC address. The next power level is HOLD, which offers higher power consumption than PARK because the device maintains its MAC address and synchronization. Having a MAC address in the HOLD state allows the device to transmit immediately after waking up and entering ACTIVE state. Although the SNIFF power state consumes the largest amount of power among the low-power states, it allows the device to listen to all the traffic on the piconet at a reduced rate. The SNIFF also allows the device to send immediately after waking up because it maintains synchronization and its piconet address.

The SWAP-CA standard also provides a mechanism for devices to manage power. If power management functions are needed, a connection point, which allows nodes to enter into power-saving modes and buffers data until they wake up, must be utilized. For isochronous (voice) nodes, each device wakes up and checks the connection point for buffered data during its assigned TDMA slot. If data are not present, the device goes back into sleep mode and waits for the next slot. For asynchronous (data) nodes, the sending node sends a wake-up flag to the sleeping receiver node. The receiving node wakes up and checks with the connection point to see if its node address wake-up flag is set. It the flag is set, the node stays on and waits for data from the sender. If the flag is not set, the node goes back to sleep for a frame period. The HomeRF Working Group's 10-mA standby power requirements are in line with Bluetooth, but the peak power needed by HomeRF devices is 300 mA.

IrDA does not have true power management features but does offer power-saving modes related to the distance between the communicating devices. The distance can vary from 0 to 1 m for standard operation, from 0 to 0.2 m when both devices are operating in low power mode, and from 0 to 0.3 m when one device is in low-power and the other is in standard-power mode. The data rate for all three scenarios can vary from 2400 bps to 4 Mbps as long as the bit error ratio is met over the range for the particular data rate desired. Power management features for IrDA AIR are similar to IrDA data. For purposes of a WPAN, Bluetooth's ability to allow any node to become a master and provide power management for the other nodes is superior to what is offered by SWAP-CA and IrDA. The SWAP-CA protocol offers an adequate method; however, the need for a connection point limits the ability to have this function available to nodes in ad hoc WPANs.

6.7.6 WLAN vs. WPAN

It is important to discuss and clarify the applications and differences between WLAN and WPAN systems because there is a long and continuous debate regarding their distinctions and whether they are competing technologies. The situation in which it is reasonable to assume WPAN and WLAN are the same is when both are wireless technologies, i.e., their role is to allow the transmission of information between devices by a radio link. This situation is also shared by devices such as cellular phones, walkie-talkies, garage door openers, cordless phones, satellite phones, etc. However, no one would assume that walkie-talkies are in competition with satellite phones because of several fundamental differences, such as range, price, abilities, primary role, power consumption, etc.

One of the most important issues is the range, and this field is often used (combined with role) to distinguish between wireless technologies. Figure 6.20 shows the various wireless technologies (except WMANs) and their suitability for a given radio coverage and the type of networks. As can be seen, WPAN and WLAN systems have completely different scopes and thus distinct applications.

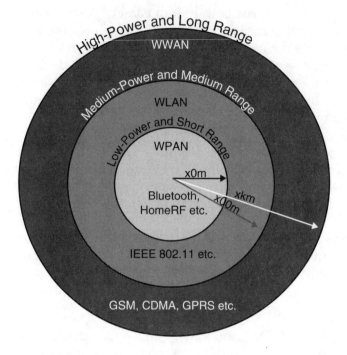

Figure 6.20   Wireless technologies.

There is no question (particularly if cost, size, and power are not the primary factors) about which technology to use for a wireless mobile network. WLANs have been designed for this environment. Yet, it is important to realize that WPAN wireless communication devices are complementary to WLANs. Designed as a cable-replacement technology and not intended as a WLAN competitor, WPAN is not likely to eclipse WLAN and vice versa. However, WPAN products often have interesting and flexible features not found in WLAN systems. This includes co-located separate (personal) networks, ad hoc networking (not present in HomeRF only), and synchronous channels that are particularly effective for voice applications and exceptionally low-power operation for wireless link members. These connections can range within group, point to point, and point to multipoint, all of which can be accomplished automatically.

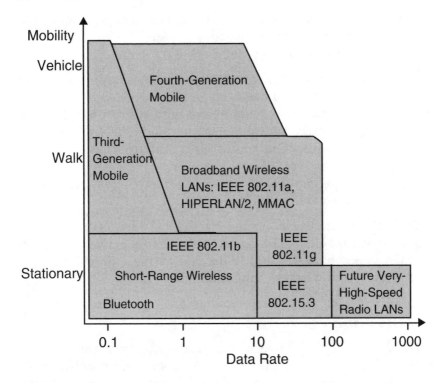

Figure 6.21 Summary of mobility and data rate for last mile wireless broadband technologies.

6.8 CONCLUSIONS AND FUTURE DIRECTIONS

The speed of change in digital communications is most prominently represented by broadband wireless access and home networking technologies. In this chapter, we have thoroughly investigated current high-speed broadband wireless communication systems, data distribution, and wireless local and personal area networking proposals for homes and offices. The technologies covered are the major players in the world of last mile broadband wireless access, given the rapid growth in demand for high-speed Internet and Web access. In the future, seamless coverage, ubiquitous computing, and operation at vehicular speeds up to 200 km/h (120 mph) will be supported in macrocells and microcells for data rates up to 64 kbps. Figure 6.21 characterizes most of the wireless technologies discussed in this chapter that provide broadband last mile access according to data rate and speed.

Important future directions of BWA could be[45]:

- Evolving BWA from mostly business applications to residential applications
- Bandwidth on demand and high spectrum utilization
- More reconfigurable systems and adaptivity to support multiband, multiple standards, and multiple carriers
- Convergence of broadband wireless access and broadband wireless mobile

With the spectrum allocated by the ITU for future fourth-generation mobile communications (4Gmobile), we expect the convergence of BWA and 4Gmobile to be a major focus of activity in the wireless communications field.

GLOSSARY

1/2/3/4G	first/second/third/fourth generation
4Gmobile	fourth-generation mobile communications
ADPCM	adaptive differential pulse code modulation
AP	access point
ARQ	automatic repeat request
ATM	asynchronous transfer mode
BPSK	binary phase shift keying
BRAN	broadband radio access networks
BS	base station
BSS	basic service set
BTS	base transceiver station
BWA	broadband wireless access
CCI	co-channel interference
CCK	complimentary code keying
CDMA	code division multiple access
C/I	carrier-to-interference ratio
C/N	carrier-to-noise ratio
COFDM	coded OFDM
CPE	customer premises equipment
CRC	cyclic redundancy check

CSMA/CA	carrier sense multiple access with collision avoidance
DAVIC	digital audio/visual council
DBPSK	differential BPSK
DBS	direct broadcast satellite
DCF	distributed coordination function
DHCP	dynamic host configuration protocol
DIFS	DCF interframe space
DQPSK	differential QPSK
DS-CDMA	direct sequence CDMA
DSL	digital subscriber line
DSSS	direct sequence spread spectrum
DVB	digital video broadcasting
EDCF	enhanced DCF
EDGE	enhanced data rates for GSM evolution
END	electrodomestic network devices
ESS	extended service set
ETSI	European Telecommunications Standards Institute
FBWA	fixed broadband wireless access
FCC	Federal Communications Commission
FDD	frequency division duplex
FEC	forward error correction
FFT	fast Fourier transform
FHSS	frequency-hopping spread spectrum
GFR	guaranteed frame rate
GFSK	Gaussian frequency-shift keying
GPRS	general packet radio service
GSM	global system for mobile communication
GTS	guaranteed time slot
HAN	home area network
HC	hybrid coordinator
HCF	hybrid coordination function
HDLC	high-level data link control
HIPERACCESS	high-performance radio access
HIPERLAN	high-performance LAN
HRFWG	HomeRF working group
HUMAN	high-speed unlicensed metropolitan area network

IAS	information access services
IBSS	independent BSS
IFFT	inverse FFT
IP	Internet protocol
IR	infrared
IrDA	infrared data association
ISI	inter-symbol interference
ISM	industrial–scientific–medical
ISO	International Standards Organization
ITU	International Telecommunication Union
LAN	local area network
LLC	logical link control
LMCS	local multipoint communication systems
LMDS	local multipoint distribution service
LOS	line-of-sight
LR-WPAN	low rate WPAN
MAC	medium access control
MAN	metropolitan area network
MMDS	multichannel multipoint distribution system
MSK	minimum shift keying
NAV	network allocation vector
NLOS	non-LOS
OFDM	orthogonal frequency-division multiplexing
PAN	personal area network
PCF	point coordination function
PDU	protocol data unit
PHY	physical
PLCP	PHY layer convergence procedure
PN	pseudonoise
PPP	point-to-point protocol
PSDU	PHY service data unit
PSTN	public switched telephone network
QAM	quadrature amplitude modulation
QBSS	QoS supporting BSS
QoS	quality of service
QPSK	quadrature phase shift keying
RF	radio frequency
RG	residential gateway

RLC	radio link control
RMS	root mean square
SC	single carrier
SDU	service data units
SIFS	short interframe space
SIG	special interest group
SNMP	simple network management protocol
SOHO	small office/home office
SS	subscriber station
SSCS	service specific convergence sublayer
STA	station
SWAP-CA	shared wireless access protocol — cordless access
TCM	trellis code modulation
TCP	transmission control protocol
TDD	time division duplex
TDMA	time division multiple access
TFTP	trivial file transfer protocol
TG	task group
UNII	unlicensed national information infrastructure
UWB	ultra-wideband
UWBM	UWB modulation
VLAN	virtual LAN
VoIP	voice over IP
WAN	wide area network
WG	working group
WLAN	wireless LAN
WLL	wireless local loop
WPAN	wireless PAN

REFERENCES

1. Infrared Data Association. IrDA standards, http://www.irda.org/standards/standards.asp, April 2000.

2. V. Erceg, L.J. Greenstein, S.Y. Tjandra, S.R. Parkoff, A. Gupta, B. Kaulic, A.A. Julius, and R. Bianchi, An empirically based path loss model for wireless channels in suburban environments, *IEEE J. Selected Areas Commun.*, 17(7), 1205–1211, July 1999.

3. V. Erceg et al., A model for the multipath delay profile of fixed wireless channels, *IEEE J. Selected Areas Commun.*, 17(3), March 1999, 399–410.

4. L.J. Greenstein et al., A new path-gain/delay-spread propagation model for digital cellular channels, *IEEE Trans. Vehicular Technol.*, 46(2), May 1997, 477–485.

5. L.J. Greenstein et al., Ricean K-factors in narrowband fixed wireless channels, *WPMC '99 Conf. Proc.*, Amsterdam, The Netherlands, September 1999.

6. V. Tarokh, N. Seshadri, and A.R. Calderbank, Space–time codes for high data rate wireless communication: performance criterion and code construction, *IEEE Trans. Info. Theory*, 44, March 1998, 744–765.

7. A.J. Paulraj and T. Kailath, Increasing capacity in wireless broadcast systems using distributed transmission/directional reception, U.S. Patent, no. 5,345,599, 1994.

8. G.J. Foschini, Layered space–time architecture for wireless communication in a fading environment when using multielement antennas, *Bell Labs Tech. J.*, Autumn 1996, 41–59.

9. G.D. Golden et al., Detection algorithm and initial laboratory results using the V-BLAST space–time communication architecture, *Elect. Lett.*, 35(1), January 1999, 14–15.

10. K. Sheikh et al., Smart antennas for broadband wireless access networks, *IEEE Commun. Mag.*, November 1999, 100–105.

11. S.R. Bullock, *Broadband Communications and Home Networking*, Noble Publishing, 2001.

12. L. Wirbel, LMDS, MMDS race for low-cost implementation, *Electron. Eng. Times*, November 30, 1999.

13. V. Tipparaju, Local multipoint distribution service (LMDS), Ohio State University, http://www.cse.ohio-state.edu/njain/cis788-99/LMDS/, December 2000.

14. C. Mason, LMDS: fixed wireless wave of the future? America's Network, Advanstar Communications, Santa Clara, CA, June 1998.

15. D.P. Agrawal and Q.-A. Zeng, *Introduction to Wireless and Mobile Systems*, Brooks/Cole Publishing, 2003.

16. ITU, *Satellite Communications Handbook, Fixed Satellite Service.* International Telecommunications Union, International Radio Consulting Committee, Geneva 1988.

17. IEEE 802.16-2001, IEEE Standard for Local and Metropolitan Area Networks — Part 16: Air Interface for Fixed Broadband Wireless Access Systems, April 8, 2002.

18. IEEE P802.16a/D3-2001: Draft Amendment to IEEE Standard for Local and Metropolitan Area Networks — Part 16: Air Interface for Fixed Wireless Access Systems — Medium Access Control Modifications and Additional Physical Layers Specifications for 2–11 GHz, March 25, 2002.

19. B. Bossard, Low power multi-function cellular television system, U.S. patent no. 4 747 160, 1988.

20. ITU-R Draft Recommendation, 3/47, rev. 1, Propagation data and prediction methods required for the design of terrestrial broadband millimeter radio access systems operating in a frequency range of about 20 to 50 GHz, Geneva, Switzerland, 1999.

21. N. Scevak, Can 802.11 become a viable last mile alternative? http://australia.internet.com/r/article/jsp/sid/11942, May 2002.

22. IEEE Std. 802-11, IEEE Standard for wireless LAN medium access control (MAC) and physical layer (PHY) specification, June 1997.

23. IEEE Std 802.11a/D7.0-1999, Part 11: wireless LAN medium access control (MAC) and physical layer (PHY) specifications: high speed physical layer in the 5-GHz band.

24. ETSI, Broadband radio access networks (BRANs); HIPERLAN type 2 technical specification; physical (PHY) layer, August 1999.

25. R. Van Nee et al., New high-rate wireless LAN standards, *IEEE Commun. Mag.*, 37(12), December 1999, 82–88.

26. A. Doufexi et al., Throughput performance of WLANs operating at 5 GHz based on link simulations with real and statistical channels, *IEEE VTC '01 Spring*.

27. G. Anastasi, L. Lenzini, and E. Mingozzi, MAC protocols for wideband wireless local access: evolution toward wireless ATM, *IEEE Pers. Commun.*, October 1998, 53–64.

28. A. Hettich and M. Schrother, IEEE 802.11a or ETSI BRAN HIPERLAN/2: who will win the race for a high-speed wireless LAN standard? *Eur. WL Conf.*, Germany, Oct. 1999, 169–174.

29. J. Khun-Jush et al., Structure and performance of HIPERLAN/2 physical layer, *IEEE VTC '99 Fall*, 2667–2671.

30. ETSI, Broadband radio access networks (BRAN); HIPERLAN Type 2; data link control (DLC) layer; part 1: basic transport functions, December 1999.

31. A. Doufexi et al., A study of the performance of HIPERLAN/2 and IEEE 802.11a physical layers, *IEEE VTC*, Spring 2001.

32. Bluetooth SIG, http://www.bluetooth.com/.

33. Spike, http://www.spikebroadband.net/.

34. K.J. Negus, A.P. Stephens, and J. Lansford, HomeRF: wireless networking for the connected home, *IEEE Personal Commun.*, February 2000, vol. 7, no. 1, pp. 20–27.

35. B.P. Crow, I. Wadjaja, J.G. Kim, and P.T. Sakai, IEEE 802.11 wireless local area networks, *IEEE Commun. Mag.*, September 1997, 116–126.

36. IEEE 802.15 Working Group for WPANs, http://grouper.ieee.org/groups/802/15/.

37. Bluetooth Special Interest Group, Bluetooth FAQ, http://www.bluetooth.com/bluetoothguide/faq/1.asp, April 2000.

38. A.S. Tanenbaum, *Computer Networks*, Prentice Hall, Upper Saddle River, NJ, 1996.

39. HomeRF Group Working Group, About HomeRF Working Group, http://www.homerf.org/about.html, April 2000.

40. HomeRF Technical Committee, Shared wireless access protocol, cordless access (SWAP-CA) specification, revision 1.2, October 1999.

41. G. Ungerboeck, Trellis coded modulation with redundant signal sets part 1: introduction, *IEEE Commun. Mag.*, 25(2), February 1987.

42. C.M. Cordeiro and D.P. Agrawal, Mobile ad hoc networking, tutorial/short course, *20th Brazilian Symp. Computer Networks*, 125–186, May 2002, http://www.ececs.uc.edu/~cordeicm/.

43. 802.16 Working Group, http://www.ieee802.org/16.

44. ITU Wireless Access System (WAS), http://www.itu.int/was.

45. W.W. Lu, Compact multidimensional broadband wireless — convergence of wireless mobile and access, *IEEE Commun. Mag.*, November 2000, 119–123.

46. J. Cai and D. Goodman, General packet radio service in GSM, *IEEE Commun. Mag.*, October 1997, vol. 35, no. 10, pp. 122–131.

47. D.L. Waring, J.W. Lechleider, and T.R. Hsing, Digital subscriber line technology facilitates a graceful transition from copper to fiber, *IEEE Commun. Mag.*, March 1991, vol. 29, no. 3, pp. 96–104.

48. G.T. Hawley, Systems considerations for the use of xDSL technology for data access, *IEEE Commun. Mag.*, March 1997, vol. 35, no. 3, pp. 56–60.

49. D. Raychaudhuri and K. Joseph, Channel access protocols for Ku-band VSAT networks: a comparative evaluation, *IEEE Commun. Mag.*, May 1988, vol. 26, no. 5, pp. 34–44.

50. T.S. Rappaport, *Wireless Communications*, Prentice Hall, Upper Saddle River, NJ, 1996.

51. S. Middleton, IEEE 802.15 WPAN Low Rate Study Group PAR, doc. number IEEE P802.15-00/248r3, submitted September 2000; http://grouper.ieee.org/groups/802/15/pub/2000/Sep00/00248r4P802-15_LRSG-PAR.doc.

52. D. Lal, R. Toshniwal, R. Radhakrishnan, D.P. Agrawal, and J. Caffery, A novel MAC layer protocol for space division multiple access in wireless ad hoc networks, *Proc. Int. Conf. Computer Commun. Networks*, Miami, FL, 2002, pp. 614–619.

53. 3GPP – 3rd Generation Partnership Project, www.3gpp.org.

54. 3GPP2 – 3rd Generation Partnership Project, www.3gpp2.org.

55. The Software Defined Radio Forum, www.sdrforum.org.

56. IEEE 802.11 WG, Draft supplement to standard for telecommunications and information exchange between systems — LAN/MAN specific requirements - Part 11: wireless medium access control (MAC) and physical layer (PHY) specifications: medium access control (MAC) enhancements for quality of service (QoS), IEEE 802.11e/D2.0, November 2001.

57. The IEEE 802.11 Working Group, http://grouper.ieee.org/groups/802/11/.

7

Satellite Technologies Serving as Last Mile Solutions

PAUL G. STEFFES AND JIM STRATIGOS

7.1 THE ROLE OF SATELLITES FOR BROADBAND ACCESS

As demand for broadband Internet access has grown, so too has the demand in areas of the globe that are underserved or poorly served by terrestrial broadband networks. With the rapid adoption of satellite direct to home (DTH) video distribution from its introduction in the mid 1990s (with over 80 million subscribers worldwide at the end of 2002), satellite operators have increasingly looked to Internet access for growth in service revenues. The cost of high-speed digital satellite receivers and other components has also declined dramatically, making it possible to design satellite broadband networks that are in many cases cost competitive with terrestrial alternatives.

Satellite-based systems provide a unique solution for broadband access for a number of reasons. First, the physical length of the last mile (whether 1.6 or 1600 km) has only a minor effect on the costs of the system. Second, in locations where the user density is low, or where usage is light or transient, the high per-user cost of a local "backbone" system is avoided. Third, a single system can cover wide geographic areas, allowing for portable, or even mobile, terminals, without the significant build-out cost required for a terrestrial wireless infrastructure. Although the accompanying cost of customer premises equipment (CPE) is higher, the overall cost per subscriber may actually be lower due the ability of a satellite link to connect widely dispersed users back to a central location.

Another advantage of satellite systems is that they can support a large range of broadband applications, including multicast (point to multipoint), unicast (multipoint to point), or single user to single user (point to point) using the same CPE, satellites, and infrastructure.

7.2 SATELLITE ORBITS AND FREQUENCY BANDS

7.2.1 Selection of Orbits

The selection of orbits strongly affects the performance and coverage of a satellite system. Orbits are described by several parameters (for example, see Morgan and Gordon[1] or Pratt et al.)[2] The first is *altitude* above the mean surface of the Earth. Because of the radial dependence of the force of gravity, the higher above the Earth a satellite flies, the more slowly it travels. Slower velocity makes the satellite easier to track because it moves more slowly across the sky; however, the increased distance reduces the power received on a communication link from the satellite and increases links delay (see Section 7.2.2).

The second orbital parameter is *eccentricity*, which describes the variation in the satellite's altitude and therefore also variations in its velocity. Because Kepler's laws require orbiting satellites to travel on a planar, elliptical path, with the Earth's center included on the plane, an orbit with no eccentricity results in a perfectly circular orbit, with the Earth at its center and with constant spacecraft velocity. The third orbital parameter is *inclination*, or the angle between the plane of the orbit and the Earth's equatorial plane. An orbit with no inclination is referred to as an *equatorial* orbit, and one with maximum inclination (90°) is referred to as a *polar* orbit. Note that the vast majority of satellites orbiting the Earth in nonpolar orbits have a significant eastward component to their motion because the Earth's rotation imparts an eastward velocity of 464 m/sec × cos[latitude of launch site] at launch. However, it is possible to expend additional launch resources to place a satellite in an orbit with significant westward motion. Such an orbit is referred to as *retrograde*.

The most commonly used orbit for modern satellite communications systems is the *geostationary* orbit. First proposed by Arthur C. Clarke in 1945,[3] the geostationary orbit is an equatorial, circular orbit (resulting in a constant spacecraft velocity), with an orbital period matching the rotation period of the Earth (23 h, 56 min, 4.09 sec). Because the orbital period (and thus velocity) is related to the altitude of the orbit, according to Kepler's third law, the radius of the orbit can be determined from the relation,

$$T^2 = (4\pi^2 a^3)/\mu \qquad (7.1)$$

where

T = the orbital period in seconds

a = the length of the semimajor axis of the orbital ellipse, which for a circular orbit is simply the radius of the orbit

μ = Kepler's constant for the Earth, which is approximately 3.986×10^5 km^3/sec^2 (see, for example, Morgan and Gordon)[2]

Thus, for a geostationary orbit, the radius of the orbit must be about 42,164 km, which corresponds to an orbital altitude of 35,786 km, assuming a mean Earth radius of 6378 km. Relating the distance traveled in one orbit to the orbital period gives a spacecraft velocity of 3.07 km/sec.

The geostationary orbit has the unique feature of making the satellite appear stationary relative to the rotating Earth. This reduces or eliminates the troublesome Doppler shift that occurs when a transmitter moves toward or away from a receiver. In fact, the satellite hovers 35,786 km above the equator at its assigned longitude. Because of the high spacecraft altitude, over 40% of the Earth's surface can be viewed by a single geostationary satellite. However, because of the high altitude of the satellite and thus the long distance from ground stations, the typical delay over a one-way link from one ground station to another (via the satellite) is around 0.25 sec. The corresponding two-way delay (or "response time") is around 0.5 sec.

Additionally, because of the required equatorial position of the satellite, latitudes above 81° are not visible to a geostationary satellite, and ground stations at high latitudes less than 81° must orient their antennas to low elevation angles in order to point toward the satellite, risking additional atmospheric attenuation or blockage by trees or buildings. Finally, because of the relatively large distance between geostationary satellites and ground stations (36,000 to 41,000 km), higher power transmitters or larger antennas are required relative to comparable links to lower altitude satellites.

In contrast, low Earth orbiting (LEO) satellites with orbital altitudes in the 200- to 2000-km range encounter significantly smaller delays and can operate with lower performance ground terminals to provide the same quality of service (QoS). However, because of the lower altitude, a single satellite provides significantly smaller percentage coverage of the Earth. For example, a satellite at an altitude of 1000 km is only visible to about 8% of the Earth's surface at any given time. Moreover, the high velocities characteristic of LEO satellites (over 7 km/sec) mean that they pass overhead quite rapidly (a single satellite might be in range for only 10 to 15 min of its 100-min orbital period) and that the uplink and downlink signals are subject to significant Doppler shift. In order to provide truly ubiquitous worldwide coverage with LEO satellites, a fleet of ~60 satellites (such as those used in the IRIDIUM system) is required.

Alternate orbital configurations include MEO (medium Earth orbit, with altitudes in the 2000- to 30,000-km range, which allows ubiquitous worldwide coverage with a smaller number of satellites) and Molniya,[4] which is a highly elliptical inclined orbit allowing the satellite to "hover" for large portions of its orbit near apogee over some nonequatorial zone such as northern Siberia or North America.

7.2.2 Link Budgets and Carrier-to-Noise Ratios

The carrier-to-noise ratio (C/N) and the accompanying energy-per-bit to noise-spectral-density ratio (E_b/N_o) are the key metrics of link performance. For uplinks and downlinks, the carrier-to-noise ratio can be computed from:

$$C/N = P_t G_t G_R /(L_p L_{ATM} kTB), \tag{7.2}$$

where[5]

P_t = the transmitted power on the link
G_t = the gain of the transmitting antenna
G_R = the gain of the receiving antenna
L_p = the "free space loss," which equals $(4\pi R/\lambda)^2$, where R is the distance between the transmitter and the receiver, and λ is the wavelength
L_{ATM} = the loss due to atmospheric gases or precipitation
k = Boltzman's constant $(1.38 \times 10^{-23}$ W/Hz-K)
T = the noise temperature of the receiving system (in Kelvins)
B = the bandwidth of the channel (in Hertz)

These quantities are expressed as linear quantities rather than in decibels. (Note that the conversion of a linear quantity to decibels is G (dB) = $10 \log_{10} G$.) If written in decibel format, the carrier-to-noise equation becomes

$$\begin{aligned}(C/N)(\text{dB}) = &\ P_t \text{ (dBw)} + G_t \text{ (dB)} + G_R \text{ (dB)} \\ &- L_p \text{ (dB)} - L_{ATM} \text{ (dB)} - k \text{ (dB-W/Hz/K)} \\ &- T \text{ (dB-K)} - B(\text{dB-Hz}), \tag{7.3}\end{aligned}$$

where the quantities are as above, except expressed in decibel form. The energy-per-bit to noise-spectral-density ratio (E_b/N_o), which is directly related to the bit error rate on a link, can be related to the carrier-to-noise ratio by

$$E_b/N_o = (C/N)/thruput \tag{7.4}$$

where *thruput* is the ratio of data rate to bandwidth (R_b/B) for the modulation type employed. For QPSK modulation, thruput can range from one to two, depending on the amount of tolerable intersymbol interference. For a QPSK signal with little or no intersymbol interference (i.e., thruput = 1), E_b/N_o

$= C/N$. The probability of a bit error for a link with a specific value of E_b/N_o and no intersymbol interference is calculated from[2]:

$$PB = 1/2 \; erfc \; [E_b/N_o]^{1/2} \qquad (7.5)$$

where *erfc* is the complementary Gaussian error function and E_b/N_o is expressed as a linear (not decibel) number. As an example, for $E_b/N_o = 13$ dB $= 20$, $PB \simeq 2 \times 10^{-10}$.

7.2.3 Band Selection

Because of technological limitations, the first satellite systems providing broad bandwidth service (bandwidths larger than 1 MHz) operated in the 4/6 GHz (downlink/uplink) band, which was shared with fixed terrestrial microwave users and is also known as the "C-band." Large dish antennas were used for ground stations because satellite transmitted powers were low and the large apertures were necessary to keep the ground station beamwidths small enough so as not cause significant interference to adjacent satellites. The half-power beamwidth of a circular antenna can be approximated by[2]:

$$\theta_{3-dB} = 70 \; \lambda/d, \qquad (7.6)$$

where θ_{3-dB} is the half-power beamwidth (in degrees); λ is the wavelength (in meters); and d is the diameter (in meters).

As technology improved at higher frequencies and bandwidth demands increased, new allocations at Ku-band in the 12-GHz range (downlink) and 14-GHz range (uplink) were exploited. This also allowed the use of smaller antennas because the necessary aperture size required to maintain a beamwidth less than 2° (the nominal spacing between satellites) is much reduced at higher frequencies (shorter wavelengths), and antenna gain is proportional to the square of frequency. For a circular aperture, the gain of an antenna can be expressed as

$$G = \eta \; (\pi d/\lambda)^2, \qquad (7.7)$$

where η is the aperture efficiency (typically between 0.5 and 0.9); d is the diameter of the antenna (in meters); and λ is the wavelength (in meters).

Thus, antennas exhibit higher gain and are more directive at higher frequencies. (Note, however, that the propagation loss, L_p, also increases as the square of frequency.) Another significant issue arising with Ku-band operation involves attenuation due to rain, which is the largest component of L_{ATM}. Rain attenuation is significantly higher at Ku-band than at C-band, requiring additional margins in the link C/N budget to allow for additional attenuation. On satellite downlinks, the noise temperature of the receiving antenna system is also compromised in the presence of rain due to thermal noise from the absorbing rain particles. This can worsen the downlink noise by up to 5 dB (a factor of ~3) in the heaviest rain, further compromising the downlink C/N.

In spite of rain attenuation, the success of Ku-band has been phenomenal. It is widely used for satellite-based data networks and for satellite broadcasting. In the U.S., the FSS Ku-band allocation provides a total of 500 MHz split into two bands: 11.7 to 12.2 GHz (downlink) and 14.0 to 14.5 GHz (uplink). The FCC has allocated orbital locations every 2°, and all locations are licensed to carriers and contain operating satellites with few exceptions. The operator of a Ku-band broadband system must lease existing continental U.S. (CONUS) capacity from the carrier or partner with it to launch a Ku-band spot beam satellite into a licensed location.

The FCC originally licensed Ku-band FSS operators for "big dish" systems — i.e., antennas of 2 m in diameter or larger. Each two-way system required a specific license from the FCC to ensure that it would not interfere with other satellites. The advent of enterprise VSAT (very small aperture terminal) products in the mid 1980s saw relief from this restriction with the advent of blanket licensing for antennas down to 1 m. Below 1 m, there is no guarantee that adjacent satellite systems will not cause interference from their transmission of high-power, low data rate carriers. Thus, the coordination challenges in launching Ku-band systems that will serve millions of sub-1-m terminals could potentially be very complex and time consuming; ultimately, this may not be possible in many orbital locations.

Table 7.1 U.S. Allocations for Fixed Satellite Service
Broadband Applications

Downlink (space–ground) (GHz)	Uplink (ground–space) (GHz)
3.7–4.2	5.925–6.425
10.7–11.7 (nongeostationary)	14.0–14.5
11.7–12.2	14.0– 4.5
18.3–20.2	29.5–31
38.6–41	47.2–50.2

At Ka-band, the FCC has assigned uplink (29.5 to 30.0 GHz) and downlink (19.7 to 20.2 GHz) allocations to various licensees in two rounds of license filings over the past 7 years. A second uplink (28.35 to 28.60 GHz; 29.25 to 29.5 GHz) and downlink (18.3 to 18.8 GHz) band may be available in some orbital locations (Table 7.1). Currently, no commercial Ka-band satellites are operating in the U.S. (the first launch, Spaceway 1, is expected in late 2005), but the FCC has specifically coordinated licenses to serve sub-1-m terminals.

An additional 250 MHz of capacity is available at most licensed locations. This other 250 MHz is shared with some terrestrial microwave services and is generally not considered available for user beams; however, it may in some cases be available for gateway beams providing access to hub stations in the networks that are directly connected to the terrestrial internet/private network. Thus, if a market exists for millions of satellite broadband subscribers, Ka-band satellites are essential at some point in order to have adequate spectrum in locations within 20° of the prime broadcast satellite service (BSS) assignments at 101, 110, and 119°.

The effect of rain on satellite signal propagation is much greater in the Ka-band than in the Ku-band. Ku-band systems typically compensate for rain fade effects by operating the satellite links with excess margin so that service is not degraded during a typical rain event. It is not practical to operate with enough margin to eliminate rain fades completely, so satellite systems are designed to provide an availability of more than 99%, but never 100%. A 3- to 4-dB margin is usually enough for a Ku-band system to provide link availabilities of 99.5% or better. A fixed link margin basically reduces the capacity of the satellite by the amount of the margin. In other words, one could double the capacity of Ku-band broadcast satellites if it did not rain.

Unfortunately, at Ka-band, the margin required to achieve 99.5% is more than 10 dB in many areas of the U.S., and the resulting loss in capacity would be untenable. Therefore, Ka-band systems generally employ a combination of power control and data rate control to compensate dynamically for rain fades. Demonstration terminals used with NASA's advanced communications technology satellite (ACTS)[6] automatically lowered the data/coding rate of the forward link signal in the presence of rain fades.

Because rain fades are a local phenomenon, the greater the number of localized beams supported by a spot beam satellite, the smaller will be the impact of rain fade mitigation on total system capacity. Although rain mitigation at Ka-band using variable rate transmission would have little impact on terminal equipment costs, it would add a significant layer of complexity to the overall system design. An alternative, hybrid approach would be to use a few transponders on a Ku-band CONUS satellite to provide services in high-rain regions (i.e., southern Florida) or in areas with low look angles for less than ideal Ka-band orbital locations. Additionally, "lower margin" services may be offered in some areas.

Although the first Ka-band systems are only now being implemented, allocations for an even higher frequency satellite band have been recently made by the ITU. V-band uses uplinks in the 50-GHz range and downlinks near 40 GHz. Bandwidths in excess of 2 GHz are expected to be available; however, the effects of even greater rain attenuation, as well as attenuation from atmospheric gases, will necessitate the use of power control as well as data rate adjustment.

7.3 SATELLITE ARCHITECTURE

7.3.1 Types of Transponders

Communications satellites have traditionally been configured to translate a received uplink passband to a lower downlink frequency and then to provide significant amplification so as to assure an acceptable carrier-to-noise ratio (C/N) on the downlink. This approach is referred to as a "bent-pipe" satellite, in that the uplink carrier and noise are repeated onto the downlink frequency. The device responsible for this function is called a *transponder*. Because the uplink noise is rebroadcast onto the downlink, the overall carrier-to-noise ratio from one ground station to another becomes

$$(C/N)_{OVERALL} = 1/[(C/N)^{-1}_{uplink} + (C/N)^{-1}_{downlink}] \qquad (7.8)$$

Thus, the overall link performance can be no better than the weakest link in the system.

An alternative with digital systems is to demodulate the uplinked carrier at the spacecraft, so that the data stream can be rebroadcast without noise. This approach, referred to as *regeneration*, results in significantly better system performance because the bit error probabilities on uplink and downlink will simply add and, given the better C/N on either individual link, the aggregate bit error probability will be significantly lower (usually by an order of magnitude or more) than that obtained with a "bent-pipe" satellite transponder. Regenerative transponders are more complex, however, and usually involve commitment to a specific modulation type and data framing before launch. This lack of flexibility has limited the number of satellites currently employing regeneration.

7.3.2   Switched-Beam and Spot-Beam Antennas

The gain of the receiving and transmitting antennas on a satellite is usually limited by the coverage area that they are designed to provide. For example, a satellite in geostationary orbit providing coverage to the continental U.S. will require a beam that is approximately $3° \times 6°$. This will limit the dimensions of the spacecraft antenna (see, for example, Equation 7.6), which will then limit the gain of the antenna (see, for example, Equation 7.7). The overall relationship between beamwidth and gain for an elliptically shaped beam of dimension θ_V by $\theta_H°$ is

$$G = \eta_{LOSS} \, (33708/ \, \theta_V \theta_H), \qquad (7.9)$$

where η_{LOSS} is the efficiency of the antenna, excluding losses from nonuniform illumination.

Thus, for a typical geostationary satellite serving the continental U.S., the gain of the receiving and transmitting antennas is approximately 1685, or 32.3 dB. Although such a gain is significant, it may not be adequate to provide the necessary quality of service for a given link, especially if the ground station performance is marginal. One approach for improving uplink or downlink performance is to increase the gain of the spacecraft antennas. Doing so, however, results in the reduced beamwidths causing loss of coverage, unless multiple beams are employed.

A multibeam satellite system can function by switching between fixed position beams or electronically steering the beams (usually by way of a beam-forming network in the feed system of the spacecraft parabolic dish). This "switchboard in the sky" concept was first demonstrated with the NASA Advanced Communications Technology Satellite (NASA/ACTS), which was the first operational Ka-band communications satellite.

The multibeam approach has the additional advantage of providing "frequency reuse." In a conventional single-beam satellite system, only two carriers can be accommodated on a given frequency (i.e., one on each of two orthogonal polarizations, such as horizontal and vertical linear polarization for domestic satellites, or right-hand circular and left-hand circular polarization for international satellites). In a multibeam system, two uplinks and two downlinks (each in orthogonal polarizations) can be carried in each beam, resulting in substantial frequency reuse. Of course, to support retransmission without interference between carriers, multiple transponders are required aboard the satellite.

7.3.3 Spacecraft/Ground Station Performance Trade-Off

Increasing satellite performance through the use of technologies such as spot beams or regeneration can result in a corresponding reduction in the cost of ground station equipment. For example, a spot-beam system can reduce the size of the required ground station antennas by a factor of four over a standard CONUS (continental U.S.) satellite beam, especially at Ku- and Ka-band, where interference to adjacent satellites is less problematic. However, spacecraft costs are extraordinarily large because of necessary requirements for space qualification and high reliability, so such large investments can only be justified if the number of ground stations is large.

With satellite systems, the largest investment is in the spacecraft elements, which must be launched before any service can be provided. Thus, accurately modeling the number of potential users as a function of satellite performance becomes crucial. As in any market model, the number of users is driven by the price of the ground equipment, the monthly service charge, and the quality of service. Therefore, the break-even point for larger investments in satellite resources can only be reliably determined if an equally reliable market model is available.

7.3.4 Modern Satellite Architecture

Satellite systems currently under development for ubiquitous delivery of broadband service are largely focused toward Ka-band operation, although some service is currently available using Ku-band (see Section 7.6). Current Ka-band satellite systems in development or operation include Spaceway (Hughes Network Systems); ANIK F2 (Telsat); Intelsat Americas 8 (Intelsat); GenStar (Northrup–Grumman/TRW); Galaxy (PanAmSat); and Pegasus (Pegasus Development Corporation). Some of these spacecraft will be bent-pipe satellites with fixed beams. Others will employ regeneration and switchable spot beams.

7.4 EARTH STATION DESIGN

A satellite Earth station intended to provide last mile access to a broadband satellite system comprises three principal components: satellite modem, RF transceiver, and antenna (see Figure 7.1).

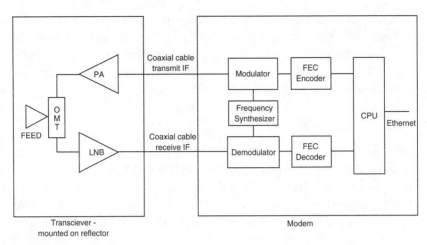

Figure 7.1 Typical block diagram of a satellite Earth station.

7.4.1 Antenna

Broadband satellite links using GEO (as opposed to LEO and MEO) satellites are the systems most likely to be viable in the near future. Parabolic reflector antennas are commonly employed for access to geosynchronous satellite systems used to distribute television broadcasts to cable systems and homes. The size and shape of these antennas and their associated feeds must be selected to yield the desired gain at the transmitting and receiving frequencies and to reject signals from adjacent satellites that may share the same frequency bands. Cross polarization rejection is also a critical design parameter because geosynchronous satellites often reuse the same frequencies on the orthogonal polarization (vertical/horizontal if linear or left/right hand if circular polarized).

Other considerations in antenna design include wind loading, aesthetics (especially important for residential broadband service), and cost. Offset-fed parabolic designs are popular for DTH and VSAT services because the shaped reflector allows the feed to be located so that blockage of the antenna is minimized, thus increasing aperture efficiency. In some cases, elliptically shaped reflectors are used to provide improved off-axis rejection while maintaining a smaller profile suitable for residential applications. Typical materials used for small aperture Ku- and Ka-band antennas include stamped steel and composite materials such as fiberglass impregnated or coated with a conductive material (see Table 7.2).

7.4.2 Transceiver

The Earth station transceiver consists of a microwave or millimeter wave transmitter integrated with the antenna feed and a low-noise, block (LNB) down-converter to minimize cost and reduce the loss of interconnection waveguide or coaxial cables. For residential last mile applications, the overriding consideration is cost. Transmit power will be the driving factor in the cost of a Ku- or Ka-band transmitter, with the final amplifier transistor or MMIC (microwave monolithic integrated circuit) comprising as much as 50% of the cost. Because uplink power is proportional to transmitted data rate, the overall system design involves interacting trade-offs that have an impact on the transmit power requirements.

Current Ku-band systems are limited to transmit data rates of 50 to 150 kb/sec in order to keep the transmit power levels below about 3 W. The use of Ka-band satellites with spot beams will provide a higher uplink G/T and higher antenna gain for a given antenna size. These systems will be able to provide uplink data rates of up to 1 Mb/sec using power levels on the order of 2 to 4 W. The transmitter on most consumer terminal designs is configured as a block up-converter (BUC), which will translate and amplify signals in the IF band (typically 950 to 1450 MHz) to the appropriate Ku- or Ka-band transmit frequency.

In addition to power output, the transmitter must meet other performance criteria, such as phase noise, frequency stability over time and temperature, etc., to ensure compatibility with gateway station receiving systems. Each of these parameters involves trade-offs in the overall system design and will have a direct impact on costs (Table 7.3).

The receiving portion of the system consists of an LNB with characteristics similar to units used for DTH video

Table 7.2 Performance Characteristics for Some Typical Antennas Designed for Satellite Broadband Applications

Material	Effective aperture (cm)	Tx/Rx frequency (GHz)	Tx/Rx gain (midband) (dBi)	Cross-Pol isolation (dB)	Sidelobe performance
Glass fiber reinforced polyester	60	13.75–14.5 10.70–12.75	37.2 35.7	>30	29–25 Logθ dBi 2.5° < θ < 20°
Glass fiber reinforced polyester	75	13.75–14.5 10.70–12.75	39.1 37.8	>30	29–25 Logθ dBi 2.5° < θ < 20°
Glass fiber reinforced polyester	67	29.5–30.0 19.7–20.2	44.5 41.0	>30	29–25 Logθ dBi 2.5° < θ < 7°
Steel	76	27.5–31.0 GHz 18.2–21.2 GHz	45.5 42.0	>30	29–25 Logθ dBi 2.5° < θ < 20°

Table 7.3 Performance Characteristics of Typical
First-Generation Ka-Band Transmitter

Output frequency	28.35–28.6 or 29.5–30.0 GHz
Compression point (P1dB)	34 dBm min
Third-order intercept	38 dBm min
RF spurious	<–55 dBc
IF frequency	1.0–1.5 or 1.5–2.0 GHz
Gain variation over temp	±1 dB
Gain flatness	±0.5 dB/30 MHz
Phase noise (SSB)	–30 dBc/Hz @10 Hz
	–60 dBc/Hz @100 Hz
	–70 dBc/Hz @1 kHz
	–80 dBc/Hz @ 10 kHz

services. Using satellite forward links compatible with the DVB standard (whether Ku or Ka) allows the use of these very low-cost, high-production volume products. They typically operate in the opposite RF direction from the BUC by translating the 500-MHz downlink frequency band into the 950- to 1450-MHz IF band.

7.4.3 Modem

Modems for satellite broadband terminals are typically housed in an enclosure connected to the antenna via coaxial cables. Depending on the choice of modulation and coding waveforms, some components of the satellite modem may be borrowed from the designs of satellite DTH receivers. In addition to the physical (PHY) and multiaccess (MAC) layer interfaces for transmit and receive links, the modem must also contain protocol translation software to improve the efficiency of TCP over the relatively long delay of the satellite channel.

Performance-enhancing proxies (PEP) are also likely to be required to improve the efficiency of http and other higher level Internet protocols. Because it is highly desirable for the satellite link to look and operate like any other TCP/IP network, TCP emulation and PEP functions must be housed in the modem, resulting in more memory and a faster processor than is typically required in a cable or DSL modem. Like DSL and cable modems, the satellite modem is typically operated in a highly asymmetrical fashion, with forward link data rates appropriate for a saturated 24- to 100-MHz wide transponder and return links on the order of 100 to 3000 kb/sec with multiple carriers per return link transponder.

7.5 NETWORK ARCHITECTURE

7.5.1 Modulation and Coding

The selection of modulation waveforms for the air interfaces in a satellite network depends on the link characteristics, such as noise and power limitations and the desired operating bit rates and error rates. Shannon's limit defines the maximum capacity of a communications link (see, for example, Couch[7]):

$$C = W \log_2 (1 + S/N), \qquad (7.10)$$

where C is the channel capacity in bits per second; W is the channel bandwidth; and S/N is the signal to noise ratio. Uncoded digital modulation techniques such as PSK or FSK are not capable of providing data throughput even approaching Shannon's limit.

Typically, geosynchronous satellite links are power limited, and thus modulation and coding techniques that trade off improved bit error rate at the expense of bandwidth are preferred. Forward error control (FEC) is used with digital satellite links to reduce the bit error rate associated with PSK modulation to acceptable levels. QPSK in combination with convolutional and block codes has been the standard practice in most VSAT products. Concatenated Reed–Solomon and Viterbi (RSV) decoding is specified in DVB and other satellite standards. With modern, higher powered transponders, 8 PSK modulation is practical and results in improved bandwidth efficiency, especially for saturated transponder (i.e., DTH) applications at the expense of additional power compared to QPSK. In recent years, turbo codes[8] have become popular due to their ability to operate very close to the Shannon limit and essentially to negate the power disadvantage of 8PSK. The use of turbo codes concatenated with Reed–Solomon codes has been recently adopted as an option to the DVB-RCS standard.[9]

7.5.2 Multiaccess Protocols

From the standpoint of data network design, the satellite channel provides many challenges that must be addressed at the MAC layer. Since the pioneering work on the ALOHA protocol by Abramson,[10] researchers have focused on the satellite multiaccess problem and developed a number of innovative solutions to the problem of efficiently sharing satellite channels among a population of terminals. As VSATs have evolved from a primarily low- to medium-speed enterprise solutions to providing broadband Internet access, so too have the capabilities required of the MAC protocols. Most commercial VSAT systems and proposed broadband Internet access systems use some variation of the reservation TDMA (R-TDMA) protocol.

The wide variation in applications and traffic patterns associated with Internet access requires a more complex MAC protocol due to the long propagation delays and asymmetric nature of the physical interface (stations cannot hear their transmissions). Multiple levels of QoS are essential, as in other RF-based systems (like cable modems) to ensure that each user is granted a fair share of the satellite return link. Simulations using Web traffic models and TCP on top of various MAC protocols have shown that protocols such as centralized priority reservation (CPR), which is used in HFC systems, can be adapted to the satellite environment.[11]

Protocols such as CPR divide the in-routes into fixed time slots and allow some slots to be used for contention access using a slotted Aloha and reserve the bulk of slots to be assigned to terminals on a reservation basis. Typically, the contention slots are only used for reservations necessitating a minimum of at least 500 msec (a round-trip delay to GEO orbit) before actual traffic packets can be sent.[12] Even with this constraint, broadband satellite networks are capable of providing response times comparable to that experienced by users with DSL and cable modems when TCP emulation and http proxies are used.

Although widely deployed in terrestrial cellular and broadband wireless networks, CDMA has rarely been used in commercial two-way satellite networks. The complications of code management and precise power control required for CDMA have not been offset by enough improvement in performance (higher throughput and lower delay). Abramson[13] has proposed a novel combination of CDMA and Aloha called spread Aloha multiple access (SAMA) that at least theoretically offers performance comparable to R-TDMA without the penalty associated with the round-trip delay for reservation request/grant.

7.5.3 Standards

The role of standards in broadband satellite systems is emerging. The DVB (digital video broadcast) standard is viewed as a critical factor in the rapid adoption of digital DTH services around the world, but it only specifies the broadcast or forward link (network to user). Although numerous international standards cover basic two-way satellite and Earth station functions have allowed vendors to produce interoperable satellite ground stations for years, standards that apply to complex broadband access capabilities were only first proposed in 1999. Two competing standards have been developed and seen limited application: DVB-RCS and DOCSIS-S.

7.5.3.1 DVB-RCS

DVB-RCS (DVB return channel system) was developed by an ETSI technical working group in an effort to specify standards for the physical, link, and network layers of a two-way broadband satellite terminal.[14] The DVB family of standards were first released in 1997 and provide a standardized method to encode, multiplex, and transmit digitized video and audio over a variety of transmission media including cable (DBV-C), terrestrial broadcast (DVB-T), and satellite (DVB-S). DVB-S, in particular, has found widespread adoption in direct-to-home satellite broadcast networks around the world.

With the growing demand for broadband Internet services and the desire among satellite operators to provide these services in conjunction with DTH services, the DVB technical committees developed a return channel standard for the subscriber-to-network path that would be applicable to residential and enterprise broadband services. The forward path (network to subscriber) is assumed to be compatible with DVB-S and may be dedicated totally to Internet access or shared with DTH broadcasting. Some of the important characteristics of DVB-RCS are:

- Use of multifrequency TDMA (MF-TDMA) at the MAC layer to allow many users to share the same return channel efficiently — MF-TDMA requires return path burst accuracy of within 1-b time to avoid the need for synchronization preambles
- ATM packetization of forward and return link traffic — a revision was later proposed allowing an option to use native MPEG packetization
- QPSK modulation with concatenated Viterbi/Reed–Solomon coding as is used in the DVB-S forward link — a revision was later proposed to allow higher order modulation (8PSK) and turbo coding

7.5.3.2 DOCSIS-S

Data over cable interface specifications (DOCSIS) was developed by Cable Labs as a standard for broadband Internet and telephony over hybrid fiber-coax (HFC) networks. In 2001, an industry group lead by Wildblue, ViaSat, and Broadcom proposed an adaptation of DOCSIS for broadband satellite applications called DOCSIS-S. The proposed standard incorporates the basic MAC return path protocol and packetization specified for HFC applications in DOCSIS with modifications of the physical layer (QPSK/8PSK and turbo coding) for use over satellite links.

One of the major advantages cited for DOCSIS-S is that equipment manufacturers could leverage the significant investment made in DOCSIS chip sets by incorporating them directly into products with ASIC glue logic to adapt the physical layer or by creating new ASICS using available DOCSIS cores. Another advantage would potentially be compatibility with the higher layer functions (such as QoS and network management) available from DOCSIS vendors and use of existing cable modem termination systems (CMTSs) in the satellite gateways. Some of the important parameters specified by DOCSIS-S are:

- QPSK or 8PSK combined with turbo coding
- Native IP encapsulation instead of ATM as used in DVB-RCS
- Native forward and return rate link control to mitigate rain fades at Ka-band

7.5.3.3 DVB-RCS vs. DOCSIS-S

Both standards require the use of a DVB-S formatted forward carrier and its associated MPEG-2 multiplexing and framing. The underlying assumption for this requirement is the ability to mix DTH video streams and IP streams in the same carriers or transponders. Both standards allow the use of standard DVB modulation waveforms (QPSK) with concatenated Reed–Solomon convolutional coding or turbo coding. DOCSIS-S, however, also specifies the use of 8PSK plus turbo coding, thus allowing higher bandwidth efficiency. DVB-RCS also covers the scenario of a satellite return link as an adjunct to a DVB-S DTH video system. Based on the status of Ku-band DTH and the evolving Ka-band Internet access systems, this is not a likely scenario and results in extra complexity not required in Internet-only access system.

To mitigate the effects of rain fades at Ka-band, the use of dynamic symbol and coding rate control is essential. In this area the two specs differ. As written, the DVB-RCS specification (ETSI EN 301 790) only states that the forward link must be DVB-S compliant. DOCSIS-S provides a very dynamic approach that varies the modulation and coding on a frame-by-frame basis to allow the link margin to increase only for terminals actually undergoing a fade. The impact of this difference in a Ka-band system will be dramatic because the DVB-RCS system assumes that the forward link margin will compensate for rain fades, resulting in a significant reduction in satellite capacity. By dynamically varying the burst parameters, the capacity of a given beam with faded terminals is only reduced during the fade as opposed to a fixed link margin, which reduces the beam capacity in clear air as well. This technique necessarily prevents the use of the same forward link for both video and IP traffic.

The most significant difference between the two standards is the requirement for ATM framing in DVB-RCS. This results in a additional overhead of about 10% compared with DOCSIS-S. The short length of an ATM cell is inefficient for encapsulating IP traffic and the variable QoS potential for ATM is not usually required in a residential broadband access product as opposed to an enterprise networking solution, for example. The other major difference is, again, the range of symbol and code rates specified by DOCSIS-S vs. DVB-RCS. DOCSIS-S appears to contemplate a more flexible, automatic process of scaling back transmitted data rate to mitigate rain fade, and DVB-RCS seems to rely more heavily on return link power control. Table 7.4 shows differences between DVB-RCS and DOCSIS-S.

7.6 EXISTING SYSTEM EXAMPLES

7.6.1 Historical Perspective

Satellites have played an important role in providing two-way digital communications links to telecom operators for more than three decades. In the early 1980s, advances in low-cost microprocessors with the power to provide protocol processing and packet switching led to the development of two-way interactive satellite terminals. These terminals were cost effective for large enterprise users to deploy in lieu of terrestrial private line services. These very small aperture terminals (VSATs), as they became known, served an important role for enterprise users whose needs for advanced data communications services were not met by the mostly monopoly terrestrial telecommunication carriers around the world. Even with the breakup of AT&T and the divesture of the local exchange carriers in the U.S. in 1984, large- and medium-sized enterprise users continued to utilize VSAT networks to provide homogenous data networks to geographically dispersed locations of

Table 7.4 Summary of Key Forward and Return Link Differences

	DOCSIS-S	DVB-RCS
Forward link framing	IP in MPEG-2	IP in MPEG-2
Forward link adaptive modulation/coding	Yes	No
Forward link modulation	QPSK/8PSK	QPSK, 8PSK optional
Forward link coding	RS/turbocode	RS/convolutional
Return link framing	IP in minislots	IP in ATM in minislots (option for MPEG-2)
Return link modulation	QPSK	QPSK
Return link coding	RS/turbocode	RS/convolutional (turbocode optional)
Forward link IF	950–2150 MHz	950–2150 MHz
Return link IF	5–65 MHz	950–3000; 950–1450; 20–60 MHz
IFL signaling	Not specified	DiSEqC, 10.7 MHz FSK

corporate facilities, especially in the retail and automotive industries. According to COMSYS,[15] over 400,000 VSATs were deployed at the end of 2001.

Most VSAT installations provided what would be termed today low- or medium-speed data network services of from 64 to 256 kb/s, because these speeds were required by mainframe-dominated corporate networks of the time. With the advent of distributed enterprise computing and the growth of Internet usage in the mid 1990s, VSATs began to evolve to provide higher speed services. During the same period, advancements in video compression technology created the opportunity for wide scale deployment of direct-to-home satellite television networks and the requisite low-cost silicon physical layer solutions.

VSATs began to evolve at this point to incorporate higher link speeds by utilizing the advanced modulation and coding techniques embodied in standards such as DVB. By leveraging higher levels of silicon integration and placing more functions in software as opposed to hardware, VSAT costs began to drop to the point at which services providers and vendors began to look at the emerging residential and small enterprise Internet access market seriously. Although this market traditionally would have been assumed to exist only in areas of the world lacking in broadband infrastructure, the exact opposite has occurred. Areas of the world rich in broadband infrastructure supported the growing demand for broadband Internet access, even if the established telecommunications service providers had significant gaps in its service areas. This created a natural opportunity for broadband satellite services, especially in areas of the world with growing populations of DTH users who were already comfortable with satellite delivery of video programming and with the aesthetics of having a 60- to 90-cm antenna on their property.

Planning for a huge demand for broadband satellite services, in the early 1990s the industry began to lobby regulators around the world for more spectra and frequencies favorable for the development of high-speed, low-cost satellite products. Because C- and Ku-band GEO satellite allocations were made assuming mostly broadcast or backbone two-way applications and the limited supply of favorable orbital locations at these frequencies, allocations for spectrum in the Ka-band (20 to 30 GHz) were heavily sought by established telecom operators and new ventures created solely to pursue the broadband market. Although notable failures (Astrolink) and setbacks (Wildblue) occurred among the early Ka-band players, the first commercial Ka-band broadband systems were expected to launch in late 2003.

7.6.2 Starband

Starband was launched in February 2000 as a joint venture among Gilat Satellite Networks, Microsoft, and EchoStar Communications Corporation. It was the first residential satellite broadband service to utilize two-way satellite terminals without the need for a terrestrial ISP connection. Utilizing VSAT products originally designed for enterprise networks by Gilat, Starband focused on rural users outside the footprint of cable modem and DSL providers who desired to have a full-time, high-speed connection to the Internet.

Rather than launching its own satellites, Starband employed the existing Ku-band geosynchronous satellite systems operated by U.S. carriers such as SES Americom and PanAmSat. Typical pricing for residential service was $499 for equipment and $69 per month for service with a 1-year contract. A Starband terminal consists of a 76-cm offset-fed elliptical reflector connected to a modem vial two coaxial cables. Limited by the G/T of CONUS Ku satellites, Starband offers uplink speeds of about 40 to 60 kb/s and downlink speeds of 500 kb/s.

7.6.3 DirecWay

As precursor to true two-way satellite broadband, Hughes Network Systems launched a hybrid product using a satellite forward link combined with a standard dial-up return link in 1995. DirecPC took advantage of the relatively high download speeds achievable over a satellite link without the need for an expensive satellite transmitter and transmit-capable satellite antenna. By taking advantage of the asymmetrical nature of Web access, Hughes was able to appeal to "early adopters" who wanted faster download speeds and to gain valuable experience with deploying and operating a satellite broadband service. As broadband Internet access became less "asymmetrical" with the popularity of peer-to-peer services and transfer of music and video files, the limitations of a hybrid system grew. Hughes launched a two-way satellite broadband service (DirecWay) in 2002 using standard Ku-band satellites and technology derived from its enterprise VSAT products. With similar pricing model and performance as the Starband service, Hughes had over 300,000 subscribers in 2004.

REFERENCES

1. Morgan, W.L. and G.D. Gordon, 1989. *Communications Satellite Handbook*. John Wiley & Sons, New York.

2. Pratt, T., C.W. Bostian, and J.E. Allnutt, 2002. *Satellite Communications*. John Wiley & Sons, New York.

3. Clarke, A.C., 1945. Extra-terrestrial relays. *Wireless World*, 305–308.

4. Johnson, N.L, 1988. Satcom in the Soviet Union. *Satellite Commun.*, 21–24, June.

5. Pratt, T. and C.W. Bostian, 1986. *Satellite Communications*. John Wiley & Sons, New York.

6. Gargione, F., R. Acosta, T. Coney, and R. Krawczyk, 1996. Advanced communications technology satellite (ACTS): design and on-orbit performance measurements. *Int. J. Satellite Commun.*, 14, 133–159.

7. Couch, L. 1997. *Digital and Analog Communication Systems*. Prentice Hall, Upper Saddle River, NJ.

8. Berrou, C., A. Glavieux, and P. Thitimajshima, Near Shannon limit error-correcting coding and decoding: turbo codes, *IEEE Proc. Int. Conf. Commun.*, Geneva, 1993.

9. ETSI, 2000. Digital video broadcasting (DVB), PR EN 301 790, interaction channel for satellite distribution systems, rev 14, Geneva.

10. Abramson, N., 1970. The ALOHA system: another alternative for computer communications, *AFIPS Conf. Proc.*, 37, 281–285, Montvale, NJ: AFIPS Press.

11. Choi, H. et al., Interactive Web service via satellite to the home, *IEEE Commun. Mag.*, Mar. 2001, pp. 182–190.

12. Hu, Y. and V. Li, 2001. Satellite-based Internet: a tutorial, *IEEE Commun. Mag.*, March, 154–162.

13. Abramson, N., 1990. VSAT data networks, *Proc. IEEE*, 78(7), 1267–1274, July.

14. ETSI, 1997. Digital video broadcasting (DVB), EN 300 421, framing structure, channel coding and modulation for 11/12 GHz satellite services, Geneva.

15. The VSAT Report, 2001. COMSYS, (Communication Systems Limited), 42 Holywell Hill St Albans, Herts, AL1 1BX U.K.

8

Management of Last Mile
Broadband Networks

MANI SUBRAMANIAN

8.1 BROADBAND NETWORK MANAGEMENT ISSUES

This chapter will treat network management of the various broadband access networks that have been covered elsewhere in the book. Broadband access network is an evolving telecommunication area that comprises many technologies to provide broadband service. It is extremely important from operations as well as business points of view to manage networks remotely. The subject will be reviewed in this chapter from the networking viewpoint and in the next chapter from the service and users' viewpoint.

The network management protocol most often used in remotely managing networks is the simple network management protocol (SNMP) developed by Internet Engineering Task Force (IETF). Another standard is the open system interface (OSI) management protocol, common management information protocol (CMIP), which will be referenced wherever applicable. The reader is referred to Subramanian[1] for details on network management protocols.

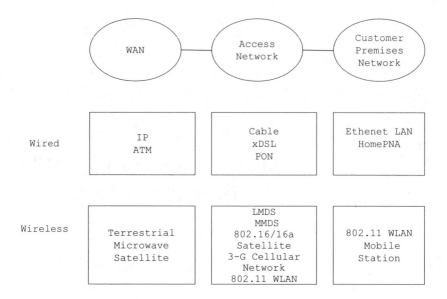

Figure 8.1 Broadband networks.

Figure 8.1 shows the various networks carrying broadband traffic using wired and wireless technologies. The wide area network (WAN) is primarily ATM (asynchronous transfer mode) or IP (Internet protocol) wired networks and terrestrial microwave or satellite wireless networks. The access networks, which will be the primary focus here are cable, various DSL (digital subscriber line), and PON (passive optical network) fixed wireless networks comprising LMDS (local multipoint distribution system), MMDS (multichannel multipoint distribution system), IEEE 802.16 and 802.16a, satellite, 3-G cellular, and 802.11 WLAN (wireless local area network) networks. The wired customer premises networks shown in Figure 8.1 are Ethernet LAN and HomePNA (Home Phone Networking Alliance). The management of familiar Ethernet LAN has been extensively dealt with in the literature.[1] The emerging HomePNA network is still in its early stage as the last meter network. The wireless customer premises networks shown in Figure 8.1 are the 802.11 WLAN and mobile station. The mobile station could be associated with wireless LAN or cellular network and is an integral part of the last/first mile/meter technology; thus, their management will be discussed in detail along with 3-G cellular network and WLAN.

The network management protocols associated with data networks are primarily concerned with the management of parameters or managed objects associated with the data link layer and above, i.e., ISO (International Organization for Standardization) protocol layers 2 and above. However, in broadband access networks, management of objects is dealt with at the physical layer (ISO protocol layer 1), which strongly depends on the access network technology used. For each technology, the management of the physical layer along with higher layers is addressed by the respective industrial consortium or standards body; this not only helps with remote management, but also ensures interoperability of equipment manufactured by various vendors. Each consortium's specifications define conformance and compliance requirements to meet its standards and certify passage of those requirements for the equipment prior to the release.

Just as in the management of data-centric WANs and LANs, remote network management of access and customer premises networks is an essential part of providing broadband service to subscribers. Contrary to the experience of telephone service, when the service provider invariably "rolled the truck" for installation and maintenance of the loop segment of the network, the broadband and Internet service providers cannot afford this practice to continue to stay in business. Also, subscribers cannot afford interruption of essential services while receiving multimedia service on the same physical medium. For example, imagine the situation of not having emergency 911 service because the telephone is out of service!

Remote network management of broadband access networks becomes a necessity for service provider and subscriber from an operational aspect. Operations support systems, such as service provisioning and bandwidth performance management systems, would be required for self-provisioning by subscribers, as well as ordering and receiving video-on-demand services on-line. This is a convenience for the subscriber and also proves economical for the service provider.

The two technologies widely deployed today are cable or hybrid-fiber-coaxial (HFC) technology and the set of digital subscriber line (DSL) technologies. Management of these two is covered in Section 8.2 and Section 8.3, respectively. In 2002, 16 million homes in the U.S. subscribed to broadband services. Of these, 62% deployed cable modems, 28.1% DSL modems, and the rest of the other technologies.[2] Fixed wireless technology is predominant among the other categories of access networks and is primarily deployed in rural areas. Use of mobile devices is spreading at a rapid rate and standards for managing mobile wireless enterprise networks are under development.

Management of fixed and mobile cellular wireless networks is discussed in Section 8.4, Section 8.5, and Section 8.6. The last mile access network to residence needs to address the network from the edge router to the residence, as well as the last meters inside the home. IEEE 802.11 WLAN is a fast growing broadband distribution system for home, enterprise, and public use. Management of WLAN is covered in Section 8.7. The new emerging optical fiber access network is the passive optical network (PON), whose management is addressed in Section 8.8. In practice, broadband access network in the last/first mile and meters is a heterogeneous network, whose management aspects are covered in Section 8.9.

The rest of this section introduces the various aspects of network and service management. Borrowing the terminology from telecommunication service providers, operation support systems (OSSs) perform four functions: operations, administration, maintenance, and provisioning (OAMP). The OSI specifications classify them as fault, configuration, accounting, performance, and security (FCAPS) management. Mapping the two terminologies, operations encompass fault, performance, security, element management, and network management. Administration addresses accounting, service, and business management. Maintenance comprises installation and maintenance of the network. Provisioning defines configuration management of network and network elements. The critical service issues associated with broadband service offering will be reviewed briefly.

8.1.1 Triple Play: Types of Services

The term "triple play" is used to describe broadband multi-media services and means that voice, video, and data are carried over the same medium. Although triple play could be implemented in the current service offerings and available in selected sites, it is not universally available. Cable systems provide predominantly video and data, whereas the asymmetric DSL (ADSL) system offers voice and data. Fixed wireless, which uses cable technology, is also limited to video and data services.

8.1.2 Reliability and Dependability

The telecommunications (telephone) network is a highly dependable and reliable global network. The expectation of telephone subscribers has been set by the quality of telephone service. The data network is not as reliable and dependable. Packets are dropped or lost and even a session could be abnormally terminated. With voice and video riding along with the data, such lapses need to be made transparent to users. Equivalence of "establishment of" and "tearing down" of a circuit in telecommunications network needs to be replicated reliably in a packet-switched multimedia network.

8.1.3 Tiered Service

Independent of the access technology, the data rate (which depends on bandwidth and modulation schemes) needs of upstream and downstream are limited. Data rate needs of subscribers are different, based on the application and class of subscribers. For example, a SOHO (small-office home-office) subscriber would require higher data rate symmetrical service, but for a residential subscriber executing e-mail and surfing the Web, low data rate asymmetrical service is adequate. In a shared medium, such as cable network, this becomes a critical problem. The network needs to be provisioned for different classes of tiered service, which define the priority and type of traffic — voice, video, or data. Because this would change with time, the network management should be capable of provisioning tiered service that could be dynamically changed. The classification of packets for real-time traffic of voice and video, and non-real-time traffic of data becomes important in the quality of service (QoS) offering by the service providers.

8.1.4 Quality of Service

The data rate, delay, and jitter characteristics of voice, video, and data services are different. In the core network, ATM has the capability to distinguish the difference in requirements for these services. It provides all three services by provisioning in the ATM switches from a stringent constant bit rate for voice to available bit rate for data service. Significant efforts are in progress to provide the desired quality of service (QoS) to switch from ATM to Internet core with IP-based routers handling broadband traffic. Management of this problem is exacerbated in the last mile going from the edge router to the customer premises. Not only is active management of the packets required, but also monitoring the quality of service achieved to meet the service level agreement (SLA) contracted between the service provider and the subscriber. QoS issues associated with each technology will be addressed as it is discussed.

8.1.5 Security Management

Security management continues to be a serious issue in emerging technologies. In wired access network technologies, security has been dealt with satisfactorily. However, the security in wireless networks still remains a concern. This is a combination of lack of standards as well as lack of implementation of security function in public and private networks. In discussing security management, it is necessary to distinguish between security management of the on-line payload data from that of security of network management information going across the network. The former is covered as each technology is discussed and the latter is addressed in Section 8.1.8.

Information faces six types of threats while it is being transported from one network entity to another:

- *Modification of information.* Some unauthorized user may modify the contents of the message while it is in transit. The modification is to the data contents, including falsifying the value of an object. It does not include changing the originating or destination address.
- *Masquerade.* Masquerade occurs when an unauthorized user assuming the identity of an authorized user sends information to another. This can be done by changing the originating address.
- *Message stream modification.* Some of the communication modes use connectionless transport service, such as UDP. This means that the message could be fragmented into packets with each packet taking a different path. The packets could arrive at the destination out of sequence and have to be reordered. The threat here is that the intruder may manipulate the message stream and maliciously reorder the data packets to change the meaning of the message. For example, the sequence of data of a table could be reordered to change the values in the table. The intruder could also delay messages so that they arrive out of sequence. The message could be interrupted, stored, and replayed at a later time by an unauthorized user.
- *Disclosure.* For disclosure of information, the unauthorized user need not intercept a message, but simply eavesdrop. For example, the message stream of accounting could be promiscuously monitored by an employee with a TCP/IP dump procedure, and then the information could be used against the establishment.
- *Denial of service.* In a denial of service situation, an authorized user is denied service by a network component, such as a router, or by a management entity. This is usually addressed by the safeguards in the protocol used.

- *Traffic analysis by an unauthorized user.* Information could be used to do a pattern analysis of the traffic and the results abused by the unauthorized user.

8.1.6 Installation and Maintenance

Cable and DSL require modems at the customer premises and a significant cost is associated with installation and maintenance of customer premises equipment (CPE), although many service providers claim that this is no longer a serious problem due to maturity in technologies. This requires self-provisioning of the modem and not all the subscribers are able to do this even if the service provider offers the feature. Service provisioning is further complicated by a video-on-demand (VoD) service offering. This requires bandwidth on demand and needs to be balanced with the provisioned SLAs of other subscribers sharing the network. Some operations support systems (OSSs) can predict the integrated usage of subscribers, which helps in the management of VoD.[3]

In telecommunications network maintenance, the access network is the loop and the demarcation between the subscriber and service provider networks is clear. Although such a demarcation point exists in broadband access networks, it is not as simple to isolate the problem between CPE and access network from a centralized management system.

8.1.7 Billing

Billing is a complicated function in providing broadband services. The service provider needs to bill based on the type of service and the tier of service offering. Furthermore, additional billing requirements are introduced with VoD service and usage-based fees that would be offered in the future. The issue of bundled services is of great importance to the service providers and could be offered as a menu of options for the broadband service, as it is for the telephone service. However, this needs a sophisticated billing system that is integrated with other OSSs.[4]

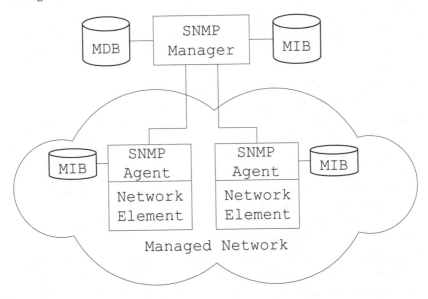

Figure 8.2 SNMP managed network architecture.

8.1.8 SNMP Management

This section will present a brief introduction to the network management protocol, SNMP. Subramanian[1] and Stallings[5] offer detailed treatments of SNMP management. SNMPv1 has the basic specifications for network management with five messages exchanged between the network management entities. SNMPv2C (community-based SNMPv2) adds two more messages in addition to several enhancements in formalization of definitions. SNMPv3 addresses the security considerations lacking in the previous two versions.

The architecture of a managed network is shown in Figure 8.2. An SNMP manager in a network management system (NMS) acquires data about a network element from an SNMP agent residing in it. The information is stored in the management database (MDB) and is displayed by the application in the NMS. Information is transmitted and received by the manager and the agent. For example, when a new network element with a built-in management agent is added to the network, the discovery process in the network manager broadcasts queries and receives positive response from the added element. The information must be interpreted semantically and syntactically by the agent and the manager.

The managed object is defined by a set of rules in the structure of management information (SMI). The managed objects are grouped and organized in a hierarchical structure, called management information base (MIB) —a virtual information database compiled in the agent and management software modules. Communication between the manager and agents is exchanged using the TCP/IP protocol suite for the transport mechanism. SNMP messages are defined in the application layer protocol.

The communication of management information among management entities is realized through exchange of seven protocol messages: five in SNMPv1 and two more added in SNMPv2. Four of these (*get-request, get-next-request, get-bulk-request*, and *set-request*) are initiated by the manager. Two messages, *get-response* (response to request messages from the manager) and *trap* (unsolicited messages), are generated by the agent. The *inform-request* is a manager-to-manager message.

As the various broadband access network technologies are discussed in the following sections, two key areas are relevant to their management of the network and services. In order to facilitate centralized and remote management using SNMP, new management objects must be defined, grouped, and structured. In other words, a MIB must be developed and agreed upon as standard by the industry at large, so that equipment developed by different vendors is interoperable and could be managed by a common network management system. Furthermore, OSSs need to be developed for the network of components of each technology and management of services, each independently, and jointly when they depend on each other. These will be addressed as emerging technology is discussed.

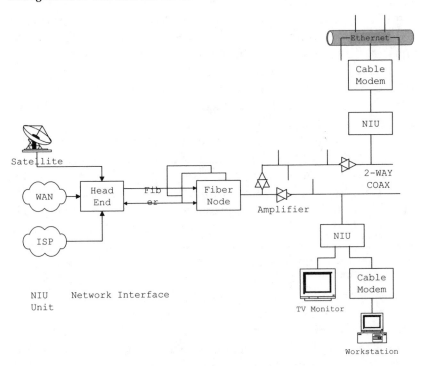

Figure 8.3 Cable access network.

8.1.9 Operations Support Systems (OSSs)

Network management applications are fault, configuration, account, performance, and security (FCAPS) management. Operations support systems, including NMS, perform these applications by collecting the data from managed objects using SNMP and/or other management protocols.

8.2 BROADBAND CABLE ACCESS NETWORK (BCAN)

Figure 8.3 shows the basic components of a broadband cable access network (BCAN), also called hybrid fiber coaxial (HFC) network. The HFC network uses a tree structure. It is based on the cable TV system and service is provided by cable service providers, also referred to as multiple system operators (MSOs). The signal is brought to a fiber node via a pair of optical fibers and then distributed via coaxial cable to the customer premises. The signal is mostly analog radio frequency (RF); each TV station is carried over a 6-MHz channel. One or more channels are used for transmitting broadband multimedia transmission in digital format. Separate channels are used for carrying the upstream and downstream traffic. These channels are of interest in BCAN management.

At the head end, signals from various sources, such as traditional satellite services, analog and digital services using WAN, and Internet service provider (ISP) services using private backbone network, are multiplexed and up-converted from electrical (radio frequency) to optical signals. The communication is one way on the optical fiber, which is a passive medium. There is a pair of optical fibers from the head end to the fiber node; each carries one-way traffic in the opposite direction. The HFC plant consists of multiple pairs of optical fibers to the fiber nodes. Each node serves typically 200 to 2000 households. The optical signal is down-converted to RF at the fiber node and travels over the coaxial cable in a duplex mode. The coaxial medium is an active medium.

The signal that goes from head end to the customer premises is called downstream signal and that going from the customer premises to head end is called upstream signal. The downstream transmission is broadcast mode. The upstream traffic from the cable modems to the head end is coordinated by the head end. Duplex mode of communication is achieved in U.S. by transmitting the downstream signal in high-frequency band (approximately from 50 to 860 MHz) and the upstream signal in the low-frequency band of 5 to 42 MHz. The downstream signal includes analog cable television. The last section of the HFC plant consists of the section from the coax running along the street to the network interface unit (NIU) in the house.

At the customer premises, NIU is the demarcation point between the customer and service provider networks. The analog signal is split at NIU. The TV signal is directed to TV and the data to cable modem. The cable modem converts the

RF signal to an Ethernet output feeding a PC or LAN. Telephone signal is also transmitted along with video and data in some cable sites.

The cable modem modulates and demodulates the digital signal from the customer equipment to the RF signal carried on the cable. A similar operation occurs at the head end equipment. A single 6-MHz channel in the downstream can support multiple data streams. Different modulation techniques support different capabilities. The more common modulation techniques used are quadrature phase shift keying (QPSK) in the upstream, which is more sensitive to ingress noise at the low end of the spectrum, and quadrature amplitude modulation (QAM) in the downstream.

All the cable modems terminate on a router called the cable modem termination system (CMTS) at the head end. The HFC connects the cable modems to the CMTS at the head end. CMTS performs two functions of routing (to the external network) or bridging (intra-access network). It is the gateway to the external network from the access network and multiplexes and demultiplexes the signals from the cable modems to interface to the external network.

The second function that CMTS performs as a bridge is frequency conversion. The upstream message to another cable modem in the broadband LAN is converted to the downstream carrier frequency by the CMTS and propagated downstream as a broadcast message. The receiving cable modem picks up the message by reading the destination address in the message. The CMTS interfaces with operations support systems that serve the function of managing the access network. It also supports the security and access control system to handle the integrity and security of the access network.

8.2.1 DOCSIS and IETF Standards

DOCSIS (data over cable interface specifications) is the industry standard developed by CableLabs, a consortium of MSOs, to ensure interoperability of cable modems and CMTSs manufactured by different vendors. DOCSIS 1.0 is the first version of the specifications and the most deployed in the field; DOCSIS

1.1 and DOCSIS 2.0 are enhancements to it. DOCSIS along with the MIBs developed by IETF address the network management requirements needed to manage the broadband cable access networks remotely.

8.2.2 DOCSIS 1.0

The broadband cable access network comprising cable modems, CMTS, and HFC operating in the RF spectrum is more complex to manage than the computer or telecommunication networks. Management of the computer network is involved with data layers, data link layer, and above. Telecommunication network management is primarily involved with physical layer management and BCAN management is involved with both. Part of the HFC link is fiber and the other part is coaxial cable. There is the complexity of frequency spectrum management. Because cable access technology is serviced by a multiple system operator, who must deal with other content and network service providers in close business relationships, service and business management need to be addressed. These are the top two layers of the telecommunications management network (TMN) defined by OSI.[1,6] Numerous OSSs have been developed to address these issues.

The top half of Figure 8.4 shows the system reference architecture of data-over-cable services and interfaces.[1] It is made up of a head end, HFC link, cable modem, and subscriber PC.[7] The head end is connected to WAN. Multiple head ends could be connected via the WAN to a regional center head end. In such a case, the local head end may be referred to as a distribution hub.

The bottom half of Figure 8.4 presents an expanded view of the head end, which comprises CMTS, switch/router, combiner, transmitter, receiver, splitter and filter, servers, operations support system/element manager, and security and access controller. The CMTS consists of a modulator, *mod*, and a demodulator, *demod*, on the HFC link side, and a network terminator, *term*, to the switch/router connecting to the wide area network. The modulator is connected to the combiner, which multiplexes the data and video signals and feeds them

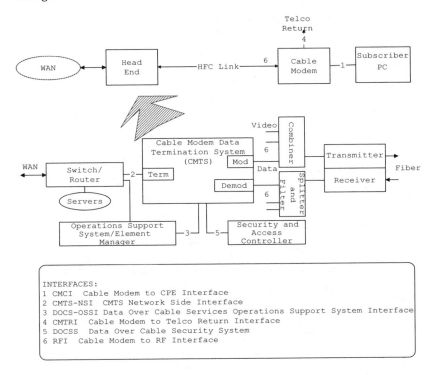

Figure 8.4 Data-over-cable system reference architecture.

to the transmitter. The RF signal is converted to optical signal in the transmitter. The receiver converts down the optical signal to the RF level and feeds it to the splitter, where the various channels are split. The demodulator in the CMTS demodulates the analog signal back to digital data.

Servers at the head end handle the applications and databases. The security function is managed by the security and access controller. The operations support system and element manager perform the functions of management at various management levels: elements, network, and service.

Six interfaces are indicated in Figure 8.4. DOCCIS categorizes these into three groups: (1) data interfaces; (2) operations support system interfaces and telephony return interface; and (3) RF and security interfaces. The documents pertaining to these categories can be downloaded from

Table 8.1 DOCSIS 1.0 Documentation

Designation	Title
SP-BPI-I02-990319	Baseline Privacy Interface Specification
SP-OSSI-I02-990113	Operations Support System Interface Specification
SP-OSSI-RFIV01-I01-990311	Operations Support System Interface Radio Frequency MIB
SP-OSSI-TR	Operations Support System Interface Specification Telephony Return MIB
SP-OSSI-BPI-I01-980331	Operations Support System Interface Baseline Privacy MIB
RFC 2669	DOCSIS Cable Device MIB
RFC 3083	Baseline Privacy MIB
TR-DOCS-OSSIW08-961016	OSSI Framework
RFC 2670	DOCSIS RF Interface MIB
draft-ietf-ipcdn-tri-mib-00.txt, July 30, 1998	Telephony-Return Interface MIB for Cable Modems and CMTS
draft-ietf-ipcdn-interface-mib-03.txt, January 1998	MCNS Interface MIB
draft-ietf-ipcdn-qos-mib-07.txt, February 1, 2003	DOCSIS Quality of Service MIB

http://www.cablemodem.com/specifications. Management-related documents are listed in Table 8.1.

8.2.3 BCAN MIBs

Cable modems and cable modem terminating systems (CMTSs) are managed using SNMP management. Different vendors implement the network management function in different ways. Some cable modems and CMTSs have built-in agents and are managed from a centralized NMS directly. However, others have the network management agent interface built into the CMTS, which acquires information on individual cable modems.

Figure 8.5 shows the MIBs associated with BCAN that are relevant to managing cable modem and CMTS. The MIBs could be grouped into three categories. The first category is the generic set of IETF MIBs, *system* {*mib-2 1*}, *interfaces* {*mib-2 2*},[8] and *ifMIB* {*mib-2 31*}, that describes interface

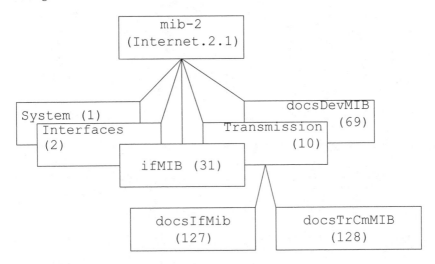

Figure 8.5 BCAN management MIBs.

types.[9] The second category comprises MIBs for the interfaces of cable modem and CMTS, *docsIfMib*. The *docsIfMib* {*mib-2.transmission 127*} is a subnode under transmission and includes objects for CMTS and cable modem {*docsIfMIBObjects*}, baseline privacy interface {*docsBpiMIB, docsBpi2MIB*}, and QoS {*docsQosMIB*}. The *docsTrCmMIB* {*mib-2.transmission 128*} specifies the telephony-return (or telco-return) interfaces for cable modem and CMTS. The third category deals with the set of objects for cable modem and CMTS. Here, only the second and third categories will be discussed; Subramanian[1] and Stallings[5] contain details on the first category.

Figure 8.6 shows the data over cable system (DOCS) interface MIB that supplements the standard SNMP interface MIBs, RFC 1213 and RFC 1573. The subnodes 1, 2, and 3 under *docsIfMIBobjects* are shown in Figure 8.7 and address the base objects common to cable modem and CMTS and the individual objects specific to cable modem and CMTS, respectively. The notification subgroups in the interface MIBs deal with traps. The baseline privacy MIB, *docsBpiMIB*, deals with the privacy requirements for the cable modem and CMTS. The specifications include definition of objects, authorization

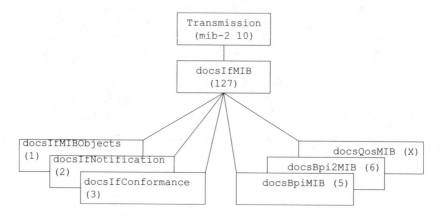

Figure 8.6   DOCS interface MIB.

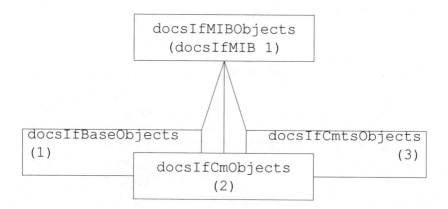

Figure 8.7   DOCS interface MIB objects.

tables, encryption keys, and multicast control tables. The *docsbpi2MIB* addresses enhancements to baseline privacy in DOCSIS 1.1. DOCSIS quality of service MIB, *docsQoSMIB,* is under draft status and is identified as node(X) under *docsIfMIB*. It describes the control of QoS features for the cable modem and CMTS.

As an example of how the network management process works, an NMS would remotely acquire a serial number of a cable modem by sending a request:

```
GetRequest 192.168.100.100 cableco {docsDevBase 4}.0
```

where 192.168.100.100 is the IP address of the cable modem, cableco is the community name (only management entities with common community name could communicate with each other), and *docsDevSerialNumber* is the fourth node under *docsDevBase*, which is a node in the branch under *docs-DevMIB* shown in Figure 8.5. The management agent residing in the cable modem would respond with GetResponse message and would provide its manufacturerís serial number. The ".0" at the end of the message indicates that the entity is scalar.

```
GetResponse 192.168.100.1 cableco {docsDevBase 4}.0
SN123456
```

Another example would be when an unauthorized source tries to access the cable modem with a wrong community name. In this situation, the management agent in the cable modem would generate an unsolicited *authenticationFailure* (trap type 4) trap message to the management system, identifying itself as xyzcable enterprise:

```
trap xyzcable 192.168.100.100 cableco {trapType 4} 12:23
```

The format in the preceding examples is conceptual and not an actual data format. It is intended to present the type of message.

8.2.4   OSS Framework and Interface Specifications

OSSs support the management of data and telecommunications networks and network components. As mentioned earlier, the telecommunications industry established the term OAMP, which stands for operations, administration, maintenance, and provisioning. These are the functions needed to support the smooth functioning of the network. ISO developed the TMN (telecommunications management network) standard that addresses the five layers of management and is shown in Figure 8.8. The lowest layer is the network element layer comprising network elements. The second layer is the element management system. Most network management systems fall under this category. The third layer is the network management layer, which manages the network and

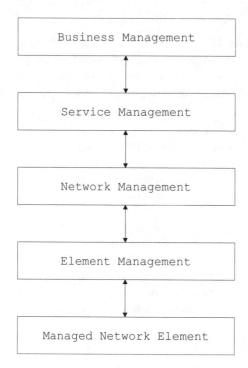

Figure 8.8 TMN architecture.

subnetworks. The service management layer is concerned with managing the services provided by the service provider — MSO in this case. The top layer, business management, deals with the business operations. The OSS is concerned with systems that deal with the top four layers; DOCSIS specifications address these four layers.

OSI specifications define management application functions — fault, configuration, accounting, performance, and security (FCAPS) that enable service providers to manage the network components remotely. OAMP defined by the telecommunications industry was mapped to FCAPS and the functions defined by TMN in Section 8.1. OSS for BCAN uses SNMP to gather data from the access network components.

TR-DOCS-OSSIW08-961016 and SP-OSSI-I02-990113 listed in Table 8.1 describe the framework and interface specifications, respectively. The former outlines assumptions about key business processes and operational scenarios of potential providers of such services, so that the DOCS-OSSIS specification may be examined in the context of plausible service and business models based on those assumptions. It defines the protocol and managed objects required to ensure interoperability of CMs and CMTSs with OSSs and business support systems (BSSs). The business processes involved are service delivery, service assurance, billing, capacity planning, and configuration management.

An overview of management requirements for DOCS is presented in the SP-OSSI-I02-990113 document. Managed objects are specified in SNMP MIBs in the documents on radio frequency interface (RFI), baseline privacy interface (BPI), and telephony return interface (TRI) listed in Table 8.1, and were discussed in Section 8.2.3. OSSI for RF interface is used for fault and performance management. Fault management comprises remote monitoring and fault detection. These are done by polling by NMS or traps generated by management agents; the fault is prioritized into eight levels. Performance management of media access control (MAC) and physical layers is done by measuring the rates of upstream traffic and collision. Bridge traffic is measured at the LLC layer. Security management includes privacy and authorization and is accomplished using BPI MIB.

8.2.5   DOCSIS 1.1

DOCSIS 1.1 has several major enhancements over DOCSIS 1.0. They are related to performance, QoS, multicast service, and security, which affect OSSs in general and network management in particular. DOCSIS 1.1 builds upon DOCSIS 1.0, and several features[10] that can be classified as QoS-related enhancements and OSS-related enhancements. The former includes service classes designated by classifier IDs; dynamic services, which can change classifier IDs dynamically; concatenation and fragmentation of packets; payload suppression; and IP multicast. The latter includes SNMPv3 implementation and FCAPS implementation specifications. The documents pertaining to DOCSIS 1.1 can be downloaded from

http://www.cablemodem.com/specifications. Network management-related enhancements will be considered here.

8.2.5.1 QoS-Related Enhancements

Several QoS-related enhancements are introduced in DOCSIS
1.1. These include packet classification and flow identification,
service flow QoS scheduling, fragmentation, and the two-
phase activation model (first, admission of resources and then
activation of resources). The packets traversing the RF MAC
layer interface are classified into service flows. A service flow
is a unidirectional flow of packets that is assigned a particular
QoS, which is provided by CM and CMTS by shaping, policing,
and prioritizing the packets. Service flows exist in upstream
and downstream directions. Each packet arriving at BCAN is
matched to a classifier that determines its priority and is
forwarded to a QoS service flow. Downstream classifiers are
applied by the CMTS and the upstream classifiers are applied
by the CM and policed by the CMTS.

The priority management is done using packet classification table. Classifiers can be added to the table using SNMP
management operations or dynamic operations using dynamic
signaling at DOCSIS MAC sublayer interface. In the latter
case, SNMP-based management can monitor the classifiers.
DOCSIS 1.1 supports IP multicast mode. Because of the compatibility requirement between versions 1.0 and 1.1, multicast
support is provided in DOCSIS 1.0 using SNMP.

8.2.5.2 OSS-Related Enhancements

DOCSIS 1.1 OSS interface specification details management
of BCAN using SNMPv3 specifications. It addresses all management applications: fault, configuration, accounting, performance, and security (FCAPS). A subset of management-
related documents for DOCSIS 1.1 is shown in Table 8.2. SP-
OSSIv1.1 lists all the IETF/IPCDN (Internet Engineering
Task Force/IP over cable data network) and RFCs applicable
to MIBs for CM and CMTS in DOCSIS 1.1.

DOCSIS 1.1 supports all three versions of SNMP:
SNMPv1, SNMPv2C, and SNMPv3. Furthermore, it is

Table 8.2 DOCSIS 1.1 Documentation on Network Management

Designation	Title
SP-BPI+	Baseline Privacy Plus Interface Specification
SP-OSSIv1.1	Operations Support System Interface Specification
Draft-ietf-ipcdn-bpiplus-mib-08	MIB for DOCSIS for CM and CMTS for Baseline Privacy Plus

required to support the coexistence of all three versions according to RFC 2576.[11] The baseline privacy interface specifications in DOCSIS 1.0 addressed the security considerations on privacy and authorization; DOCSIS 1.1 extends it to authentication using SNMPv3 in baseline privacy plus interface specifications. Cryptographic suites pairing data encryption standard (DES) and cyclic block check (CBC) authentications have been defined. It also supports key management using the Diffie–Hellman exchange mechanism for user-based security model (USM) specified in SNMPv3.[12] A BPI+ security associations ID is defined that ensures secure communication between the CMTS and CMs in upstream and downstream directions for unicast and multicast traffic.

8.2.6 DOCSIS 2.0

DOCSIS 2.0 enhancements are heavily focused on improving the performance of upstream traffic. This is achieved by implementation of more efficient modulation techniques using two different protocols: synchronous CDMA (code division multiple access) and asynchronous TDMA (time division multiple access). The upstream channel bandwidth is increased from 3.2 MHz in DOCSIS 1.0 and 1.1 to 6.4 MHz in DOCSIS 2.0. Maximum data rate per channel is 5.12 Mbps in DOCSIS 1.0, 10.24 Mbps in DOCSIS 1.1, and 30.72 Mbps in DOCSIS 2.0.

Network management requirements to support a DOCSIS 2.0 are specified in detail in DOCSIS operations support system interface specification, SP-OSSIv2.0-I03-021218. The SNMPv3 protocol is required to coexist with SNMP v1/v2. The RFCs and MIB requirements are detailed, including interface numbering, filtering, and event notifications. Network management applications, FCAPS, are incorporated in this specification for better understanding of managing a high-speed cable modem environment. The requirements on performance management are made more stringent due to the high data rate requirement in the upstream in the presence of ingress noise.

8.3 BROADBAND DSL ACCESS NETWORK (BDAN)

The broadband digital subscriber line access network (BDAN) technology for broadband services is motivated by the preexistence of local loop facilities to most households. An unloaded twisted pair of copper wire from central office to a residence can carry a digital signal of >1 Mbps up to 18,000 ft compared to a single 4-kHz analog voice signal (digital equivalent of 56-kbps data rate) in POTS (plain old telephone system). This is the basic concept behind DSL (digital subscriber line) technology.

The distance can be increased for analog telephony if loaded cables (not only passive copper) that compensate for loss and dispersion are used. However, they cannot support the digital subscriber loop as the loaded coils attenuate high frequencies. Many modern communities have been cabled with fiber coming to the curb with digital multiplexer at the end of the fiber. The length limitation of copper cable in this configuration is practically eliminated. Multiplexing is done at the termination of fiber loop by using DSLAM (DSL access multiplexer).

The basic xDSL architecture consists of an unloaded pair of wires connected between a transceiver unit at the central office and a transceiver unit at the customer premises. This transceiver multiplexes and demultiplexes voice and data and converts the signal to the format suitable for transmission on the xDSL link. Of the various forms of xDSL, asymmetric digital subscriber line (ADSL) is the most deployed and very high data rate digital subscriber line (VDSL) is being introduced; both operate asymmetrically. As in BCAN, the downstream signal has a higher data rate and the upstream has

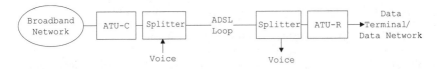

Figure 8.9 A simplified ADSL access network.

a lower data rate. The functional difference between ADSL and VDSL is that VDSL operates at a higher data rate than ADSL and over shorter lines.

A simplified access network using ADSL is shown in Figure 8.9 and consists of an ADSL transmission unit (ATU) and a splitter at each end of the ADSL line. The ATU at the central office is the ATU-C and the one at the customer residence is the ATU-R. The data and video signals from the broadband network are converted to an analog signal by the ATU-C and multiplexed and demultiplexed. The splitter at the central office combines the POTS voice signal and the broadband signal. The reverse process occurs at the splitter and ATU-R at the customer premises (residence).[1]

8.3.1 DSL Forum-Related Standards

The standards for BDAN are addressed by the various standards organizations. ANSI (American National Standards Institute) T1.413 is ANSI standard for XDSL at the physical layer protocol level. In order to accelerate the interoperability and implementation of ADSL, the industry has established a consortium, the DSL Forum. This group is developing specifications on issues associated with end-to-end system operation, management, and security. IETF has developed MIBs specific to DSL technology for OAMP/FCAPS of BDAN. Some of the key documents related to management of BDAN are given in Table 8.3.[13] More documents on the subject are currently in development; see McCloghrie and Kastenholz[14] for the latest additions.

The general framework for ADSL management is described in ADSL Forum TR-005. TR-027 refers to RFC 2662 and draft-ietf-adslmib-adsl2-00.txt, which present SNMP-based

Table 8.3 BDAN Management-Related Documents

TR-001	ADSL Forum System Reference Model
TR-005	ADSL Network Element Management System
TR-014	DMT Line Code Specific MIB for T1.413 DMT Coded Lines
TR-015	CAP Line Code Specific MIB
TR-016	CMIP-Based Network Management Framework
TR-022	The Operation of ADSL-Based Networks
TR-024	DMT Line Code Specific MIB for ITU G.992.1 and G.992.2
TR-027	SNMP-Based ADSL Line MIB
TR-030	ADSL Element Management System (EMS) to a Network Layer Interdomain Network Management System (NMS)
TR-034	Alternative OAM Communications Channel across the U Interface
TR-035	Protocol Independent Object Model for ADSL EMS-NMS Interface
TR-041	CORBA Specification for ADSL EMS-NMS Interface
TR-047	DSL Service Flow-Through Fulfillment Management Interface
TR-050	CORBAv2 for ADSL EMS-NMS Interface
TR-051	DSL Specific Conventions for the ITU-T Q.822.1 Performance Management Bulk Data File Structure
TR-052	DSL Anywhere Addendum to DSL Service Flow-Through Fulfillment Management Interface
TR-054	DSL Service Flow-Through Fulfillment Management Overview
RFC 2662	Definitions of Managed Objects for the ADSL Lines

ADSL line MIB. TR-016 contains a CMIP specification for ADSL network element management. TR-014 and TR-024 document DMT line code specific MIB and TR-015 specify CAP line code specific MIB. The management documentation is specific to ADSL and is a supplement to standard management MIB. The element management system and network management system specifications are covered in TRs 030, 035, 041, 050, and 051. Service considerations are covered in TRs 047, 052, and 054.

Table 8.3 shows the DSL Forum management system reference model used in BDAN. The management functions addressed in the ADSL-specific documents address the physical layer functions. Management of data layers is addressed by the conventional network management system. The

V-interface is between the ATU-C and the access node and the U-interfaces are off the splitters; the T-interfaces are associated with ATU-R and home network. The POTS interface is between the low pass filter and the telephone set. See Subrumanian[1] and the TRs in Table 8.3 for details on the various interfaces and the components of the model. The management of BDAN is accomplished solely through V–C-interface.

8.3.2 ADSL Network Element Management System

ADSL network element management addresses the parameters, operations, and protocols associated with configuration, fault, and performance management of an ADSL network. Management of this network involves the following five network elements:

- Management communications protocol across the network management subinterface of the V-interface
- Management communications protocol across the U-interfaces between ATU-C and ATU-R
- Parameters and operations with the ATU-C
- Parameters and operations within the ATU-R
- ATU-R side of the T-interface

All management functions in the ADSL network are accomplished via the V–C-interface shown in Figure 8.10. Thus, the management of all elements is accomplished via the V-interface.

The physical layer involves three entities: physical channel, fast channel, and interleaved channel (see Section 8.3.2.1). The fast and interleaved channels need to be managed separately. These two use the same physical transmission medium, which also needs to be managed. In addition to management of the physical links and the channel parameters, the parameters associated with the type of line coding need to be monitored.

As mentioned earlier, the ADSL Forum addresses configuration, fault, and performance aspects of OAMP/FCAPS for ADSL network. These are primarily concerned with the physical layer of ADSL. The higher layer management is left to

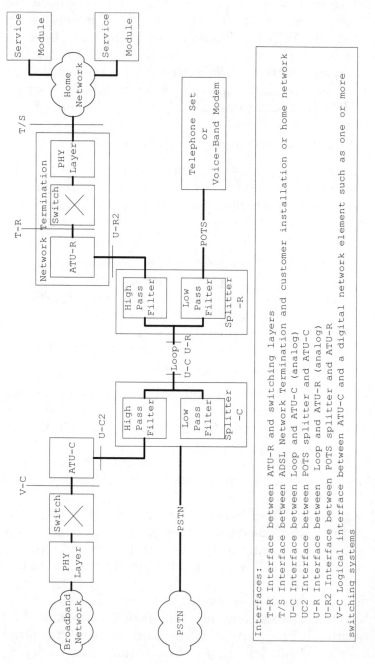

Figure 8.10 DSL forum management system reference model.

the conventional network management system. The element management system (EMS) will be addressed in this section and the EMS–NMS interface aspects in Section 8.3.4.

8.3.2.1 Configuration Management

The various parameters that need to be managed for configuration are listed in DSL Forum TR-005 and are shown in Table 8.4. The table lists the component with which the element is associated, as well as whether it pertains to the physical line or fast or interleaved channel. A brief description of each parameter is given in the last column. Line/channel in column 3 is identified as physical link (Phy) or either fast (F) or interleaved (I) channel.

The link between the ATU-C and ATU-R could be configured for upstream and downstream traffic into seven channels, referred to as bearer channels. Furthermore, each channel could be configured as a fast channel to handle real-time traffic or interleaved channel to handle data traffic (see Figure 8.11). Four are dedicated downstream channels and the other three are duplex channels that could be used for upstream and downstream traffic. The channels could be configured using SNMP MIB described in Section 8.3.3 as one of five options: no separation of channels, fast, interleaved, either, or both. These are the values for the line types listed in Table 8.4. ADSL line coding is the type of modulation scheme used, DMT, which is the standard or CAP supported as legacy system.

In Table 8.4, the noise margin elements — target, maximum, minimum, upshift, and downshift — define five levels of noise thresholds from the highest, defined by the maximum noise margin, to the lowest, defined by the minimum noise margin. These levels are shown in Figure 8.12. The transmitted power of the modem is decreased or increased based on these thresholds. The transmission rate can be increased if the noise margin goes above a threshold level that is beneath the maximum noise margin threshold. Similarly, the transmission rate should be decreased if the noise margin falls below a certain threshold that is higher than the minimum

Table 8.4 Configuration Management Elements

Element	Component	Line/channel	Description
ADSL line type	ADSL Line	N/A	Five types: no channel; fast; interleaved; either; or both
ADSL line coding	ADSL Line	N/A	ADSL coding type
Target noise margin	ATU-C/R	Phy	Noise margin under steady state (BER = $<10^{-7}$)
Max. noise margin	ATU-C/R	Phy	Modem reduces power above this threshold
Min. noise margin	ATU-C/R	Phy	Modem increases power below this margin
Rate adaptation mode	ATU-C/R	Phy	Mode 1: manual Mode 2: select at start-up Mode 3: dynamic
Upshift noise margin	ATU-C/R	Phy	Threshold for modem increases data rate
Min. time interval for upshift rate adaptation	ATU-C/R	Phy	Time interval to upshift
Downshift noise margin	ATU-C/R	Phy	Threshold for modem decreases data rate
Min. time interval for downshift rate adaptation	ATU-C/R	Phy	Time interval to downshift
Desired max. rate	ATU-C/R	F/I	Max rates for ATU-C/R
Desired min. rate	ATU-C/R	F/I	Min. rates for ATU-C/R
Rate adaptation ratio	ATU-C/R	Phy	Distribution ratio between fast and interleaved channels for available excess bit rate
Max. interleave delay	ATU-C/R	F/I	Max. transmission delay allowed by interleaving process
Alarm thresholds	ATU-C/R	Phy	15-min count threshold on loss of signal, frame, poser, and error-seconds
Rate-up threshold	ATU-C/R	F/I	Rate-up change alarm
Rate-down threshold	ATU-C/R	F/I	Rate-down change alarm
Vendor ID	ATU-C/R	Phy	Vendor ID assigned by T1E1.4
Version no.	ATU-C/R	Phy	Vendor-specific version
Serial no.	ATU-C/R	Phy	Vendor-specific serial no.

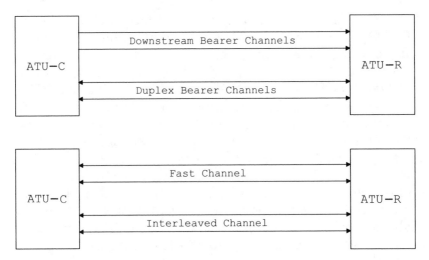

Figure 8.11 ADSL channel configuration.

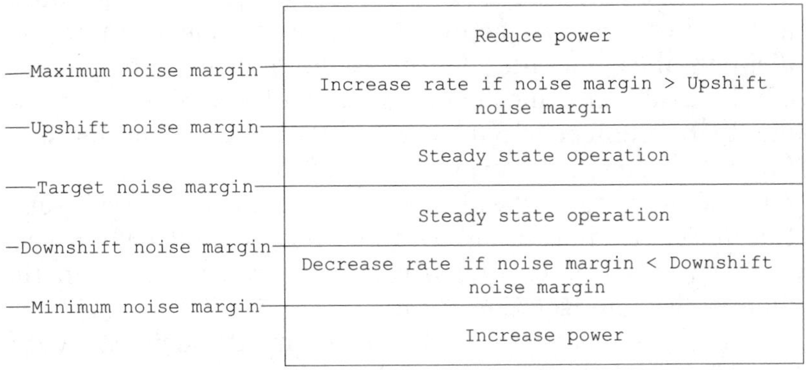

	Reduce power
—Maximum noise margin—	Increase rate if noise margin > Upshift noise margin
—Upshift noise margin—	Steady state operation
—Target noise margin—	Steady state operation
—Downshift noise margin—	Decrease rate if noise margin < Downshift noise margin
—Minimum noise margin—	Increase power

Figure 8.12 ADSL noise margins.

noise margin. Target noise margin is the value of the normal steady state operation. These values are configured as part of configuration management, and the system automatically and dynamically adjusts the power level and the rate based on the signal-to-noise ratio in the link.

A DSLAM at an ADSL access node has hundreds of ATU-Cs and is configured to link to an equal number of ATU-Rs. It would be impractical to provision all the parameters for each ATU-C individually. A MIB table, *adslLineConfProfileTable {adslMibObjects.14}* discussed in Section 8.3.3, addresses this configuration issue. The table contains the information on the ADSL line configuration shown in Table 8.4. One or more ADSL lines may be configured to share common profile information and can be changed in a dynamic mode. A second mode of configuration specifies the static mode of setting up an ADSL configuration profile. Each ADSL line interface has a static profile and a configuration profile. The alarm profile element value listed in Table 8.4 could also be structured in a manner similar to the configuration profile explained earlier.

8.3.2.2   Fault Management

DSL Forum specifications define fault management as the process of identifying fault condition, determining its cause, and taking corrective action. A network management system displays alarm on fault identification. Because of the practice of telephone companies in managing faults by alarm management, the number of alarms could be large and needs to be filtered carefully. For example, an ATU-R getting turned on and off could generate alarms, but should be ignored. After the automatic indication of faults, ATU-C and ATU-R self-tests as specified in T1.413 could be used to assist in the diagnostics and isolation of fault.

Table 8.5 lists the fault management elements. ADSL line status shows the current state of the line as to whether it is operational or a loss of any of the parameters on frame, signal, power, or link has occurred. It also indicates initialization errors. Alarms are generated when the preset counter reading exceeds 15 minutes on loss of signal, frame, power, link, and error-seconds. No distinction between major and minor alarms is made. The alarm threshold is set to a value in a counter, which if exceeded in a 15-minute interval would generate an alarm. Unable to initialize, ATU-R will generate

Table 8.5   Fault Management Elements

Element	Component	Line/ channel	Description
ADSL line status	ADSL line	Phy	Indicates operational and various types of failures of the link
Alarms thresholds	ATU-C/R	Phy	Generates alarms on failures or crossing of thresholds
Unable to initialize ATU-R	ATU-C/R	Phy	Initialization failure of ATU-R from ATU-C
Rate change	ATU-C/R	Phy	Event generation when rate changes while crossing shift margins in upstream and downstream

an alarm. Alarms are generated on rate change caused by noise margin thresholds.

8.3.2.3   Performance Management

Table 8.6 shows the elements associated with ADSL performance management. Each ATU's performance in terms of line attenuation, noise margin, total output power, current and previous data rate, along with the maximum attainable rate, channel data block length (on which the CRC check is done), and interleave delay, can be monitored. In addition, statistics are gathered for 15-minute and 1-day intervals on the error-seconds statistics. Two counters are maintained by each ATU to measure each error condition. The error statistics are maintained for loss of signal seconds, frame seconds, power seconds, link seconds, erred seconds, transmit blocks, receive blocks, corrected blocks, and uncorrectable blocks.

8.3.3   SNMP Management

The element management system could be implemented using SNMP or OSI/CMIP protocol. SNMP (TR-027)- and CMIP (TR-016)-based specifications have been developed for ADSL.

Table 8.6 Performance Management Elements

Element	Component	Line	Description
Line attenuation	ATU-C/R	Phy	Measured power loss in dB from transmitter to receiver ATU
Noise margin	ATU-C/R	Phy	Noise margin in dB of the ATU with respect to received signal
Total output power	ATU-C/R	Phy	Total output power from the modem
Max. attainable rate	ATU-C/R	Phy	Max. currently attainable data rate by the modem
Current rate	ATU-C/R	F/I	Current transmit rate to which the modem is adapted
Previous rate	ATU-C/R	F/I	Rate of the modem before the last change
Channel data block length	ATU-C/R	F/I	Data block on which CRC check is done
Interleave delay	ATU-C/R	F/I	Transmit delay introduced by the interleaving process
Statistics	ATU-C/R	Phy; F/I	15 min/1 day failure statistics

The SNMP-based MIB specified in RFC 2662 will be discussed here. ADSL SNMP MIB is presented in Figure 8.13. It is 94th node under *transmission*. Three nodes are defined under *adslLineMib*. Figure 8.13 contains objects under *adslMibObjects* and *adslTraps* that pertain to the configuration management parameters defined in Table 8.4, fault management parameters defined in Table 8.5, and performance management parameters in Table 8.6. The entity *adslConformance* {*adslLineMib 3*} ensures the conformance and compliance aspects of ADSL-managed objects and groups.

RFC 2662 details the *adslMibObjects* {*adslLineMib.1*}. The elements defined under the ADSL element management system are specified in the *adslMibObjects and adslTraps* {*adslLineMib 2*}. Node 16 {*adslLCSMib*} under *adslMibObjects* specifies the complementary line code specific objects for the physical link in terms of each of the DMT and CAP modes of operation. For example, corresponding to *adslAtucPhysTable*, is

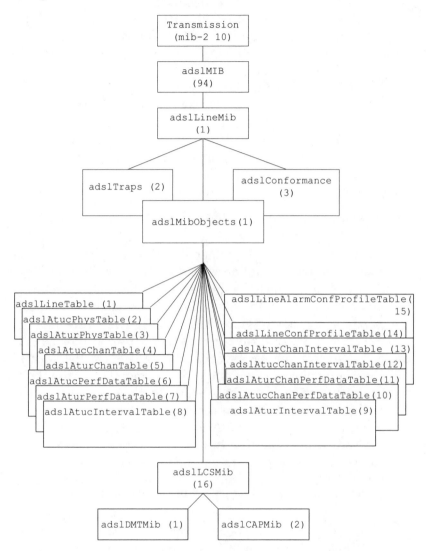

Figure 8.13 ADSL simple network management protocol MIB.

an entity, *adslAtucDMTPhysTable*, under *adslDMTMib*, and an entity, *adslAtucCAPPhysTable*, under *adslCAPMib*. The former specifies a table with each row containing the physical layer parameters associated with the link on an interface. The latter two specify tables in which each row contains parameters associated with a channel on that interface. These interfaces are defined based on RFC 1213 for *interfaces* {*mib-2.2*} and RFC 1573 for *ifMIB* {*mib-2.31*}.

DSL Forum specifications have integrated ADSL MIB with the standard IETF interfaces group of MIB-II[14] by defining *ifType* in the following manner: adsl(94) for asymmetric digital subscriber loop, adslInterleave(124) for ADSL interleave channel, and adsl(125) for ADSL fast channel.

8.3.4 ADSL EMS–NMS Management

The last section, which addressed the management of the physical layer parameters, discussed ADSL EMS. Earlier, it was mentioned that the management of higher layers will be handled by the traditional network management system. DSL Forum TR 030 defines the higher level functional requirements of such an NMS, and TR 035 specifies a protocol-independent object model for ADSL EMS–NMS interface. It is important to observe that the transport protocol for BDAN could be synchronous or asynchronous; the ATM network is the most predominantly deployed scheme. Thus, the DSL Forum specifications for EMS–NMS interface are focused on ATM as the higher layer protocol.

Figure 8.8 shows the TMN layered architecture, in which the NMS is the next higher layer to EMS. A top-down network view of a broadband NMS managing multiple EMSs is shown in Figure 8.14. This is an adaptation of a network management view of ATM network[1] and TR-030. The left-hand side of the subnetwork is a multi-supplier network and the right-hand side is a single-supplier network. Each subnetwork has an EMS managing the ADSL network elements (NE) in it. The M4 interface is defined by ATM Forum as the interface between public NMS and public network.[1]

TR-030 specifies the implementation requirements for configuration, fault, and performance management. TR-035 specifies the management information model to be used at the interface between EMS and NMS. The model specification

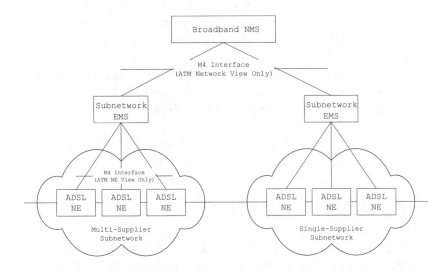

Figure 8.14   Network view of NMS-EMS physical configuration.

is independent of management protocol and includes an appendix that outlines the applicability of the ATM Forum M4 network view model to EMS–NMS interface. The TR documents in Table 8.3 deal with business aspects of service flow. These enable automation of the DSL service business-to-business interfaces between various business entities involved in providing DSL service. Other documents in the table deal with CORBA specifications for the EMS–NMS interface communication. The TRs contain details on these.

8.4   BROADBAND MOBILE AND WIRELESS ACCESS NETWORKS

It is important to distinguish clearly between mobile networks and wireless networks before addressing their management. A mobile network has the ability to perform computing anytime and anywhere. The devices on a mobile network may or may not have wireless interface and the network may not use a wireless transmission medium. There are two types of mobility: (1) a cellular network in which the devices are always on as it moves from one cell to another, such as a cell telephone; and (2) a nomadic network, in which the session in the computing device is not active while in motion, such as a laptop connected to LANs. A wireless network has a wireless interface to computing devices and may interface with wired or wireless networks. It is deployed for networking fixed and mobile users.

The network management of broadband mobile and wireless networks is management of integrated wired and wireless networks, as well as fixed and mobile networks, that carry voice, video, and data traffic. This section will consider the management of various fixed wireless networks — MMDS (multichannel multipoint distribution service), LMDS (local multipoint distribution service), IEEE 802.16 fixed wireless system, and IEEE 802.16a wireless MAN (metropolitan area network). Satellite wireless networks are currently used primarily for video broadcasting service but, in the future, will be expanded to broadband service. The section will then treat mobile wireless network primarily focusing on 3-G broadband cellular network and will finally consider the fast growing IEEE 802.11 LAN.

8.5 FIXED WIRELESS NETWORKS

The early deployment of last mile coverage of fixed broadband wireless access network (BWAN) currently uses cable modem technology with terrestrial wireless replacing HFC plant. This is shown in Figure 8.15 and is the basis for MMDS. IEEE 802.16 is a broader set of specifications for fixed wireless networks; however, this is not deployed anywhere commercially in North America. The transceiver at either end of a wireless local loop performs the up-conversion and down-conversion of the RF signal to the spectral band of the wireless local loop. IEEE 802.16 operates at a frequency spectrum starting at 11 GHz and requires line-of-sight transmission. Currently, work is in progress for IEEE 802.16a that operates

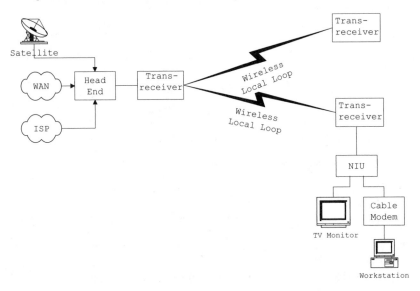

Figure 8.15 Fixed wireless access network for MMDS.

at 2 to 11 GHZ and does not require line-of-sight transmission. The application for this is wireless MAN.

Figure 8.16 shows the broadband wireless network architecture with multiple base station units (BSUs) serving residential and commercial subscribers. The core network shown is a SONET/ATM network, although, in practice, it could be any other transport scheme, such as T1 or IP network.

8.5.1 Multichannel Multipoint Distribution Services

Two versions of fixed wireless systems have been deployed: LMDS (local multipoint distribution system) and MMDS (multichannel multipoint distribution system); the difference between the two is the spectrum and range of their operations. LMDS operates over two frequency bands at 27,500 to 28,350 MHz and 31,000 to 31,300 MHz. The LMDS range is about 3 miles. MMDS operates at 2500- to 2686-MHz band and serves a longer range of up to 35 miles.

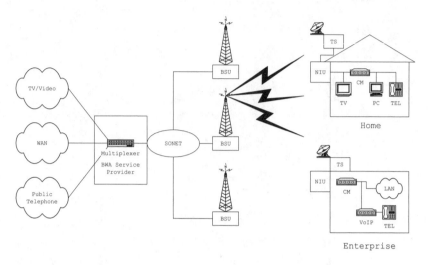

Figure 8.16 Broadband wireless network architecture.

The management of the cable modem technology part of the fixed wireless access network for MMDS follows the same procedures as discussed in Section 8.2. The wireless medium part needs additional management parameters for the base station units comprising edge routers and radio frequency transceivers, and for the transceivers at customer premises. The routers could be managed by an SNMP-based management system. The physical layer management of the radio frequency link depends on the type of transmission. For a SONET/ATM (synchronous optical network/ATM), the ATM Forum has developed specifications for management. The management procedures could be proprietary or standard based, depending on the type of transportation.

8.5.2 Local Multipoint Distribution Services

LMDS deployment, which is minimal, has been implemented by various users using different technologies; no standards are in practice. However, management of fault, configuration, accounting, performance, and security (FCAPS) can be implemented using standards such as SNMP based on the terminating end equipment used at the base station and at customer premises.

8.5.3   IEEE 802.16 Fixed Wireless Network

In spite of the lack of deployment of fixed wireless IEEE 802.16 as of now, extensive specifications have been developed on standards for implementation. The operation is at 11- to 66-GHz band with data rate of 2 to 155 Mbps with flexible asymmetry. The downstream transmission is TDMA (time division multiple access) in broadcast mode and the upstream is TDMA with DAMA (demand assigned multiple access). The components to be managed are subscriber station, base station, wireless link, and RF spectrum. The IEEE 802.16 Work Group has recommended the adoption of 802 standards framework for LAN/MAN management ISO/IEC 15802-2(E). Managed objects are to be generated using OSI GDMO (guidelines for definition of managed objects). As an alternative, SNMP could also be used for management.

Implementation of IEEE 802.16.1 specifications enables meeting SLA commitments, as well as accounting and auditing functions. The security specifications address two levels of authentication — between subscriber station and base station at the MAC level and between subscriber and the broadband wireless access system for authorization of services and privacy.

IEEE 802.16.1 specifications support classes of service with various QoS for bearer services, bandwidth negotiation for connectionless service, state information for connection-oriented service, and various ATM traffic categories: constant bit rate, variable bit rate real time, variable bit rate non-real time, and adjustable bit rate. IETF traffic categories of integrated services and differentiated services are also supported by the specifications.

8.5.4   IEEE 802.16d WirelessMAN/WiMax

IEEE 802.16d, a modified version of 802.16a and 802.16c also known as WiMax, is an extension of 802.16. The operation is in 2- to 11-GHz band and is primarily intended as an MAN. The network management considerations specified in 802.16 may be applied to it. Quality of service and performance are maintained high by implementing TDMA downstream and

TDMA/DAMA upstream. Detailed specifications on network management are yet to be developed for 802.16a.

8.6 CELLULAR ACCESS NETWORKS

Mobile wireless is primarily used for voice now using 2-G based GSM (global standard for mobile communications) and GPRS (general packet radio service) cell systems. It is migrating toward 3-G and 4-G IP-based systems, which could be accomplished by the existing cell network architecture or rapidly evolving wireless LANs. In either case, the mobile IP would play a significant part in managing the mobile elements.

In addition to the management of new elements and mobile IP parameters, the management of mobile wireless includes management of mobility, resources, and power. Mobility management addresses the smooth handling of the mobile node from one foreign agent to another while roaming. Resource management deals with call admission and control based on the availability of resources. Power management is concerned with ensuring the maximum power of transmitters to mitigate the interference of a signal between cells.

8.6.1 Mobile IP

Mobile IP is analogous to call forwarding in a telephone network, except the forwarding address is mobile while roaming with a mobile unit (station). Figure 8.17 shows cell network architecture with home agent in home network communicating with a mobile unit through a foreign agent and a mobile agent residing in the mobile unit. As the mobile unit roams from one cell to the next, it registers with the foreign agent in that cell and the communication link is transparently transferred to the foreign agent. Each mobile unit is given a mobile IP address associated with a home network and the communication link is always through the home network.

The additional functional entities to be managed in the mobile cellular network are the mobile node, home agent, and foreign agent. The mobile node is a host or router that changes

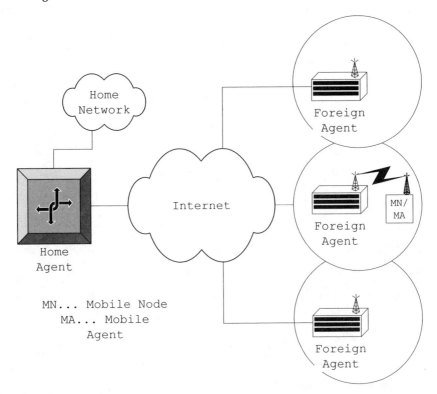

Figure 8.17 Cell network architecture.

point of attachment from one network or subnet to another. The home agent is a router or mobile node's home network that tunnels packets to and from the mobile node via a foreign agent. The foreign agent is a router on a mobile node's visited network that provides services to the mobile node. Figure 8.18 and Table 8.7[15] present the MIB and overall structure of MIB groups associated with the management of Mobile IP.

These groups are under *mipObjects*, which is the 44th node under *mib-2*. The three columns are marked with the groups associated with mobile node, foreign agent, and home agent. The system group (*mipSystemGroup*), security association group (*mipSecAssociationGroup*), and security violation group (*mipSecViolationGroup*) are associated with all three

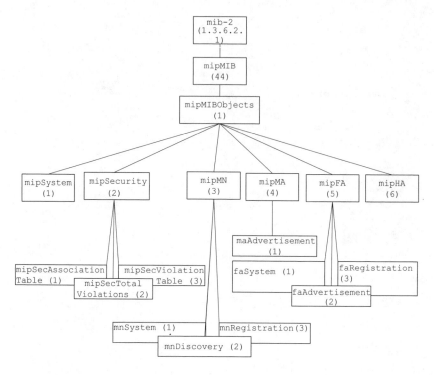

Figure 8.18 Mobile IP MIB.

Table 8.7 Mobile IP MIB Groups

Groups	Mobile node	Foreign agent	Home agent
mipSystemGroup	X	X	X
mipSecAssociationGroup	X	X	X
mipSecViolationGroup	X	X	X
mnSystemGroup	X		
mnDiscoveryGroup	X		
mnRegistrationGroup	X		
maAdvertisementGroup		X	X
faSystemGroup		X	
faAdvertisementGroup		X	
faRegistrationGroup		X	
haRegistrationGroup			X
haRegNodeCountersGroup			X

elements. The prefixes associated with the elements are *mn* for mobile node, *ma* for mobile agent in the mobile node, *fa* for foreign agent, and *ha* for home agent. The registration groups are managed objects associated with registration of mobile nodes with foreign agents. The advertisement groups are managed objects associated with the self-advertisement of the elements in the discovery and registration process. The *haRegNodeCountersGroup* contains statistics kept on the home agent of all events that occurred between the mobile nodes and home agent.

8.6.2 Mobility Management

Mobility management keeps track of a mobile node and executes procedures that include identification and authentication of the mobile subscriber, security, access to wireless services, transfer of subscriber data among network nodes, location updating, and registration.

There are two classes of schemes to locate a mobile node or station. The global positioning system (GPS) is a worldwide radio-navigation system that utilizes a constellation of 24 satellites and their ground stations to locate mobile stations with a spatial resolution of a few meters. The second approach is to use the cellular network to locate the mobile node by measuring the distance and direction from the base stations. Mobility management plays a significant role in the resource QoS management.

According to the OSI model, mobility management acts as a network layer. It works as a signaling and control entity for the mobile and network nodes with which it communicates. A subscriber can access network services only if the mobility management functions are successfully completed, the network has granted permission, and (in some cases) the network has authenticated the user.

Many designs for 3-G wireless networks have been proposed by IMT (the international mobile telecommunication standards defined by the International Telecommunications Union), some of which are very popular and widely used. Among these are PLMN (public land mobile network)-based wireless networks, mobile IP for CDMA2000 (code division multiple access), and satellite-based networks.[16-18]

8.6.3 Resource and Power Management

Resource management deals with scheduling and call admission control, load balancing between access networks, and power management. This has an impact on the management of bandwidth to provide multimedia broadband service as well as hand-off between cells so as to achieve quality of service.

In mobile cellular networks, the service area is divided into cells, each of which is allocated a certain amount of radio spectrum. The size of the cell is managed by the power management system. A mobile user moves from one cell to a different cell during the service and the "hand-off" of the user must take place without dropping the session. The process that controls whether an incoming call can be admitted is called the call admission control (CAC). The resource control mechanism must allocate the limited bandwidth resources to users in an efficient way in order to guarantee the users' QoS requirements. If the hand-off target cell does not have enough bandwidth to support this call, the call will be forced to terminate.

Hand-off calls are commonly given a higher priority in accessing the bandwidth resources in order to provide a seamless connection for users. The call dropping during hand-off is mitigated by reserving bandwidth specifically for that function in each cell.[19] For broadband service, the hand-off management gets more complicated. In addition to hand-off in the same class of service, the system needs to support multiple classes of service in which each class presents different QoS requirements.

8.6.4 Quality of Service Management

Third-generation (3-G) mobile networks and beyond are required to transmit voice, video, and data under the specifications of the universal mobile telecommunications system (UMTS) developed by the Third-Generation Partnership Project (3GPP). The equivalent of 3GPP for the U.S. is 3GPP2.

The UMTS infrastructure is expected to carry various types of applications on the same medium while meeting the QoS objectives. Four QoS classes (TS 23.107) have been defined by 3GPP: conversational, streaming, interactive, and background. These are shown in Table 8.8, along with the required delay, delay variation, and bit error rate (BER). Conversation has the most stringent requirements and background tasks such as e-mail have the least stringent requirements.

The previous two sections mentioned the dependency of QoS on mobility and resource management. The QoS discussed here is applicable only between the base station and the mobile stations. The QoS on the core network is achieved using wired network methodologies such as *intserv* (integrated services) or *diffserv* (differentiated services).

8.6.5 Security Management

Security is a major issue in mobile wireless communication. Wireless application protocol (WAP) and secured socket shell (SSL) are the two approaches commonly used for secured wireless communication. Although SSL is used extensively in wired networks, WAP is the common implementation in mobile wireless networks. It is based on transport layer security (TLS) protocol and performs the normal security functions, which include authentication, authorization, privacy, and address integrity.

Figure 8.19 shows WAP architecture.[20] The mobile node is connected to the WAP gateway through network operator control and remote access server. The security protocol used is wireless TLS (WTLS). TLS protocol is used between the WAP gateway and the remote server that the mobile subscriber is trying to access. Many security failures happen in the transition between the two. Robust security can be achieved by careful design and implementation of the WAP gateway using the available security tools.

WTLS could be implemented for the wireless leg of the link and SSL could be implemented between the WAP gateway and remote server. In this case, the WAP gateway module must perform the decryption and encryption as well as the

Table 8.8 UMTS QoS Specifications

QoS class	Transfer delay	Transfer delay variation	Low BER	Guaranteed bit rate	Example
Conversational	Stringent	Stringent	No	Yes	VoIP, video- and audio-conferencing
Streaming	Constrained	Constrained	No	Yes	Broadcast service, news, sport
Interactive	Looser	No	Yes	No	Web browsing, interactive chat, games
Background	No	No	Yes	No	E-mail, SMS, TFP transactions

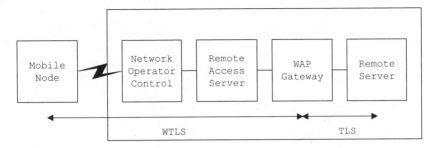

Figure 8.19 WAP architecture.

Table 8.9 802.11 Standards and Amendments

802.11a	54-Mbps data rate 5.15 MHz to 5.35 and 5.4 MHz to 5.825 MHz
802.11b	11-Mbps data rate at 2.4 GHz
802.11e	Addresses QoS issues
802.11f	Addresses multivendor AP interoperability
802.11g	Higher data rate extension to 54 Mbps in the 2.4 GHz
802.11h	Dynamic frequency selection and transmit power control for operation of 5-GHz products
802.11i	Addresses security issues
802.11j	Addresses channelization in Japan's 4.9-GHz band
802.11k	Enables medium and network resources more efficiently

protocol conversion. Plans to implement IP networks are underway in 3GPP and 3GPP2. Under that situation, open standard SNMPv3, which has security built in, could be used for mobile network management.

8.7 IEEE 802.11 WIRELESS LAN

Wireless LAN (WLAN) growth has been very rapid and is being deployed at home, in enterprise, and at public places. WiFi, as it is popularly known, is an IEEE 802.11 protocol. LAN 802.11b and 802.11g operate at 2.4-GHz band and 802.11a operates at 5-GHz band. The IEEE 802.11 working groups have amended 802.11 to address scalability, provisioning, performance, QoS, and security issues. These amendments are listed in Table 8.9. The scalability issue is

Table 8.10 IEEE 802.11 MIB Groups

Entity	OID	Description
ieee802dot11	{iso 2 840 10036}	MIB module for IEEE 802.11 entities
dot11smt	ieee802dot11 1	Station management attributes: WEP security, power, transmission
dot11mac	ieee802dot11 2	Mac attributes
dot11res	ieee802dot11 3	Resource type attributes
dot11phy	ieee802dot11 4	Physical attributes

addressed by having a "wireless" switch or a hub that serves many access points. Many of the access point functions, including some of the management functions, could be centralized and thus the "thick" access points become "thin" access points.

8.7.1 IEEE 802.11 Management Information Base

The 802.11 MIB is still in development by the various IEEE work groups. It is being developed under IEEE 802.11 group in the MIB tree under "iso" root.[21] The subgroups under it, station management, MAC attributes, resource attributes, and physical attributes, are listed in Table 8.10. The SMT object class provides the necessary support at the station to manage the processes in the station. The MAC object class provides the necessary support for the access control, generation, and verification of frame check sequences, and proper delivery of valid data to upper layers. The PHY object class provides the necessary support for required PHY operational information.

8.7.2 Quality of Service Management

An access point has multiple stations and thus acts as a bridge. Unlike wired LAN, in which the transmission between stations is a distributed process, every transmission between stations and the external interface goes through the access point and thus could affect the performance. Also, an access point handling multimedia traffic should be able to satisfy the QoS for each class of service: real-time voice, streaming video, and non-real-time data. The current access points with 802.11a/b/g protocols do not meet the QoS needs of broadband service. IEEE 802.11e is being developed to satisfy the baseband requirements.

The QoS parameters that need to be managed in the MAC and PHY layers are data rate, upper bound on delay, and jitter. The two types of MAC are DCF (distributed control function) and PCF (point control function). In DCF, each station in the LAN accesses the medium on a random access basis. In PCF, each station is polled by the access point and allocated a time slot to transmit. Neither one is satisfactory to achieve the QoS needed for broadband services. IEEE 802.11e has developed a hybrid control function (HCF) that is backward compatible with PCH and enhanced DCF (EDCF).

QoS also depends on the range of transmission from the access point. As the client distance from the access point increases, the data rate not only decreases, but also fluctuates with time significantly. Several antenna technologies can extend the range. Diversity and MIMO (multiple input multiple output) antenna schemes are two approaches to extend the range of the access point. In the diversity antenna scheme, two antennas are used at the access point and at the client. The receiver chooses the better of the signals from the two antennas or a combination of both. In the MIMO antenna scheme, multiple antennas are located at the access point and at the client. With the MIMO algorithm, the range of the access point is significantly increased.

Power management is another approach to extending the range. The maximum power that can be transmitted by a system depends on the frequency in 802.11a. IEEE 802.11h defines the specifications for dynamic frequency selection and transmitter power control.

8.7.3 Security Management

The first attempt to implement security for WLAN is 802.11 WEP (wired equivalent privacy). It has 40-b weak and static

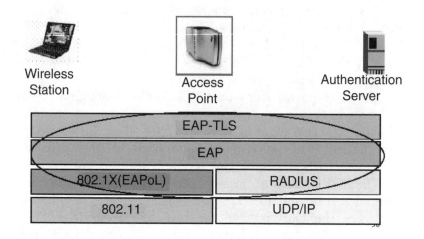

Figure 8.20   IEEE 802.11i authentication components.

encryption keys based on RC4 algorithm and thus easy to break. The WiFi Alliance has tightened the security by announcing that it would certify access points only with WPA (WiFi protected access) — a standard based on the IEEE 802.11i standard that is under development. WPA comprises dynamic temporal key integrity protocol (TKIP) and 802.1x mutual authentication mechanisms. WAP2 is the next phase of security for the WiFi Alliance and is expected to be the same as the IEEE 802.11i standard.

IEEE 802.11i data protocol includes confidentiality, data origin authenticity, and replay protection. The security architecture ties keys to authentication. There are pair-wise keys and group keys. Every session requires a new key. Key management delivers keys to use as authorization tokens after the channel is authorized.

Figure 8.20 shows the authentication components of IEEE 802.11i.[22] The wireless station and the authentication server mutually authenticate each other through the access point. The IEEE 802.1x standard is a robust authentication protocol between a client, which in this case is a wireless station, and authentication server. Several implementation types of 802.1x exist and the interoperability between a client

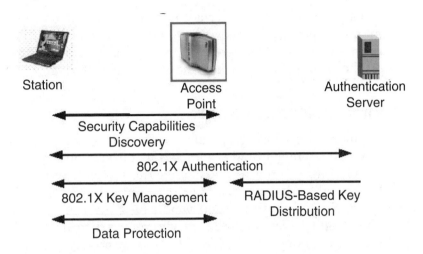

Station   Access   Authentication
Point   Server

Security Capabilities
Discovery

802.1X Authentication

802.1X Key Management   RADIUS-Based Key
Distribution

Data Protection

Figure 8.21 IEEE 802.11i operational phases.

and an access point is achieved by extensible authentication protocol (EAP). There are obviously several versions of 802.1x EAP, such as EAP over TLS (EAP-TLS), protected EAP (PEAP), etc. Shown in Figure 8.20, RADIUS (remote authentication dial-in user service) is a commonly used authentication server.

The operational phases of 802.11i are shown in Figure 8.21. In the discovery phase, the station identifies the access point. They exchange request/response transactions over 802.1x/EAP for mutual authentication. The information on station is passed on to the authentication server by the access point. Then, the authentication server and workstation perform mutual authentication using 802.1x/EAP. Pair-wise keys are exchanged between station and access point, and between access point and authentication servers. The data transfer between station and access point is protected. At the end of authentication, a concrete EAP process is validated and then a direct, mutually authenticated session is established between the station and the authentication server. Encrypted data are now transferred between the two. Robust AES (advanced encryption standard) encryption is based on the Rjindael algorithm with 128-, 192-, or 256-b key.

Another aspect of security is the detection of intrusion by foreign access point, also referred to as rogue access point. The several methods of implementing this depend on coordination among the station, access point, and the wireless switch.

8.8 OPTICAL ACCESS NETWORK

Development of providing the last mile service using fiber all the way to customer premises, known as passive optical network (PON), is in progress. Although the technology is currently available, it has not been economically feasible to deploy PON. The protocol used would be Ethernet or ATM from head end to customer premises, referred to as EPON (Ethernet PON) or APON (ATM PON), respectively.

The optical section of a local access network system could be a point-to-point, active, or passive point-to-multipoint architecture. Figure 8.22 shows the architectures defined by ITU G.983.1, which range from fiber to the home (FTTH) through fiber to the building/curb (FTTB/C) to fiber to the cabinet (FTTCab). APON and EPON are implemented as point-to-multipoint configuration. IEEE 802.3ah GigE (gigabit Ethernet) first mile access network is EPON at the gigabit data rate.

8.8.1 OAM in EFM

An IEEE workgroup, IEEE 802.3ah Work Group, is working on the development of link monitoring, remote failure indication, and fault isolation by loop back scheme for EFM (Ethernet in the first mile).[23] The OAM (operations, administration, and maintenance) for EPON will be considered in this section. EPON architecture is a point-to-multipoint configuration in Figure 8.22. The downstream traffic is broadcast mode and the upstream is based on the allocation of time slot. The functional OAM objectives are:

- Remote fault indication at the PHY (logic) level
- Link and performance monitoring of CRC error packets at MAC level, code violation at PHY level, and PHY-PMD signal level

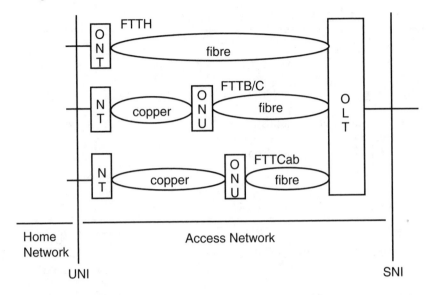

T1528110-98

ONU   Optical Network Unit
ONT   Optical Network Termination
OLT   Optical Line Termination
NT   Network Termination

Figure 8.22   ITU G.983.1 network architecture.

- Remote loopback
- Management channel for remote OAM

8.8.2   EPON MIB

The EPON MIB is in draft status[24] and comprises two sets of MIBs. The first is the EFM EPON MIBs and the second is the EPON device MIBs. The EFM EPON MIB group, dot3EfmEpon, defines the objects used for configuration and description of the (802.ah) P2MP (point-to-multipoint) section. This MIB group is shown in Table 8.11. Table 8.12 presents the MIBs defining the objects used for configuration and description of management objects for EPON-compliant devices.

Table 8.11   EFM EPON MIB Group

Entity	OID	Description
dot3EfmEpon	Mib-2 XXX	Configuration and description of EFM
dot3Mpcp	dot3EfmEpon *1*	Multipoint control protocol attributes
dot3OmpEmulation	dot3EfmEpon 2	Point to point emulation attributes
dot3EponMau	dot3EfmEpon 3	EPON MAU managed object

Table 8.12   EPON Device MIB

Entity	OID	Description
eponDeviceMIB	Mib-2 XXX	Objects for managing EPON devices
eponDeviceObjectMIB	eponDeviceMIB 1	EPON-compliant devices
eponDeviceObjects	eponDeviceObjectMIB 1	Table of objects for EPON devices
eponDeviceConformance	eponDeviceObjectMIB 2	EPON device conformance object group

8.9   HETEROGENEOUS LAST/FIRST MILE/METER NETWORK MANAGEMENT

In practice, the broadband service at the edge network, which is from the edge router to the customer premises equipment (also referred to as last/first mile and last/first meter access networks), will comprise heterogeneous networks. For example, in Figure 8.22, the transmission to the ONU over fiber could be EPON or APON, and the transmission over copper could be VDSL. Again in the configuration of Figure 8.22, the FTTH could be EPON or APON, and the distribution in the customer premises could be implemented using 802.11 wireless technology or HomePNA. In metropolitan areas, 802.16a could be implemented in conjunction with 802.11 for distribution for the last meters at the customer premises. In this situation, even though the two technologies are wireless, they are different.

Fortunately, network management using SNMP technology enables management of heterogeneous networks and presents a unified picture for handling OAMP. In the OSI terminology, this would be FCAPS management. Various technologies are handled using the MIB schema associated with each, and the common presentation is done using the graphic user interface that many of the enterprise network management systems offer. The MIBs are usually standard MIBs associated with the technology. Even in a case in which no standardization is implemented, e.g., many of the cell networks, private MIBs are developed by the equipment vendors. A private MIB could be integrated with standard MIBs in a network management system by implementing a proxy server for the private MIB.[1]

8.10 SUMMARY

We have presented in this chapter an overview of management of various broadband access networks associated with the last mile. We have dealt with the popular cable and ADSL access network management considerations and the MIBs that have been developed for them. CableLabs, a consortium of MSOs, is continuing the enhancement of BCAN to handle all multimedia services and a similar effort is continuing on DSL technology by DSL Forum. Fixed wireless technology is being used in rural areas and uses cable modem technology by replacing the HFC with wireless. We have addressed the two implementations of LMDS and MMDS of fixed wireless access network and then the specifications for the broader category of IEEE 802.16. We have briefly addressed the extension of 802.16 to 802.16a MAN. Mobile and wireless network technologies of 3-G cellular network and IEEE 802.11 WLAN have been presented. PON is an emerging technology and we have detailed the current management standards work in progress for EPON. Finally, we have addressed the management of heterogeneous access networks.

REFERENCES

1. M. Subramanian, *Network Management: Principles and Practice*, Reading, MA,: Addison-Wesley, 2000.

2. Last-mile lament, *Dallas Business Journal*, High Tech Internet, February 7, 2003.

3. M. Subramanian, A software-based elastic broadband cable access network, *Proc. Manual*, Cable-Tec Expo 2002, San Antonio, Texas, June 2002.

4. M. Subramanian and L. Lewis, QoS and bandwidth management in broadband cable access network, special issue on network management, *Computer Networks*, 43(1), 16 September 2003.

5. W. Stallings, *SNMP, SNMPv2 SNMPv3, and RMON 1 and 2*, Reading, MA, Addison–Wesley, 1999.

6. L.G. Raman, *Fundamentals of Telecommunications Network Management*, New York, IEEE Press, 1999.

7. *Cable Data Modems — A Primer for Nontechnical Readers*, CableLabs, April 1996.

8. M. Rose, Management information base for network management of TC/IP-based Internets: MIB-II, RFC 1213, March 1991.

9. K. McCloghrie and F. Kaastenholz, Evolution of the interfaces group of MIB-II, RFC 1573, January 1994.

10. http://www.cablemodem.com/faq/#FAQ17

11. R. Frye, D. Levi, S. Routhier, and B. Wijnen, Coexistence between version 1, version 2, and version 3 of the Internet standard network management standard, RFC 2576, March 2000.

12. M. St. Johns, Diffie–Hellman key management information base and textual convention, RFC 2786, March 2000.

13. www.dslforum.org

14. K. McCloghrie and F. Kastenholz, The interface group MIB using SMIv2, RFC 2233, November 1997.

15. D. Cong, M. Hamlen, and C. Perkins, The definitions of managed objects for IP mobility support using SMIv2, RFC 2006, October 1996.

16. M.M. Deore and A.K. Mathur, Mobility management for wireless networks, IBM developerWorks, 15 July 2003.

17. J.J. Caffery, Jr. and G.L. Stuber, Overview of radiolocation in CDMA cellular systems, *IEEE Commun. Mag.*, 38–45, April 1998.

18. I.F. Akyildiz, J. McNair, J.S.M. Ho, H. Uzunalioglu, and W. Wang, Mobility management in next generation wireless system, *IEEE Proc.*, 87(8), 1347–1385, August 1999.

19. J. Ye, J. Hou, and S. Papavassiliou, A comprehensive resource management framework for next generation wireless networks, *IEEE Trans. Mobile Computing*, 1(4), 249–264, Oct.–Dec. 2002.

20. R. Howell, WAP Security, www.vbxml.com/wap/articles/wap_security

21. http://lists.skills-1st.co.uk/mharc/html/nv-l/2001-12/txt00001.txt

22. N. Cam-Winget, T. Moore, D. Stanley, and J. Walker, IEEE 802.11i overview csrc.nist.gov/wireless/S10_802.11i%20Overview-jw1.pdf

23. H. Suzuki, H. Barras, P. Kelly, R. Muir, and B. Berret, OAM in EFM, IEEE 802.3 EFM SG, May 2001.

24. L. Khermosh, Managed objects for the Ethernet passive optical networks, <draft-ietf-hubmib-efm-epon-mib-00.txt>

9

Emerging Broadband Services Solutions

MANI SUBRAMANIAN

9.1 INTRODUCTION

The subject of broadband multimedia services to home as they relate to the last mile and last meters of broadband networks has been presented. Chapter 1 introduced the broad overview of applications with special emphasis on the role of media compression. It also introduced the operating conditions of media over broadband networks. Chapter 2 set the stage for background of the last mile network as it relates to the edge and core networks. Chapter 3 through Chapter 7 discussed each of the specific access technologies of the last mile and last meters network. In the previous chapter, the subject of network and services management was discussed. This chapter attempts to look ahead to where broadband services are heading in terms of applications and technologies. This is, as could be expected, a difficult task because the multidimensional aspects of technologies and user needs, as well as economical and political considerations, influence the outcome.

After a slow start of almost a decade, the broadband service to home has picked up speed in deployment in 2003.

Table 9.1   U.S. Residential Broadband Household Forecast

	In millions				
	2003	2004	2005	2006	2007
Total residential broadband households	22.5	28.9	35.4	42.0	48.2
Cable modem households	14.6	18.7	22.9	26.5	30.2
DSL households	7.4	9.6	11.7	14.0	16.3
Satellite broadband households	0.25	0.3	0.5	0.9	1.1
Broadband wireless households (MMDS, unlicensed spectrum)	0.06	0.09	0.16	0.25	0.29
Other (FTTH, powerline, T1, etc.)	0.1	0.15	0.19	0.25	0.3

Source: D. Ainscough, The Yankee Group Research Report, 27 June 2003.

At the end of the second quarter of 2003, 26.7 million homes in North America and 77.1 million homes worldwide were using broadband service. There were 10.6 million DSL (digital subscriber line) subscribers in North America and 46.7 million subscribers worldwide. Corresponding statistics for cable are 16.1 million and 30.4 million, respectively.[1] As can be observed, the number of homes using cable technology in the U.S. is 60% higher than that of DSL. However, DSL technology is more heavily deployed worldwide compared to cable technology.

The Yankee Group forecasts that U.S. residential broadband subscribers will number more than 48 million by the end of 2007 (Table 9.1).[2] Cable modems are expected to maintain a strong lead over DSL. Cable operators will preserve their advantage in broadband availability in the near term. Beyond 2005, cable operators will sustain subscriber growth through triple-play bundles of video, data, and telephony. The subscriber base of satellite broadband may be limited through 2007. Opportunities for satellite service in rural markets exist, but broadband optical and wireless alternatives could be strong competitors. It is also possible that wireless MAN (metropolitan wireless network) could be a potential competitor for metropolitan broadband market for cable and DSL.

Cable and DSL technologies that have been deployed provide primarily only two of the three broadband services. Cable service providers offer video and data, and DSL service

Table 9.2 Growth of Networked Homes in the U.S.

	Actual	Projected				
	2002	2003	2004	2005	2006	2007
Networked homes (in millions)	6.7	11.0	15.7	21.3	26.5	32.3
Annual growth (%)	60	65	43	35	24	22

providers offer voice and data services. Some cable service providers have introduced telephony over cable recently in some regions. With the unpredictability of the rapidly changing broadband technologies, this chapter will address possible trends in various emerging technologies. The proven technologies are cable and DSL and the emerging technologies are wireless (fixed and wireless) and optical networks.

The last mile network emphasis has recently been extended to the last meters of broadband service to the home. Home networking is one of the areas that will see a high level of activity in the near future. With numerous technologies capable of bringing broadband service to customer premises, the challenge has shifted to its distribution within the house. Strong advances are occurring in wired and wireless LAN technologies, USB (universal serial bus), HomePNA (Home Phone Networking Alliance), HomePlug, and 1394b. Table 9.2 shows the Yankee Group projection of networked homes in U.S. This is expected to grow to 32 million homes by 2007.

Section 9.2 addresses the perspectives of broadband services from the point of view of applications. The emerging trends in IP-based services are viewed from the voice-over-IP and video-over-IP perspectives. The embedded technologies are introduced and specific applications in each category are discussed. The other emerging category of applications is residential applications that include home entertainment, data applications, and home- and small-business applications.

Section 9.3 discusses the needs of subscribers and Section 9.4 discusses the service issues in meeting those needs. The requirements of future broadband services are listed in Section 9.5. Section 9.6 addresses the future trend in technologies and networks in the last mile and last meters networks in meeting the future applications, broadband service requirements, and subscriber needs. Section 9.7 outlines the trends in OAMP (operations, administration, maintenance, and provisioning) to keep in step with broadband service offerings.

9.2 BROADBAND SERVICE PERSPECTIVES

As mentioned in Chapter 1 and in *Broadband: Bringing Home the Bits*,[3] offering broadband services has multiple dimensions. The infrastructure perspective addresses attributes such as the speed, latency, jitter, and symmetry between the uplink and downlink capacities. These attributes depend on the technology used for the access network. Chapter 2 though Chapter 6 in the book have been devoted to the various technologies and their characteristics. Chapter 7 addressed the management considerations of the infrastructure comprising the various technologies.

When the subject of broadband service is mentioned, it is invariably implied that this refers to broadband service to the home because that it is where the last mile and the last meter challenges currently exist in bringing the voice, video, and data simultaneously over the same medium. The problem is much less severe in the enterprise environment because it is mostly an IP and Ethernet environment that can handle large bandwidth multimedia requirements. However, there are issues that affect voice and video over IP, especially when it is related to telephony and videoconferencing. Internet protocol (IP) will be playing a significant role as an enabler technology in broadband service offerings for enterprise and residential services. The IP as an enabling technology on the future of telephony and broadcast video services will be addressed in Section 9.2.1.

There are two market drivers for the large scale deployment of broadband services. The first is "killer application(s)" that would drive residential or enterprise subscribers to cross the "tipping point." It is generally believed that broadband service will take off with the discovery of killer applications, even though it is hard to predict what those would be. Section

9.2.2 and Section 9.2.3 discuss several of the potential residential, small office home office (SOHO), and SME (small and medium enterprise) applications that could contribute to this explosion.

The second driver is associated with the economics of broadband services. The demand for broadband services has so far been more in the residential market. The pricing of bundling of services among service providers of different technologies would play a significant role in the rapid deployment of broadband services to home in metropolitan and rural areas. Although this is an important subject, due to the complexity associated with federal regulations and other noneconomic factors, it will not be dealt with here.

9.2.1 IP-Based Services

Internet protocol-based services have until now been predominantly data oriented, such as e-mail, Web surfing, and e-commerce. IP has been an enormously successful protocol for non-real-time traffic. With converging broadband multimedia services, the use of IP is being extended for real-time traffic of voice and video. Numerous IP-based voice and video applications are currently offered on the Internet over the various media. Several technology enablers, such as compression technologies (see Chapter 1), high-speed transcoding, Internet group management protocol (IGMP), signaling protocols for control of sessions, real-time protocol (RTP) for transport of streaming data, etc., have brought these closer to commercial deployment.

9.2.1.1 Voice over IP

Juniper Research predicts that voice over IP (VoIP) revenue will exceed $20 billion by 2009.[4] VoIP will account for approximately 75% of world voice services by 2007 (Frost and Sullivan, 3/2002).[5] Over the next 5 years, 90% of enterprises with multiple locations will start switching to IP systems for voice (Phillips Group, via Aspect, 6/2001).[5] All these quantitative predictions could vary in the future. However, they all tend to indicate the direction in which the transport of voice across the public Internet is progressing: namely, VoIP. However, the quality and economics of IP telephony must overcome those of the current offering by PSTN (packet switched telephone network).

9.2.1.1.1   Signaling Protocols

One of the several factors that would determine the rapid adoption of VoIP solutions for commercial and residential broadband services would be signaling protocol. This protocol controls all facets of the connections — setting up, managing, and releasing — across a telecommunications network, creating a fixed and reliable link over the entire session of a conversation.

A disjoint family of signaling protocols is being used now for implementation of voice and video over IP.[6,7] H.323 is the most widely deployed VoIP signaling protocol in converged networks today. Session initiation protocol (SIP) is an emerging VoIP signaling protocol under development by the IETF (Internet Engineering Task Force) as the standard signaling protocol over the Internet.[8–12] A third way of VoIP implementation is with media access gateway using any one of the standard VoIP gateway control protocols. Control protocol (MGCP) is a standard VoIP control protocol that controls gateway devices in a VoIP network.[13] MEGACO, another gateway-control protocol standardized by IETF, is based on ITU-T H.248 protocol standard and has improved features over MGCP.[14,15] It is a master–slave, transaction-oriented protocol in which media gateway controllers (MGCs) control the operation of media gateways (MGs). However, the first two approaches are the leading contenders for VoIP implementation and are described here.

Figure 9.1 shows the H.323 VoIP network configuration based on H.323 signaling protocol. The four components to the H.323-based VoIP network are terminals, gateways, gatekeepers, and multipoint control units (MCUs). A terminal is an IP telephone or a PBX serving regular telephones. Gateway/router functions as a protocol translator between the PBX, which appears as a PSTN network, and Internet, or it

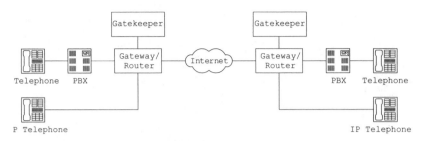

Figure 9.1 H.323-based VoIP configurations.

acts simply as a router for IP telephones. Gatekeepers provide call control functions such as address translation between telephone number and network address, admission control based on bandwidth availability, etc. MCUs (not shown in Figure 9.1) provide conference facilities for users who want to conference three or more end points together. Once the connection is established, audio stream is exchanged using RTP (see Section 9.2.1.1.2).

SIP is the standard for multimedia conferencing over IP of the Internet Engineering Task Force (IETF). SIP is an ASCII-based, application-layer control protocol that addresses the functions of signaling and session management within a packet telephony network. Signaling allows call information to be carried across network boundaries. Session management provides the ability to control the attributes of an end-to-end call.

A VoIP network configuration based on SIP signaling protocol is shown in Figure 9.2. It illustrates the information transfer over the Internet as well as the legendary PSTN. The gateway/router functions as a gateway when it interfaces with PSTN or as a router when it interfaces with the Internet. SIP is a peer-to-peer protocol and functions in a client/server mode. The peers in a session are called user agents (UAs). A user agent functions as a client and as a server. A client application in UA at the source initiates the SIP request and a server at the destination UA returns a response. The proxy server performs functions such as registration, authentication,

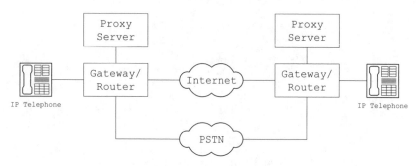

Figure 9.2   SIP-based VoIP network configurations.

authorization, network access control, and network security. It also finds the next-hop routing information based on the received or translated destination address. Again, as stated in the H.323-based network, the information transfer is implemented using RTP.

The SIP protocol provides name translation and user location, thus ensuring that the call reaches the destination irrespective of the called party's location. SIP address is similar to e-mail address and thus each user has a unique address. This enables implementation of unified messaging for users to receive a telephone message, e-mail, and fax anywhere, anytime (see Section 9.2.1.1.4).

9.2.1.1.2   Data Transport Protocol

For IP to be used in a real-time application, it is supported by reliable TCP (transmission control protocol) for signaling and UDP (user datagram protocol) for data transfer across media. RTP is used at the application level over UDP. Use of RTP is essential for performance. Section 9.2.1.1.1 discussed the first phase of multimedia session, signaling protocols, that deals with user location, session initiation, call setup and teardown, media negotiation, and conference control. Now, the second component of the multimedia session, namely, the data transport protocol will be discussed.[16]

RTP is the protocol used to transfer data for real-time traffic of voice and video over IP. IETF established the Audio/

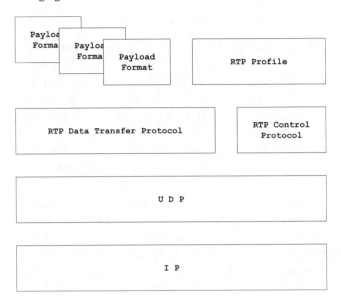

Figure 9.3 RTP components.

Video Transport (avt) Working Group to specify a protocol for real-time transmission of audio and video over unicast and multicast UDP/IP. Among other goals, the working group activities include developing specifications for secured RTP (SRTP) and an RTP MIB (management information base) for management of elements and networks.

Figure 9.3 shows the components of RTP.[16] The RTP data transfer protocol transfers data streams, one stream for each media type. It is responsible for source and destination identification, media sequencing, and timing recovery. Each RTP data flow has associated with it a control flow that addresses time-base management, quality of service feedback, and membership management. Several payloads, such as H.323, MPEG (Moving Picture Experts Group), etc., can be handled by RTP. The RTP profile contains the various parameters for the particular application. RTP is a robust, flexible, and secure protocol that has forward error correction, ability to handle multiple codecs, and secured authentication and privacy.

9.2.1.1.3   PSTN and IP Telephony

The legacy network-based telephony using public switched telephone network (PSTN), is highly matured and is offered at a competitive economical rate to subscribers by various carriers. It is reliable and the quality of service required for real-time voice is easily met with the circuit-switched network. Thus, any alternative to PSTN, such as public Internet, must offer enough market drivers to replace the existing circuit-switched PSTN. Some of these market drivers will be reviewed now.

One of the major driving forces to migrate to Internet telephony is the convergence of voice and data, i.e., telephone, fax, and e-mail accessed via a single medium. Also, the SIP protocol discussed in Section 9.2.1.1.1 enables a single unique and ubiquitous personal address for each subscriber. This would naturally lead to unified messaging with a single personalized address (see Section 9.2.1.1.4). Voice telephony applications do not require large bandwidth. The data rate is only 64 kbps and, with compression techniques, it is even less. This should help significantly in achieving the quality of service that needs to be met. In addition, the economic factor is a driving force. As mentioned in Reference 3, the cost reduction is due to several factors. Residential subscribers pay a flat charge for Internet connection and thus subscribers do not pay the per-minute charge. Long-distance calls can be placed to a local ISP number on dial-up calls and thus will be considered local calls without per-minute charging.

9.2.1.1.4   Unified Messaging

The International Engineering Consortium in its tutorial on unified messaging[17] defines unified messaging as the integration of several different communications media so that users will be able to retrieve and send voice, fax, and e-mail messages from a single interface, whether it is a wireline phone, wireless phone, PC, or Internet-enabled PC. Technologies exist that enhance the integration of voice mail and e-mail, such as text-to-speech software that converts e-mail into spoken words. Likewise, voice recognition technology converts voice into data for voice-activated telephony. As mentioned in Section 9.2.1.1.1, SIP is the enabler to implement unified messaging.

9.2.1.1.5 Instant Messaging

Instant messaging is an interactive process that allows one to maintain a list of persons (often called a *buddy list* or *contact list*) and send messages to anyone on the list, as long as that person is online. Most of the popular instant-messaging programs provide a variety of features such as send notes back and forth on the Internet, create custom chat room, etc. An IETF working group is working on the application of the SIP to the suite of services collectively known as instant messaging and presence (IMP). Because most common services for which SIP is used share quite a bit in common with IMP, the adaptation of SIP to IMP seems a natural choice, given the widespread support for (and relative maturity of) the SIP standard.

9.2.1.1.6 Call Centers

Session initiation protocol is ideally suited for the operation of call centers. It has several advantages over H.323 signaling protocol for this application. First, when a user logs onto a network using SIP, his or her presence on the network is declared, much like the example of a "buddy" appearing as available on an instant-messaging window. Second, SIP signals presence (and therefore availability) to a multitude of applications supporting voice calls, data calls, video calls, and instant messaging. Third, SIP permits concurrent sessions even if the sessions use different media. For example, SIP users can log onto their call center's systems and "announce" to the network that they are available to receive a Web chat session, an instant message session, e-mail, and voice calls simultaneously. All these features of SIP signaling protocol could be taken advantage of in the application of a call center as described next.

Consider the scenario of a customer e-mailing a customer support center with a particular problem (adapted from Taylor and Hettick[19]). At the customer support center, the call is redirected by the user agent server to the next available support engineer, based on the engineer's announced "presence" at the customer support center. The customer's profile and his instant messaging address are displayed on the support engineer's screen. Using instant messaging, the engineer sends a message to the customer asking him or her to send over instant messaging a "Web push" page containing the error message that the customer had reported. As the chat over the instant messaging continues, it becomes apparent that a phone conversation is in order. Therefore, the engineer asks the customer to press the "push to talk" icon on his instant messaging screen to initiate a real-time voice call. The call, along with the Web chat and error message attachment, allows the support engineer to identify and resolve the customer's problem.

9.2.1.1.7   VoIP over Cellular Network

Application of VoIP in its use in cellular networks is shown in Figure 9.4. The voice data in a digital cellular network are

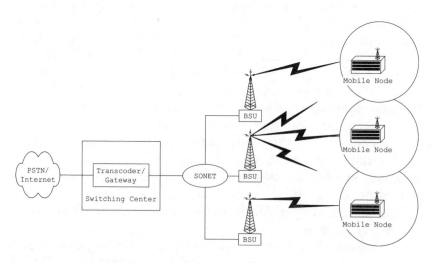

Figure 9.4   VoIP over cellular networks.

generally compressed and packetized for transmission over the air by the cellular phone/mobile node and BSU (base station unit). Packet networks can then transmit the compressed cellular voice packet, saving a tremendous amount of bandwidth. The transcoding function is accomplished by the transcoding gateway in the central office to convert the cellular voice data to the format required by the public switched telephone network (PSTN) or Internet in the future. The current separate message center concept for voice mail and e-mail will be integrated into a single unified message server with applications. The message will be automatically sent to or retrieved by the subscriber.

9.2.1.2   Video over IP

Some of the application drivers for the deployment of video over IP are video conferencing, digital video broadcasting, and video on demand. Devices such as on-line gaming consoles, stereos, and personal video recorders could connect to home networks and would add to the rapid deployment of video on IP. They depend on the multimedia and multicast capabilities of the network with real-time streaming protocol and session control protocol. The technology enablers for multicast multimedia network will be discussed, followed by the video conferencing application for illustration.

9.2.1.2.1   Technology Enablers

As mentioned at the beginning of Section 9.2.1, several technology enablers also help implement video over IP, just as they do for VoIP. These include compression technologies, high-speed transcoding and transrating, IGMP, SIP and H.323 signaling protocols, real-time protocol RTP, and quality of service (QoS) improvements. Many of these were addressed earlier.

IP multicast and IGMP are important technology enablers for group video conferencing. IGMP[20] is used by IP hosts to report their multicast group memberships to any immediately neighboring multicast routers. Routers that are members of multicast groups broadcast their membership to other routers. Thus, they behave as hosts as well as routers.

Video conferencing over IP can also be implemented using the H.323 protocol. Contrasting IGMP with H.323, the former is a one-to-many relationship and the latter is a one-to-one unicast protocol. Thus, using H.323 for group video conferencing requires an additional component such as a multistation conferencing unit (MCU) acting as a server on the LAN that controls the session. MCU receives unicast packets from each client and then replicates or "reflects" them to each other participant. This is very inefficient and would heavily load the network. IGMP is efficient in that it behaves in a multicast mode. IGMP also handles multiple data rates and thus all the clients do not need to operate at the same media speed.

9.2.1.2.2   Video Conferencing

Although video conferencing using IGMP and SIP is the wave of the future, it has been slow in deployment for several reasons.[21] One of them is meeting the quality of service for throughput level for video and voice over the IP data network. An overlay (parallel) network may be required instead of the converged network to accomplish this, which would add cost. Although the security issue is being resolved by firewalls and network address translator (NAT) in enterprises, they would have an impact on connectivity and performance and may need to be reconfigured or upgraded. Another reason that multicast-based multimedia conferencing has been slow to enter the commercial mainstream is that much of the focus has been placed on refining the H.323 standard for multimedia conferencing and in modifying proprietary unicast-based products to be interoperable with H.323. As mentioned earlier, H.323 is a unicast protocol and MCUs are needed to extend this to group video conferencing efficiently.

Current implementation of multimedia multicast for video conferencing is done using MBone networks. MBone stands for multicast backbone, a virtual network that originated from an IETF effort to multicast audio and video

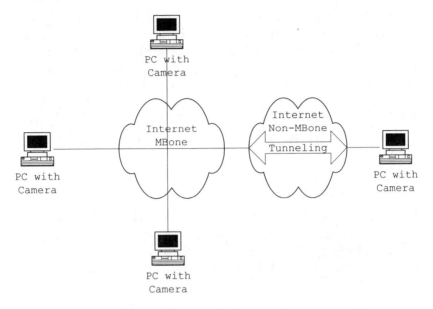

Figure 9.5 MBone multimedia conferencing.

streams over the Internet. It shares the same physical media as the Internet, with routers that can support multicast. Tunneling is used to send the packets through routers that do not support multicast. MBone applications use RTP protocol for streaming video and audio (see Figure 9.5). For a review of SIP, IGMP, and MBone the reader is referred to Banikazemi[22] and Gómez–Skarmeta et al.[23]

Several MBone applications, such as sdr, vic, vat, rat, wb, and nt, have been developed that support video conferencing and other multimedia applications. The session directory (sdr) is a tool for announcing and scheduling multimedia conferences on the MBone. Users can use sdr to see which conferences are available and to join them. They can also use it to announce conferences and to specify timing, media, and other details. Sessions can be public or private and may optionally link to further information on the Web. The video conferencing (vic) tool is the video portion of the multicast and is an application for sending/receiving streaming video in real time over IP networks. The visual audio tool (vat) and robust audio tool (rat) are audio conferencing applications that can be used with the vic and/or whiteboard (wb) software tool to produce a multimedia conference. The software tool, wb, is a shared "whiteboard" drawing surface that can also be used as a writable screen on which to view and annotate ASCII text or PostScript files. The text annotations made with the network text (nt) utility can be used to lend chat session capability to wb.

9.2.2   Residential Applications

Residential applications may be split into three classes: home entertainment, data applications, and business applications. These applications are discussed in this section and home networks that serve these applications are discussed in Section 9.6.10.

9.2.2.1   Home Entertainment

Home entertainment is one of the primary market drivers of broadband to home. Analog TV is migrating to digital TV. Figure 9.6 shows the projected growth of digital TV in the U.S. and worldwide.[24] The U.S. government has made the decision to turn off shipment of analog TVs by 2006 and

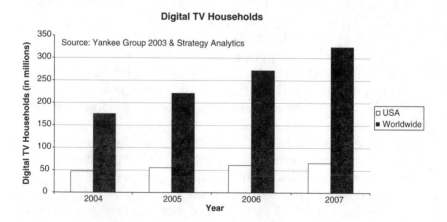

Figure 9.6   Growth of digital TV households.

Canada will do so by 2007. A significant percentage of digital TV is projected to be high-definition TV, which requires at the current time a data rate close to 20 Mbps per channel. This demand along with data applications is expected to accelerate the broadband market to home.

Digital TV growth is expected to come primarily from two applications: broadcast TV and video on demand. Most of the service providers are currently offering a significant number of digital TV channels, several of which are HDTV (high-definition TV) channels. HDTV is available to over 55 million U.S. households, but only about 2 million households were viewers of HDTV in 2003. HDTV households, however, are expected to grow to 16.4 million by 2006[24] — about one third of the digital TVs in U.S. households.

Direct broadcast satellite (DBS) TV provides exclusively digital TV channels and is gnawing away the market share from cable TV, which is mostly analog. Efforts to offer totally IP-based digital TV on cable in the U.S. have been made, but the progress on this is very slow. Worldwide, the cable market share of the broadband market is much less compared to the U.S., and satellite-based digital TV would be competing with the DSL offering, especially with the lack of success of switched TV over DSL until now.

The other major application driver for residential broadband is the gaming market. Interactive games within the house and over mobile devices and the Internet will be a strong driver for the broadband market, especially in the youth and young adult markets.

9.2.2.2 Residential Data Applications

In addition to e-mail, chat, and Web surfing, intrahome multimedia applications would demand large-bandwidth and high-performance home networks. The personal video recorder (PVR) would be a standard home appliance to record broadcast TV programs, downloaded streaming audio, and video clips. Not only highly efficient, large storage devices are required for this purpose, but also a good intrahome communication system with the ability to display programs in any room from centralized servers.

9.2.2.3   Residential Business Applications

Internet shopping and banking are increasing year by year and this e-commerce would add to the residential broadband data needs. Commercially available in the market today are introductory products and solutions that form an interconnected web of personal applications, such as calendar, things to do, appointments, address book, and telephone directory stored in multiple computing devices that need to be kept in sync on a real-time basis. Such an interconnected web of applications is expected to increase in the future, acting as another driver for data applications.

9.2.3   SOHO and SME Applications

With the enterprise market heavily leaning toward entrepreneurial efforts, SOHO and SME will be located in residences and residential neighborhoods. Applications based on video conferencing and residential business servers would put a large demand on large two-way data rate requirements in contrast to the asymmetrical bandwidth needs of today. Also, a large number of SOHO and SME applications require handheld field devices to access back-end office servers. For example, a real-estate broker showing a house to a potential customer may need all sorts of information on the house, such as its history, local schools, local commercial market, and its availability. This information is dynamic and needs constant update to and retrieval from back-end office servers from the field. Numerous types of such systems and solutions are currently offered in the market as multimedia applications. Wireless networks discussed in several places in the book, along with security considerations, play a significant role in the future broadband service of these types of applications.

9.3   SUBSCRIBERS' NEEDS

Subscribers would like to have multimedia service anywhere, anytime, and on any terminal at an economical offering. Technologies and pricing structures are far from meeting these requirements.

9.3.1 Economical Service

There is a large gap between the cost of providing broadband service to home and the fee that the subscriber is willing to pay for the service. The broadband data service currently offered in the U.S. is about $40 per month for a standard offering of data, and less for a lower data rate. The majority of telecommunications subscribers are extremely pleased with the voice service provided by the telephone service provider and the video (primarily television) service provided by the cable service provider. The reliability and fees are acceptable to subscribers.

The data needs of subscribers vary. The majority of subscribers have low data transmission requirements because their applications are primarily e-mail and, secondarily, Web surfing. At the other end of the spectrum are the SOHO and SME subscribers, whose needs are high bandwidth in the upstream and downstream directions. Furthermore, the diurnal needs of these two sets of subscribers are significantly different. To satisfy this wide range of subscribers, the service provider needs to offer usage-based data fees, similar to the long-distance telephone service fees. This requires broadband systems that can monitor the traffic and dynamically control its flow, and operations support systems (OSSs) that support with dynamic provisioning and usage-based billing.

9.3.2 Broadband Data Traffic

As mentioned earlier, residential subscribers mostly use e-mail and Web surfing applications. The bandwidth requirement of these users is asymmetrical with low bandwidth in the upstream and high bandwidth in the downstream. This is satisfied by current offerings. However, the requirements of SOHO and SME are not only large bandwidth pipes, but also high usage in upstream and downstream directions.

A typical daily pattern of downstream broadband traffic is shown in Figure 9.7.[25] These results were obtained in a cable access network of several thousand subscribers. The upstream pattern follows a similar pattern. As could be expected, the traffic is high during evening and night hours

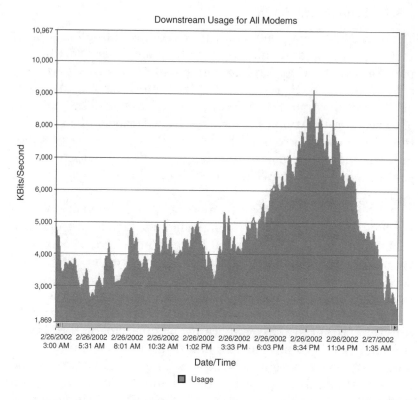

Figure 9.7 Daily broadband traffic pattern.

and low during the day. Depending on the number of users
in a shared medium such as a cable access network, users
tend to experience poorer performance at night; this needs to
be minimized.

The usage pattern of subscribers shown in Figure 9.7
follows the traditional 90/10 or 80/20 rule; that is, 80 to 90%
of the bandwidth is consumed by 20 to 10% of the subscribers.
It is logical to induce that the minority of heavy users belong
to the category of SOHO or SME. If an "asymmetry index" is
defined as upstream-to-downstream bandwidth of a subscriber, the distribution of subscribers as a function of asymmetric index is shown in Figure 9.8.[26] It can be observed that
the upstream traffic generated by 10% of the subscribers

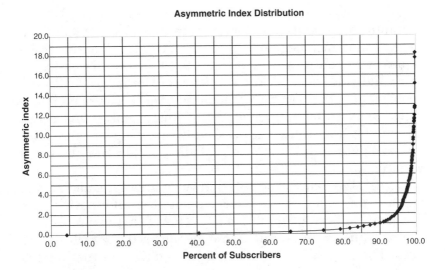

Figure 9.8 Traffic asymmetric index.

exceeds their downstream traffic. The demand on the reverse traffic pattern of such subscribers needs to be met by the broadband access network. This once again emphasizes the need for usage-based fees.

9.3.3 Quality of Service

Quality of service is multidimensional and could be interpreted in technical terms in many different ways. In addition to the traditional network quality of service, it is necessary to address other categories of QoS, such as the quality of information when much information is retrieved by a user, and content-specific QoS.

From the subscriber's point of view, broadband is a leftover culture from the telecommunications service. People are used to reliable, dependable, and good-quality telephone service. Anyone can make a telephone call at any time from any place to anywhere else in the world and can expect a good and dependable connection. Similarly, subscribers are used to reliable, dependable, and good-quality television reception service. They expect the same quality of service when tele-communication and cable services are extended to broadband service that includes voice, video, and data. Some of these issues were addressed in Section 9.2.1 in the discussion of voice and video over IP. The service provider is faced with the problem of satisfying such a demand of subscribers economically. Networking technology also must resolve this issue, prioritizing real-time voice and video traffic over store-and-forward data traffic and also providing end-to-end quality of service.

With the help of the Internet, Web browsers, and search engines, one can retrieve a large amount of information on a specified subject instantaneously. However, the problem is that the quality of information depends on the search engine used. The search engine acts as the middleware to filter the relevant information for quality presentation to the user. Another example of a problem of quality of information is with multicast transmission, in which a single data stream could be transmitted from a single source to multiple destinations. In the reverse situation, multiple streams could arrive at a single computing device. With resource limitations on the hardware as well as embedded applications, the devices would not be able to handle the incoming data. Robinson and Schwan[27] propose solving this problem by dynamically managing the quality of information at the pervasive wireless devices.

The last category of quality of service affecting the end user as well as the source driver is the content-driven quality of service. Consider Web advertising using push technology. By profiling the Web user based on various parameters such as geography, URLs accessed, etc., the right type of advertisements could be dispatched to the user, thus increasing the value of the advertisements. A similar approach once again could be used to profile TV viewers and offer products and services that match the profiles of the subscribers. Another example of content-based QoS is Web content hosting, in which a Web server stores and provides Web access to documents for different customers. Due to the variety of customers (corporate, individuals, etc.), providing differentiated levels of service is often an important issue for the hosts.

Cisco provides a solution based on the use of content networking to provide quality of service.[28] Enterprises that deploy an array of applications that support a wide variety of business processes, including workforce optimization, supply chain management, e-commerce, and customer care, must be dynamically able to recognize each application and provide the associated set of services. The network needs to provide QoS based on the application.

9.4 SERVICE ISSUES WITH CURRENT TECHNOLOGIES

Broadband services offered at the present time are an extension of current technologies to offer data service. Multiple system operators (MSOs) have extended one-way cable TV to two-way systems and have added data service offerings. The telecommunication service providers have added data to the voice service by developing DSL technology. Neither one is providing the full capability of triple play, i.e., voice, video, and data.

9.4.1 Lack of Comprehensive Service

The lack of comprehensive service can be attributed to various reasons. First, it is not possible to provide economical large bandwidth/data rate service using either technology. Even if it could be provided, "killer applications" that require large two-way bandwidth pipe to home are lacking. The most used Web-based applications currently require large and bursty bandwidth downstream and low bandwidth upstream. Furthermore, the cost of adding the third service in cable or in DSL technology is expensive.

9.4.2 Heterogeneous Service Offerings

The broadband service offered today is multitechnology, multivendor, multistandard (sometimes proprietary) with no end-to-end coordinated quality of service. Does this mean that one could some day expect a unified approach to fix all these issues? The answer is "probably not." However, one should realistically expect that the subscriber is not going to be satisfied until he or she gets a good end-to-end quality of service on a unified interface medium at a price that an average household subscriber can afford. The multiple technologies should progress and are progressing with plug-and-play open architecture that satisfies the subscribers' needs. Some of these emerging technologies will be examined next.

9.5 NEXT-GENERATION BROADBAND SERVICES

First, some of the basic requirements of the next generation of broadband service will be discussed.

9.5.1 Triple Play

Triple play is the feature of broadband service that pertains to the subscriber receiving voice (telephone), video (television, video on demand, videoconferencing, etc.), and data (Internet connection, Web service, etc.) on a single transmission medium at the customer premises. This service should be as reliable and dependable as the telephone and TV service to which the subscriber has been accustomed until now. It is likely that multiple Internet service providers could be available.

9.5.2 User-Friendly Installation and Operations

The installation of the broadband device should be easy for the subscriber and the service provider. The subscriber device should be able to self-provision. The on-going operation of broadband devices should have user-friendly interfaces.

9.5.3 On-Demand Services

Video on demand is one of many on-demand services that need to be met by the broadband service provider. The service could be requested at any time, without advance reservation. This requires dynamic service provisioning of needed additional bandwidth to the requested subscriber without significantly degrading the QoS of others in a shared resource environment.

9.5.4   Usage-Based Fee

Bundling services and a unified billing scheme is the choice of subscribers. Also, the fee should be based on the amount of usage.

9.5.5   Universal User ID

With the proliferation of mobile devices, providing service to untethered devices becomes more challenging. With multiple devices, a portable universal addressing scheme becomes a requirement.

9.5.6   Quality of Service

Subscribers are unaware of the infrastructure that provides the broadband service, as it should be. They expect good quality of service. For real-time applications of voice and video, the delay and jitteriness should be imperceptible. Service should be highly dependable (always available) and reliable (quality is consistent).

9.5.7   Security and Privacy of Information

With universal ID and multiple service providers delivering multiple services on multiple media, the security and privacy of information become a primary concern of subscribers. This is especially critical with e-business over the Internet. In addition to implementing security and privacy — authentication, authorization, and encryption — a cultural change in the perception of the subscribers is needed to convince them that the information link is secure.

9.6   NEXT-GENERATION BROADBAND LAST MILE NETWORKS

This section discusses the emergence of the last and first mile access networks, as well as the last and first meters home network, to meet the service requirements of the broadband discussed in the previous section. Although it is beyond the

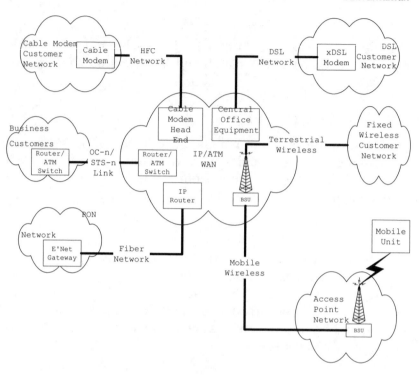

Figure 9.9   Broadband access networks.

scope of this chapter and the book to address the considerations of WAN that affect end-to-end QoS, it is a subject to
be pursued for broadband service to be successful.

Figure 9.9 shows six types of broadband access networks
that provide broadband service to homes, SOHO/SME, and
enterprises. The core network is IP or ATM WAN. The link
from the head end or edge router to business customers is
shown as OC-n link, although it could be any other transport
scheme. The HFC (cable) and DSL networks are the matured
access networks. Fixed wireless is being offered in rural areas
and some trials extending it to metropolitan areas are in
progress. Mobile wireless could be offered using 3-G technology or wireless LAN. The former has the limitation on data
rate and the latter on range. Fiber network as passive optical
network (PON) is still in the embryonic stage for economic
reasons.

Although each access network is presented as based on a single technology, Section 9.6.8 will discuss the hybrid access networks being deployed.

9.6.1 Cable

The long-term perspective of broadband technology is predicted to be in the realm of optics and wireless. However, the investments made in the cable technology and outside plant, as well as its large-scale deployment, will sustain cable broadband services for the foreseeable future. Furthermore, the enhancement to cable plant is analogous to enhancements to Ethernet LAN that have successfully fought the ATM technology for LAN. Enhancements continue to be made to cable technology, which could prolong its life for a considerable number of years.

Some of the future directions of broadband cable are the offering of VoIP, extending the cable plant to beyond the current limit of 1 GHz, all-digital offering over the cable to facilitate the fast deployment of digital TV, and the enhanced home distribution networks such as the one being developed by Multimedia over Coax Alliance (MoCA) consortium or the wireless LAN network.

Currently, several MSOs offer digital TV at the expense of analog channels. The goal in the distant future is to offer digital over the entire cable RF spectrum. In the short term, vendors are offering Ethernet transmission capability beyond 860 MHz for two-way digital transmission of 100 Mbps.

Proposals have also been made to increase the bandwidth in the reverse direction, as well as to provide two-way digital service at the high end by overlaying minifiber nodes at the coaxial amplifiers.[29] This architecture also extends the coaxial spectral range beyond 860 MHz.

PacketCable™ is a project conducted by CableLabs to support packet-based voice and video products over cable systems.[30] PacketCable is a set of protocols and associated element functional requirements developed by CableLabs to provide packet voice and packet video communications services to the home using packetized data transmission technology over the cable television hybrid fiber/coax (HFC) data network.

9.6.2 Digital Subscriber Line

The DSL Forum is working on the next-generation DSL. The focus is on increasing the data rate upstream and downstream, as well as extending the range. Considerable efforts are being spent on video over IP using xDSL. From subscriber need, the video broadcast is a major issue. At present, service providers are offering video service to subscribers as a bundled offering using terrestrial or satellite wireless, which is only a temporary solution. With the advances in data compression techniques and VDSL (very high data rate DSL) technology, IP over video offering is likely in the future. However, the switched video operational problems need to be resolved.

9.6.3 Fixed Wireless

MAN (metropolitan LAN) based on IEEE 802.16d, also known as WiMax, is one of the future directions of the fixed wireless serving the metropolitan areas for broadband. This is modified IEEE 802.16a and IEEE 802.16c standards. However, it is in competition with well-established cable and DSL and it is hard to predict its future success at this time. Cellular network will be another competitor in this space.

Fixed wireless as broadband access network is currently deployed on a limited basis in rural areas. There are two implementation schemes of fixed wireless technology. Multichannel multipoint distribution service (MMDS) operates at 2500- to 2686-MHz band. It serves a long range of up to 25 km. Local multipoint distribution service (LMDS) operates over two frequency bands: 27,500 to 28,350 MHz and 31,000 to 31,300 MHz. The LMDS range is about 5 km. Both these technologies use cable modem technology at either end and replace the HFC medium with wireless.

The standards for fixed wireless have been specified in IEEE 802.16.1, which covers from 11 to 66 GHz. IEEE 802.16d occupies the 2- to 11-MHz spectrum (licensed and unlicensed) and is based on OFDM (orthogonal frequency division multiplexing) technology to enable no-line-of-sight propagation. It is a generalized standard for point-to-multipoint architecture. Downstream transmission is TDMA (time division multiple access) and the upstream transmission is DAMA (demand assigned multiple access). The base station allocates bandwidth requested by the subscriber station to meet QoS. The specifications support various QoS for bearer services — ATM (various bit rates for real-time and non-real-time traffic) and IP (integrated and differentiated services).

9.6.4   Cellular Networks

Mobile wireless in the last and first mile broadband network could be implemented using the traditional 3-G cell network or 802.11 wireless LAN (WLAN). The former uses the licensed spectrum and the latter uses the unlicensed spectrum at 2.4 GHz (802.11b/g) and 5 GHz (802.11a). The IP-based third- and fourth-generation cellular network is an attractive vehicle for broadband service distribution. With the well established global outside plant networks, UMTS (universal mobile telecommunications system) Forum is focused on achieving this. Standards for 3-G are still under development[31] and research activities on 4-G are in progress.

9.6.5   Wireless LAN

This is the fastest growing area at the present time with hot spots established in public places such as airports and private locations such as hotels and restaurants. Mobility management and billing due to roaming is being addressed to make the hot spots seamless. Hot spots provide a convenient way to connect to the Internet for business travelers. The cost of wireless access points and wireless clients has been coming down, which is accelerating this process.

The deployment of wireless LAN has been growing at a rapid rate for last mile access as well as for home networking. It appears that this may become the *de facto* standard. 802.11b with 11 Mbps is deployed most in home networking and in hot spots. Although 802.11a has a higher data rate of 54 Mbps than 802.11b, 802.11g operating in the same spectrum as 802.11b with 54 Mbps using OPDM technology is being introduced into the market and is gaining more popularity.

The secret of success of WLAN is its meeting the "3-R" requirements: range (distance), rate (bandwidth), and robustness (constant signal strength). Development work is progressing on many fronts addressing these issues. FCC Title 47 limits the power of wireless transmitters, which limits the range. The power radiated depends on the frequency spectrum and technology used, such as frequency hopping, direct sequence, etc.

The throughput, or the effective data rate, decreases as the distance of the receiver increases from the transmitter, as well as when the direction is off the direct line of sight in the case of directional antenna. For good QoS, these need to be taken into account in designing the system. Signal strength received by a subscriber station varies with time as the profile of the terrain changes. Fading occurs due to attenuation in the direct path and due to multiple path signals arriving at the receiver. Attention needs to be paid to antenna design and configuration to mitigate this fading phenomenon.

The range, rate, and robustness can be enhanced by varying the antenna configuration at the base station. At this station, also referred to as access point, high-gain antennas are used to concentrate omnidirectional energy into narrow lobes extending the range and rate. Space-diversity antennas are improving the robustness. Antennas using MIMO (multiple input multiple output) technology have helped improve the range, rate, and robustness. However, the antenna configurations are limited at the client or workstation. Future research and development will be heavily focused in this area to meet the 3-R requirements. It is conceivable that WLAN technology will be concatenated with other technologies to extend the range, rate, and robustness.

Interference is a serious issue in the unlicensed spectral band of 2400 to 2483 MHz and 5725- to 5850-MHz bands. Interference takes place not only between stations using the same protocol, but also between dissimilar protocols such as 802.11b and Bluetooth. IETF workgroups are working on the specifications to mitigate this interference and the equipment manufacturers are developing technology to minimize the interference.

Security is another serious issue in the implementation of WLAN. The fear started by the service providers not using it in the early implementation of available security, such as WEP (wired equivalent privacy), has exacerbated this problem. Authentication and privacy need to be ensured for the users to feel comfortable. IEEE 802.1x has gone a long way in resolving this for the common use.

9.6.6 Passive Optical Networks

Along with wireless, fiber is a medium that offers great promise for the future direction of broadband access network. WAN is currently reaching close to home on fiber medium, FTTC (fiber to the curb), or FTTP (fiber to the premises) as it is called. In some sites, the fiber is brought all the way to the home.

Using fiber offers numerous advantages. It is immune from the ingress noise of electromagnetic radiation that affects other modes of transmission. The signal-to-noise ratio at the receiver is high and the bandwidth of transmission is virtually unlimited in the fiber medium, as well as through the optical components. Thus, the QoS desired for broadband service could be accomplished much more easily using this technology. Economics and the lack of applications that need very high bandwidth are holding back deployment of this technology.

The definition of PON should strictly be limited to networks that have no active components in the loop. However, the term is generally used even when an ONU (optical network unit) is in the FTTC pedestal and the signal is carried to the home using copper or fiber. In this situation, PON extends only from the head end to the FTTC.

Using Ethernet protocol over a long loop of several miles or tens of miles of PON, EPON has attracted much attention due to the simplicity of system design and cost. Another popular terminology used for this subject is "Ethernet on the first mile" (EFM). If this is combined with the IP core network, an all-IP network would be possible. Significant research is being done on this subject. There is considerable discussion in the literature on QoS that can be obtained in an all-IP broadband network[32] compared to the network using ATM in the core. The ATM network has matured in its ability to handle the multirate information of real- and non-real-time signals. Specifications for the IP network define end-to-end QoS using differentiated services (RFC 2430 and 2475) or integrated services (RFC 1633), but these are yet to be proven in the real-world environment.

Migration to PON could possibly occur by converting part of the access network to PON and the rest using legacy networks. The hybrid access networks will be discussed in Section 9.6.8.

9.6.7 Power Line Network

Interest in using power lines for Internet access has been growing.[33] The attraction to this is the existing infrastructure that reaches more residences than any other medium. However, the technology and its commercial viability are viewed with skepticism due to technical problems and regulatory issues.

9.6.8 Hybrid Technologies

It is difficult to predict how the infrastructure of a matured broadband network would look. Whatever it is, the path to that would be marked using hybrid technologies. Based on the environment and applications, the broadband network design needs to be based on plug-and-play approach.

The choices in access networks and home networks are shown in Figure 9.10. From the edge router, the alternatives in broadband access networks are the same as those shown in Figure 9.9 except for the addition of the satellite network. Satellite is primarily used for video broadcast and its role in the broadband access network is still unclear. The access network is connected to the home network by a residential gateway, which is addressed in Section 9.6.9. This gateway should be able to interface with a multitude of networks on the access side, as well as on the CPE (customer premises equipment) network side. A brief description of each of the

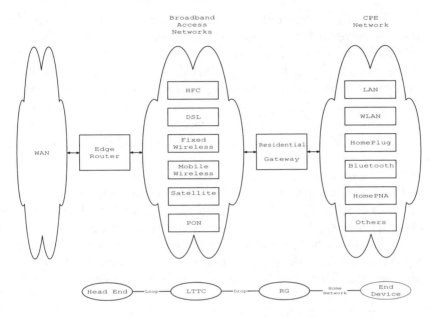

Figure 9.10 Last mile and last meters networks.

numerous alternatives in the distribution of information in the house is offered in Section 9.6.10.

The bottom of Figure 9.10 presents a segmented view of the network from the edge router at the head end to the end device, at home or as a mobile unit. The transmission of information between the head end and the residential gateway (RG) is divided into two segments: loop and drop. The loop in the cable access network is the cable that runs along the street (or buried) and terminates at LTTC (loop to the curb). The drop is the connection from the LTTC to the network interface unit at home (NIU in Figure 9.2) — what used to be called tap-to-TV. In the case of DSL, the loop is the link from the central office or remote terminal to a pedestal that serves one or more homes. There is a drop from the pedestal to each home. In both schemes, the residential gateway is a stand-alone modem with Ethernet output or a modem embedded in a residential gateway, which is generally a router.

The drop segment could be typically a few hundred meters. Several issues are associated with the drop. There is attenuation of the signal, more prone to residential ingress noise impact and expensive to install. Even the loop could be expensive to install in rural areas. In such cases, with the re-emergence of wireless technology, a hybrid approach is used. For example, EPON is used for the loop carrying the signal in the fiber to LTTC, referred to as ONU, and copper is used for the drop. Another recent example is that of DSL linked to public telephone booths, in which DSL subscribers could access the Internet on a laptop within a few hundred feet. In rural areas, fixed wireless is used for long-distance loop, and other technologies are deployed for local transmission. More of such hybrid access networks will be seen in the future. In all such cases, the QoS issue over the interface of cross-standards of different technologies becomes important.

9.6.9   Residential Gateway

The residential gateway shown in Figure 9.10 is the entry point into the customer premises. In general, it is expected to have multiple access network interfaces and multiple home networking interfaces. Telecommunication equipment manufacturers are making residential gateway a communication device, as well as one that could manage applications that run devices and appliances at home and mobile units. This trend in making the legacy set-top box into a smart residential gateway is expected to continue at a rapid pace.

9.6.10 Home Networking

Recent statistics indicate that 70% of home networks are Ethernet based.[34] However, this needs CAT5 cable. It is expected that wireless LAN based on IEEE 802.11a and 802.11g will soon surpass wired LAN. Although HomePNA and HomePlug are still a small percentage of residential networks, they claim the advantage of using the existing telephone and power-line wiring in the house, respectively. Coaxial cable that is currently used at home serves to distribute data and video. The coax network has taken advantage

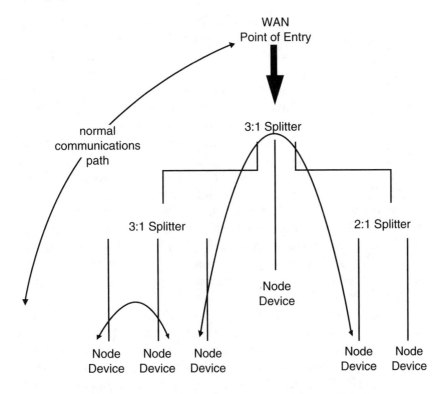

Figure 9.11   Home network based on c.LINK technology.

to serve as intrahome network for data by using special adapters by MoCA, which uses the higher end of the RF spectrum.[35,36] It is based on the c.LINK technology developed by Entropic Communications that enables data jump over the splitters in the coaxial cable network in the home and forms an intrahome data and video network.[36] This is shown in Figure 9.11 and could play a significant role because most homes have been and are wired with coaxial cable for television.

    The last-and-first-mile issue in telecommunications is extending to last and first meters. Section 9.6.8 mentioned wireless playing a role in the drop segment of access network. Home and mobile networks also play a significant role in the last and first meters in broadband service. It is not uncommon to see an Ethernet LAN or WLAN as part of the home set up for many users who have Internet connection.

Figure 9.10 shows some of the alternatives available for home networking. LAN in the figure implies wired LAN, which is 802.3 Ethernet protocol. Many commercial routers are available for this. The limitation of using this is that wiring in most houses cannot carry Ethernet and thus the computers are limited to certain specific areas in the house. Wireless LAN overcomes this problem. The 802.11b protocol is the most popular version that operates at 2400 to 2483 MHz. The trend is to move to 802.11g, which operates in the same spectral band, but provides a 54-Mbps data rate compared to the 11-Mbps data rate of 802.11b. Based on the configuration of the house and where the base station is located, WLAN could experience a reception problem. Extensive work is being done to overcome this problem with specialized antennas and repeaters.

Bluetooth is another wireless technology standard that works in the same unlicensed spectrum as 802.11b. Because these two work in the same spectrum, interference between them occurs. Research is in progress to mitigate this interference. An IETF workgroup, IEEE P802.15, is developing specifications for these two to coexist.

In addition to Ethernet and wireless LANs, a third contender for home networking is power line communication (PLC) LAN.[37] The Home Plug Power Line Alliance was formed in the U.S. in 2000 and has captured 10% of the market share in LANs.[38] In Europe, the European Home System (EHS) consortium and International Telecommunication Union (ITU) have defined specifications for power line communications. The HomePlug 1.0 specification defines a 10-Mbps home network based on orthogonal frequency division multiplexing (OFDM). HomePlug AV is being developed for a 100-Mbps home network over power line. This could easily distribute broadband signals with no additional wiring. To be highly competitive, the cost in using the technology needs to be competitive.

Consumer electronics bus (CEBus) and X-10 have emerged as standards and commercial products have been produced in PLC. CEBus has an industry council (CIC) whose mission is to provide information to the design and development community about CEBus. Home Plug and Play (HPnP) is an industry specification describing the way in which consumer products will cooperate with other products and work with CEBus standards.

The fourth category of LAN for home networking is to use the home infrastructure of telephone lines. HomePNA is a high-speed, reliable networking (LAN) technology that uses the existing phone wires in the home to share a single Internet connection with several PCs. HomePNA 2.0 claims data rate in excess of 10 Mbps and is targeting 100 Mbps in its next generation.

In addition to the LANs just discussed, one also has the choices of Firewire and USB networks for connecting appliances. These are the home networking trends of the future.

9.6.11 Integrated End-to-End Network

We have discussed implementation alternatives in core, access, and home networking. As the trend towards deployment of broadband services increases, the migratory path could be expected as an end-to-end network that would be a plug-and-play of technologies. However, this should be seamless to subscribers and should meet the requirements described in Section 9.3. This is the challenge that the broadband industry faces in the future.

9.7 OAMP

Progress in the various technologies in providing broadband service needs to go hand in hand with operations support functions: operations, administration, maintenance, and provisioning (OAMP), as they are referred to in the telecommunications industry. Equivalent to these in open system interface (OSI) terminology are fault, configuration, administration, performance, and security (FCAPS) management. These are accomplished using OSSs.

OSSs at the present time are stand-alone systems and not integrated. This is true of regular data and telecommunication

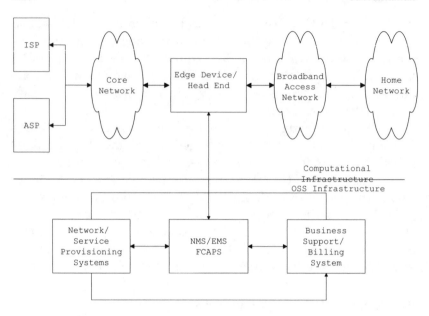

Figure 9.12   Operations support systems.

services, as well as broadband services. When broadband services are offered over the same media, it is important that the OSS functions be integrated. For example, when a subscriber requests video-on-demand service, the service provider needs to determine whether enough bandwidth resource is available over shared resources before allocating the bandwidth using a provisioning system. For this, the provisioning system may need to acquire the information from a network management system that does the FCAPS functions. Also, once the video-on-demand service is completed, the billing system needs to be updated to charge the subscriber for the service provided.

The bottom half of Figure 9.12 shows the integrated OSS infrastructure containing the network management system (NMS)/element management system (EMS), network/service provisioning system, and business support/billing system — all of which communicate to the edge device at the head end of the access network. Through the edge device, the integrated

OSS could manage the core network, access network, and the home network. The top half of the figure shows the computational infrastructure. The Internet service provider (ISP) and the application service provider (ASP) are the communication and content service providers.

OSSs play a key role in achieving satisfactory economic and QoS performance in a broadband service offering. Not only the network parameters in the various segments need to be monitored and managed in a coordinated manner, but also the coordination of the administrative functions provided by the multiple service providers. The traditional network management would be extended to cover telecommunication management network (TMN) functions of management of element, network, service, and business,[39] as well as QoS, service level agreement, and security. Thus, development of intelligent and dynamically adapting OSSs, some of whose functions would be embedded in the network, would assume a greater role in the future trend of broadband services.

REFERENCES

1. World Broadband Statistics: Q2 2003, Point Topic Ltd., September 2003.

2. D. Ainscough, The Yankee Group Research Report, 27 June 2003.

3. Committee on Broadband Last Mile Technology, *Broadband: Bringing Home the Bits*, National Academy Press, Washington D.C., 2002.

4. http://www.silicon.com/networks/broadband/ 0.39024661,39119273,00.htm.

5. http://www.techweb.com/article/com200210155002

6. M. Flannagan and J. Sinclair, Configuring Cisco Voice Over IP, 2nd ed., Cisco Systems.

7. Guide to Cisco Systems VoIP infrastructure solution for SIP Version 1.0, Cisco white paper text part number: OL-1002-01, 2000.

8. J. Rosenberg, H. Schulzrinne, G. Camarillo, A. Johnston, J. Peterson, R. Sparks, M. Handley, and E. Schooler, RFC 3261: SIP: session initiation protocol, June 2002.

9. J. Rosenberg and H. Schulzrinne, RFC 3262: reliability of provisional responses in session initiation protocol (SIP), June 2002.

10. J. Rosenberg and H. Schulzrinne, RFC 3263: session initiation protocol (SIP): locating SIP servers, June 2002.

11. J. Rosenberg and H. Schulzrinne, RFC 3264: an offer/answer model with session description protocol (SDP), June 2002.

12. A.B. Roach, RFC 3265: session initiation protocol (SIP) — specific event notification, June 2002.

13. F. Andreasen and B. Foster, RFC 3435: media gateway control protocol (MGCP) version 1.0, January 2003.

14. B. Foster and F. Andreasen, RFC 3660: basic media gateway control protocol (MGCP) packages, December 2003.

15. C. Groves, M. Pantaleo, T. Anderson, and T. Taylor, Eds., RFC 3525: gateway control protocol version 1, June 2003.

16. C. Perkins, RTP: multimedia streaming over IP, USC Information Sciences Institute, http://www.east.isi.edu/projects/NMAA/presentations/2002-06-18-Video-Services-Forum.pdf.

17. Unified Messaging, tutorial, International Engineering Consortium, 2003.

18. A. Roychowdhury and S. Moyer, Instant messaging and presence for SIP enabled networked appliances, http://www.iptel.org/2001/pg/final_program/22.pdf.

19. S. Taylor and L. Hettick, SIP in the call center, *Network World Convergence Newsletter*, 08/27/03,http://www.nwfusion.com/newsletters/converg/2003/0825converge2.html.

20. W. Fenner, RFC 2236: Internet group management protocol, version 2, November 1997.

21. S. Jang, E.B. Kelly, and A.W. Davis, Voice and video over IP FAQ, Wainhouse Research, www.wainhouse.com, January 2003.

22. M. Banikazemi, IP multicasting: concepts, algorithms, and protocols: IGMP, RPM, CBT, DVMRP, MOSPF, PIM, MBONE, http://www.cse.ohio-state.edu/~jain/cis788-97/ftp/ip_multicast.pdf.

23. A.F. Gómez-Skarmeta, A.L. Mateo, and P.M. Ruiz, Multimedia services over the IP multicast network, http://ants.dif.um.es/staff/pedrom/papers/INFORMATIK-MBone.pdf.

24. Digital television 2003: the emergence of advanced TV services, *eMarketer*, August 2003.

25. M. Subramanian, A software-based elastic broadband cable access network, *Proc. Manual Cable-Tech Expo 2002*, San Antonio, TX, June 2002.

26. M. Subramanian and L. Lewis, QoS and bandwidth management in broadband cable access network, special issue on network management, *Computer Networks*, 43(1), 16 September 2003.

27. D.A. Robinson and K. Schwan, Dynamic management of cooperating peers in embedded wireless systems: an approach driven by information quality, http://www.cercs.gatech.edu/tech-reports/tr2003/git-cercs-03-18.pdf.

28. Using content networking to provide quality of service, Cisco white paper 1999, http://www.cisco.com/offer/tdm_home/pdfs/content_network/CN_QoS.pdf.

29. O. Sniezko et.al., HFC architecture in the making, *Commun. Eng. Design*, July 1999.

30. Packet cable specifications documents in www.packetcable.com.

31. http://www.three-g.net/3g_standards.html.

32. Topics on IP-oriented QoS: technologies and standards for broadband applications, *IEEE Commun.*, 40(12), December 2002.

33. Broadband is power: Internet access through the power line network, topics in Internet technology, *IEEE Commun.*, 41(5), May 2003.

34. K. Scherf, The home network market: data and multimedia connectivity, white paper, Parks Associates, 2004.

alliance.org/.

entropic-communications.com/pages/technology.html.

issue on power line local area networking, *IEEE Com-cations Magazine,* 41(4), April 2003.

.v. Pavlidou, A.J. Han Vinck, J. Yazdani, and B. Honary, Power line communications: state of the art and future trends, *IEEE Communications Magazine,* 41(4), April 2003.

39. M. Subramanian, *Network Management: Principles and Practice,* Addison-Wesley, Reading, MA, 2000.

Index

Note: Italicized page numbers refer to illustrations and to tables

Numbers

10BASE5 standard, 346–348
10 GbE standard, 346–348
1000BASE-T, 355
100BSE-FX Ethernet, 348
2G (Second-generation) networks, 417
2.5G networks, 417
256 QAM cable modem, 12
3Com, 455
3GPP (Third-Generation Partnership Project), 568–569
3G (Third-generation) networks, 417, 567–568
4G (Fourth-generation) networks, 390, 418
8 PSK modulation, 511

A

AAC (MPEG–2 advanced audio coding), 30–34
Absolute category rating (ACR), 39
Access network frequency plan (ANFP), 240
Access points (AP), 14, 126–27, 438
ACL (Asynchronous connectionless link), 458–459

ACR (Absolute category rating), 39
ACTS (Advanced communications technology satellite), 503, 505
Adaptive differential pulse code modulation (ADPCM), 25
Adaptive media playout (AMP), 73
Adaptive modulation, 403–404
Additive increase multiplicative decrease (AIMD), 67–68
Additive white Gaussian noise (AWGN), 181
ADPCM (Adaptive differential pulse code modulation), 25
ADSL2, 163–64
    frequency band usage, 224
    operational modes, 223
ADSL2plus, 164
    bandwidth, 223–224
    ingress, 199
    twisted-pair lines, 355
ADSL (Asymmetric digital subscriber lines), 163–164; see also DSL; VDSL
    access network, 547
    bandwidth, 222–224
    in broadband DSL access network (BDAN), 546–547
    channel configuration, 553

        ement,

       .on, 207
      .nagement system,
     ɔ9
    .nanagement, 554–555
   quency band usage, *226*
  .ietwork element management
   system, 549–555
 noise margins, 553
 operational modes, 223
 performance management, 555
 simple network management
  protocol (SNMP), 555–558
 standards, 547–549
 subchannels, 222–224
Advanced communications
 technology satellite (ACTS),
 503, 505
Advanced infrared (AIR) protocol,
 461–462
Advanced time division multiple
 access (A-TDMA), 287–291
AIMD (Additive increase
 multiplicative decrease),
 67–68
AI-nodes (Asynchronous-
 isochronous nodes), 463
AIR (Advanced infrared) protocol,
 461–462
ALF (Application level framing),
 75–76
AlGaAsSb mirrors, 361–362
AlGaInAs, 361
Alien near-end crosstalk, 190
AllWave fiber, 357
Alohanet, 346–348
ALOHA protocol, 512–513
Altitude, 496
Amateur radio, 199–200
American wire gauge (AWG),
 179–180
AMP (Adaptive media playout), 73
Analysis-by-synthesis, 25, *31*
ANFP (Access network frequency
 plan), 240

ANIK F2 satellite, 506
A-nodes (Asynchronous nodes), 463
Antennas, 507–508
AP (Access points), 14, 126–127, 438
APONs (ATM passive optical
 networks), 132; *see also* PONs
 architecture of, 343–344
 broadcast downstream overlay,
  345–346
 downstream transport, 344–345
 upstream transport, 344–345
Application level framing (ALF),
 75–76
Application rate, 10
Application service providers
 (ASPs), 621
ARQ (Automatic repeat request), 10
Arrayed wavelength grating (AWG),
 140–141
ASPs (Application service
 providers), 621
Astrolink, 519
Asymmetric digital subscriber lines,
 see ADSL
Asynchronous connectionless link
 (ACL), 458–459
Asynchronous-isochronous nodes
 (AI-nodes), 463
Asynchronous nodes (A-nodes), 463
A-TDMA (Advanced time division
 multiple access), 287–291
ATM convergence sublayer, 422
ATM passive optical networks, *see*
 APONs
Attenuation, 180–181
Audio compression, 26–27, 30–34
Audio files, download speed, *8*
Audio/Video Transport Working
 Group, 590–591
Automatic repeat request (ARQ), 10
AWG (American wire gauge),
 179–180
AWG (Arrayed wavelength grating),
 140–141
AWGN (Additive white Gaussian
 noise), 181

B

Backbone networks, 5
  edge interface, 141–142
  optical label switching, 146–147
  packet switching, 143–144
  routers, 143–144
  SONET/SDH, 142–143
  tag switching, 144–145
  in telephony network, 168–169
Back channels, 48
Bandwidth (BW), 12–13
  and bit rate, 10
  Ethernet for the first mile (EFM),
    121
  frequency-division duplexing
    (FDD), 202
  and media rate, 69–71
  and rate control, 67–69
  in video streaming, 62–63
Baseband, 458
Baseline Privacy Plus, 302
Base transceiver station (BTS), 397,
    407–408
Basic service set (BSS), 437
BBM (Broadband margin), 9–10, 18
BCAN (Broadband cable access
    network), 533–546
  architecture of, *533*
  DOCSIS 1.0 standard, 536–538
  DOCSIS 1.1 standard, 543–545
  DOCSIS 2.0 standard, 545–546
  management information base
    (MIB), 538–541
  next-generation, 609
  operations support systems
    (OSSs), 541–543
BDAN (Broadband DSL access
    network), 546–559
Beamwidth, 500
BE (Best effort), 52–53
BellSouth, 334, 339, 341
Bent-pipe satellites, 504
Best effort (BE), 52–53
B-frames, 29

B-frames (Bidirectionally predicted
    frames), 29
Bidirectionally predicted frames
    (B-frames), 29
Bidirectional transport, 369–372
Billing, 530
Bit rate, 23–24
  digital subscriber lines (DSL),
    160–162
  integrated services digital
    network (ISDN), 160
  multiple-file switching, 43–44
  scalable compression, 44–45
  transcoding, 42
Bitstream synchronization, loss of,
    74–75
Bluetooth, 454–459; *see also* WPANs
  bandwidth, 456
  brief history of, 454–455
  core, 456
  data rate, 478
  and IEEE 802.15.3 standard,
    468
  links, 458–459
  next-generation, 618
  piconets, 457
  power management, 482–483
  profile, 456
  protocols, 457–458
  range, 476–478
  scatternets, 457
  support for LAN integration,
    480–481
  support for voice, 479–480
Bottlenecks, 336
BPON (Broadband passive optical
    networks), 342
BPSK modulation, 445
BRAN (Broadband radio access
    networks), 433
Bridge taps, 183–188
British Telecom, 200, 334, 339
Broadband, 1–2
  applications of, 7–9
  functional and elastic definition
    of, 6

multiple dimensions of, 5–6
options, 2–3
Broadband: Bringing Home the Bits
    (report), 2–3
Broadband cable access network, *see*
    BCAN
Broadband DSL access network
    (BDAN), 546–559
Broadband home access, 390–393
    electrodomestic network devices
        (ENDs), 392–393
    home area network, 392
    local loop, 391–392
    residential gateway (RG), 392
Broadband margin (BBM), 9–10, 18
Broadband networks, 46–47
    broadcast communications, 48
    constant- vs. variable-bit-rate
        channels, 51–52
    constant- vs. variable-bit-rate
        coding, 51
    content delivery networks, 88–92
    end-to-end security, 80–83
    future applications, 92–93
    hybrid technologies, 614–616
    interactive vs. noninteractive
        media applications, 49–50
    Internet telephony, 56–57
    joint source/channel coding in,
        79–80
    latency, 53–55
    midnetwork media transcoding,
        80–83
    multicast communications,
        48–49
    packet-switched vs. circuit-
        switched, 50
    peer-to-peer networking, 57–60
    point-to-point communications,
        48
    quality of service, 52–53
    real-time encoding vs. pre-
        encoded media, 49
    residential subscribers, 583–584
    static vs. dynamic channels,
        50–51

Broadband optical access, *see*
    Optical access
Broadband passive optical networks
    (BPON), 342
Broadband radio access networks
    (BRAN), 433
Broadband services, 583–586
    capabilities and applications, 8
    data traffic, 601–603
    economics of, 587
    fees, 601, 607
    home entertainment, 598–599
    hybrid technologies, 614–616
    installation of, 606
    Internet protocol-based, 587–598
    market drivers for, 586–587
    next-generation, 606–607
    on-demand, 606
    operations, administration,
        maintenance, and
        provisioning (OAMP),
        619–621
    quality of service, 603–605, 607
    residential applications, 598–600
    residential data applications,
        599–600
    residential subscribers, 584
    security and privacy of
        information, 607
    service issues, 605–606
    SOHO and SME applications,
        600
    Subscriber's needs, 600–605
    universal user ID, 607
Broadband wireless access, *see* BWA
Broadcast communications, 48
BSS (Basic service set), 437
BTS (Base transceiver station), 397,
    407–408
Buddy list, 593
Buffering, 71–73
Burst downlinks, 425
Burst noise, 275–276
Burst profiles, 304
BW, *see* Bandwidth

BWA (Broadband wireless access), 121–122
architecture of, *433*
basic principles, 394
challenges in, 397–398
channel coding, 403
channels, 397–400
cost reduction and reliability in, 122–123
delay spread, 399
deployment scenarios, 396–398
future directions of, 486
infrared (IR), 395
K-factor, 399–400
local multipoint distribution service (LMDS), 410–415
MAC layer, 405–406
macrocells, 397
microcells, 397
multichannel multipoint distribution system (MMDS), 408–410
multihop wireless topologies, 123–128
multiple antennas in, 406–408
path loss, 398–399
physical layer, 401–405
radio frequency (RF), 395–396
satellite communications, 415–416
services, 396
single-carrier modulation with equalization, 401
supercells, 397

C

CableLabs, 609
Cable modems, 105–106
downstream modulation schemes, 277–279
ingress, 275
upstream modulation schemes, 279

Cable modem termination system, *see* CMTS
CAC (Call admission control), 568
Caching, 88–89
Call admission control (CAC), 568
Call centers, 593–594
CAP (Carrierless amplitude/phase) modulation, 236
CAP (Contention access period), 466
Carrierless amplitude/phase (CAP) modulation, 236
Carrier sense multiple access with collision detection (CSMA/CD), 346–348, 438
Carrier-to-interference (C/I) ratio, 397
Carrier-to-noise (C/N) ratio, 397, 499–500
CAT 5e cable, 355
CATV (Community antenna television), 251–252
C-band, 500
CBC (Cyclic block check), 545
CBR (Constant-bit-rate) channels, 51–52
CBR (Constant-bit-rate) coding, 51
CC, *see* Channel coding
CCI (Cochannel interference), 397
CCK (Complementary code keying), 443–444
CDMA (Code-division multiple access), 404
CDNs (Content delivery networks), 88–89
components of, 89–90
functionalities, 90–92
overview, 87–88
CEBus (Consumer electronics bus), 618–619
Cellular access networks, 564–571
Mobile IP, 564–567
mobility management, 567–568
next-generation, 611
quality of service management, 568–569

resource and power
   management, 568
security management, 569–571
voice over IP, 594–595
Cellular telephony, 416–418
   modem efficiencies in, 13
   speech coders, *32*
Cellular Vision, 410
Centralized priority reservation
   (CPR), 300, 512
Central office (CO), 159
   DSL deployment from, 171–175
   modems, 173–174
   in telephony network, 168–170
Channel coding (CC), 15–16; *see also*
   Modulation coding (MC);
   Network coding (NC)
   broadband wireless access
      (BWA), 403
   in IP protocol stack, *11*
   and source coding, 79–80
Channels
   analog, 12, 278
   BWA, 398–400
   digital TV, 599
   optical, 132
   RF, 106, 257
   upstream and downstream, 136,
      202, 206, 213
C/I (Carrier-to-interference) ratio,
   397
Circuit-switched networks, 50, 417
Cisco Systems, 605
Clear to send (CTS) frames, 440
c.LINK technology, 617
CMIP (Common management
   information protocol), 523
CMTS (Cable modem termination
   system), 111
   adaptation approach, *323*
   experimental setup, 111–113
   functions of, 535
   in high-speed Internet service,
      257
   scheduler, 116–117

upstream and downstream
   performances of, 113–116,
   311–313
C/N (Carrier-to-noise) ratio, 397,
   499–500
Coarse wavelength division
   multiplexing (CWDM), 134,
   361, 364–365
Coaxial cables
   conductive losses, 259
   dielectric losses, 259
   elements of, *260*
   impedance of, 261
   loss vs. frequency curve, *260*
   propagation velocities, 264–265
   in tree and branch network
      architecture, *253*
Coaxial networks, 251–252, 256
Cochannel interference (CCI), 397
Coded frames, 28–29
Code-division multiple access
   (CDMA), 404
Coded orthogonal frequency division
   multiplexing (CODFM), 409
CODFM (Coded orthogonal
   frequency division
   multiplexing), 409
Combiners, 367
Committee T1 Standard T1.413, 163
Common management information
   protocol (CMIP), 523
Common open policy services
   (COPS), 110
Common path distortion (CPD), 266,
   271, 320
Community antenna television
   (CATV), 251–252
Complementary code keying (CCK),
   443–444
Complementary sequences, 444
Composite second order (CSO), 266,
   271–272
Composite triple beat (CTB), 266,
   271–272
Compression, *see* Media
   compression

Concatenation, 315
Conductive losses, 259
Configuration management, 551–554
Connection point beacon (CPB), 464
Connection point (CP), 463
Connection speed, 5–6
Constant-bit-rate (CBR) channels, 51–52
Constant-bit-rate (CBR) coding, 51
Constant-quality variable-rate (CQVR), 24
Constant-rate variable-quality (CRVQ), 24
Consumer electronics bus (CEBus), 618–619
Contact list, 593
Content delivery networks, *see* CDNs
Contention access period (CAP), 466
Continuous wave (CW) ingress, 290–291
CONUS satellite, 503
Converged networks, 338, 416
COPS (Common open policy services), 110
Couplers, 367
CPB (Connection point beacon), 464
CP (Connection point), 463
CPD (Common path distortion), 266, 271, 320
CPE (Customer premises equipment), 171, 407–408, 530
CPR (Centralized priority reservation), 300, 512
CQVR (Constant-quality variable-rate), 24
CRC (Cyclic redundancy check), 423
Cresting effect, 271
Crosstalk, 188–189
    adding from different sources, 197–198
    cancellation of, 242–243
    far-end, 192–197
    near-end, 189–192

CRVQ (Constant-rate variable-quality), 24
CSMA/CD (Carrier sense multiple access with collision detection), 346–348, 438
CSO (Composite second order), 266, 271–272
CTB (Composite triple beat), 266, 271–272
CTS (Clear to send) frames, 440
Curb-switched networks, *131*
Customer modem, 171
Customer premises equipment (CPE), 171, 407–408, 530
CW (Continuous wave) ingress, 290–291
CWDM (Coarse wavelength division multiplexing), 134, 361, 364–365
Cyclical extension, 446
Cyclic block check (CBC), 545
Cyclic prefix, 230, 243–244, 402
Cyclic redundancy check (CRC), 423
Cyclic suffix, 229–230, 243–244

D

DAMA (Demand assigned multiple access), 563, 611
Data ACK, 310
Data delivery, 18
Data encryption standard (DES), 302, 545
Data link layer, 470–473
Data packets, 15–16
Data rate, 478, 527, 612
Data rate ratios, 212–213
Data transport protocol, 590–591
DAVIC (Digital Audio/Visual Council), 413
DBPC-REQ (Downlink burst profile change request), 427
DBPC-RSP (Downlink burst profile change response), 427
DBPSK phase modulation, 442

DBS (Direct broadcast satellite), 416, 599

DCC (Dynamic channel change), 312

DCF (Distributed coordination function), 438, 573

DCF interframe space (DIFS), 439

DCR (Degradation category rating), 40

DCT (Discrete cosine transform), 27–29

Decibels, 181

Decision feedback equalizer (DFE), 378, 401

Decoders, 37–39

Degradation category rating (DCR), 40

Degradation mean opinion scores (DMOSs), 40

Delay, 23–24

Delay jitter, 63, 71–73

Delay spread, 399

Delta compression, 150
   as cross-layer technology, 153–154
   performance, 150–153

Demand assigned multiple access (DAMA), 563, 611

Denial of service, 529

Dense wavelength division multiplexing (DWDM), 134

DES (Data encryption standard), 302, 545

DFE (Decision feedback equalizer), 378, 401

DFT (Discrete Fourier transform), 214

DHCP (Dynamic host configuration protocol), 111, 428

Dielectric losses, 259

Differential mode delay (DMD), 358, 373–375

DiffServ, 109

DIFS (DCF interframe space), 439

Digital Audio/Visual Council (DAVIC), 413

Digital devices, 435

Digital loop carrier (DLC), 169–170

Digital subscriber lines, *see* DSL

Digital television (DTV), 37, 48, 598–599

Digital video broadcasting (DVB), 413

Digital video disc (DVD), 49

DirecPC, 519–520

Direct broadcast satellite (DBS), 416, 599

Direct-sequence code-division multiple access (DS-CDMA), 401

Dirty packets, 64

Disclosure of information, 529

Discrete cosine transform (DCT), 27–29

Discrete Fourier transform (DFT), 214

Discrete multi-tone, *see* DMT

Distributed coordination function (DCF), 438, 573

DIUC (Downlink interval usage code), 425

DLC (Digital loop carrier), 169–170

DL-MAP (Downlink MAP), 425, 430–431

DMD (Differential mode delay), 358, 373–375

DMOSs (Degradation mean opinion scores), 40

DMT (Discrete multi-tone), 215–222
   frequency-division duplexing (FDD), 213
   transmitter and receiver block diagrams, *222*
   very high-bit-rate DSL (VDSL), 227–228

DOCSIS 1.0, 280–283, 536–538

DOCSIS 1.1, 543–545

DOCSIS 2.0, 283–286
   advanced TDMA in, 287–291
   network management, 545–546

synchronous code division multiple access (S-CDMA), 291–298
DOCSIS cable network, 110–111
bandwidth allocations, *274*
data-over cable reference architecture, *258*
downstream error control coding, *278*
experimental setup, 111–113
MAC protocols, 300–318
symbol rates, *280*
upstream and downstream performances of, 113–116, 262–263
upstream burst receiver, *282*
wireless, 128
DOCSIS-S standard, 514–516
Doppler shift, 497
Downlink burst profile change request (DBPC-REQ), 427
Downlink burst profile change response (DBPC-RSP), 427
Downlink interval usage code (DIUC), 425
Downlink MAP (DL-MAP), 425, 430–431
Downloads, *8*
latency, 53–54
peer-to-peer networking, 149–150
Downstream MAC protocols, 306–310
Downstream modulation, 277–279
Downstream signals, 534
DQPSK phase modulation, 442
Drop, 615–616
Drop fibers, 340–341
DS-CDMA (Direct-sequence code-division multiple access), 401
DSL (Digital subscriber lines), 105–106
access multiplier, 174–176
basic concept in, 546
bit rates, 160–162

vs. copper lines, 3
deployment from remote terminals, 175–177
deployment over plain telephone services (POTS), 171–178
duplexing, 201–214
forum management system reference model, *550*
vs. integrated services digital network (ISDN), 160–161
interference, 161
line codes, 214–221
loop reach, *168*
modem efficiencies in, 12–13
next-generation, 610
residential subscribers, 584
splitters, 171–173
transmission, 201–221
types of, 162–166
use of phone lines in, 161
DSL Forum, 610
DTV (Digital television), 37, 48, 598–599
Duplexing, 201
complexity of, 213–214
data rate ratios, 212–213
echo cancellation, 206–207
frequency-division, 201–206
power consumption, 214
symmetric and asymmetric services, 210–212
synchronization requirements, 214
time-division, 207–209
DVB (Digital video broadcasting), 413
DVB return channel system (DVB-RSC), 513–516
DVB-RSC (DVB return channel system), 513–516
DVD (Digital video disc), 49
DWDM (Dense wavelength division multiplexing), 134
Dynamic channel change (DCC), 312

Dynamic channels, 50–51
Dynamic host configuration protocol
    (DHCP), 111, 428

E

EAP (Extensible authentication
    protocol), 575
Earth stations, 416
    antennas, 507–508
    block diagram, *507*
    modems, 510–511
    transceivers, 508–510
Eavesdroppers, 80–81
Eccentricity, 496
EC (Echo cancellation), 206–207
Echo cancellation (EC), 206–207
EchoStar Communications, 519
E-commerce, 416
EDC (Electronic dispersion
    compensation), 367
EDCF (Enhanced DCF), 447
EDF (Erbium doped fiber), 360
Edge-emitting laser (EEL), 369
EDGE (enhanced data rates for
    GSM evolution), 417
EEL (Edge-emitting laser), 369
EEP (Equal error protection), 17
EFM (Ethernet for the first mile),
    117–119
    copper, 119
    fiber, 119–120
    next-generation, 613
    operations, administration and
        maintenance (OAM), 576–577
    passive optical network, 120–121
    Quality of Service (QoS), 121
    topologies, 119
Egress, 200, 232–233
EHS (European Home System), 618
Electrical delay, 264–265
Electrodomestic network devices
    (ENDs), 392–393
Electronic dispersion compensation
    (EDC), 367

Electronic signal processing,
    372–373
Element management system
    (EMS), 558–559
E-mail, 416
Emitters, 370–371
EMS (Element management
    system), 558–559
Encoders, 37–39
ENDs (Electrodomestic network
    devices), 392–393
End-to-end delay, 63–65
Energy-per-bit to noise-spectral-
    density ratio, 499–500
enhanced data rates for GSM
    evolution (EDGE), 417
Enhanced DCF (EDCF), 447
Entropic Communications, 617
EPONs (Ethernet passive optical
    networks), 346–350; *see also*
    PONs
    10 GbE standard, 346–348
    architectures, 348–350, *577*
    downstream traffic, *350*
    Gigabit Ethernet, 346–348
    management information base
        (MIB), 577–578
    network management, 576–577
    standards, 132, 342
    upstream traffic, *350*
Equal error protection (EEP), 17
Equalization, 375–378, 401
Equatorial orbit, 496–497
Erbium doped fiber (EDF), 360
Ericsson, 455
Error concealment, 45–46
Error control, 73–80
    application level framing, 75–76
    bitstream synchronization,
        74–75
    error-resilient video coding, 74
    intracoding, 77
Error propagation, 77
Error recovery, 71
ESS (Extended service set), 438
Ethernet, 335, 346–348

Ethernet for the first mile, *see* EFM
Ethernet passive optical networks,
    see EPONs
European Home System (EHS), 618
Excess bandwidth, 326
Extended service set (ESS), 438
Extensible authentication protocol
    (EAP), 575

F

Fabry-Perot lasers, 363, 369
Faraday effect, 368
Faraday rotators, 368
Far-end crosstalk, *see* FEXT
Fault management, 554–555
FBWA (Fixed broadband wireless
    access), 393
FDD (Frequency-division
    duplexing), 201–206; *see also*
    Duplexing
  coded orthogonal, 409
  optimal frequency plans,
    202–206
  symmetric and asymmetric
    services, 210–212
  vs. time-division multiplexing,
    404–405
FDM (Frequency-division
    multiplexing), 255–256, 402
FEC (Forward error correction),
    15–16
  coding, 429–431
  error concealment, 71–72
  in GbE standard, 372
  IEEE 802.11a standard, 445
  satellite networks, 511
  and transmission control protocol
    (TCP), 68
FEQ (Frequency-domain equalizer),
    221
FEXT (Far-end crosstalk), 192–197;
    see also Crosstalk; NEXT
  cancellation of, 243

in very high-bit-rate DSL
    (VDSL), 240–241
FHSS (Frequency-hopping spread
    spectrum), 396, 442
Fiber, 357–360
Fiber to the curb (FTTC) networks,
    613
Fiber to the home networks, *see*
    FTTH
Fiber to the premises (FTTP)
    networks, 339–340, 613
Filters, 367–368
First mile broadband, *see* Last mile
    broadband
Fixed broadband wireless access
    (FBWA), 393
Fixed wireless networks, 397–398,
    560–561, 610–611
Foreign access points, 576
Forward error correction, *see* FEC
Forward link, *517*
Fourth-generation (4G) networks,
    390, 418
France Telecom, 334
Frequency-division duplexing, *see*
    FDD
Frequency-division multiplexing
    (FDM), 255–256, 402
Frequency-domain equalizer (FEQ),
    221
Frequency domains, 26
Frequency-hopping spread
    spectrum (FHSS), 396, 442
Frequency plans, 202–206
  for asymmetrical services, *205*
  for symmetrical services, *205*
  very high-bit-rate DSL (VDSL),
    233–234
Frequency spreading, 292–293
FSAN (Full services access
    network), 339
FTTC (Fiber to the curb) networks,
    613
FTTH (Fiber to the home) networks,
    333–335
  architectures, 335–339

bottlenecks, 336
download time, *353*
early trials, 334
installation costs, 334
passive optical networks (PONs), 339–343
FTTP (Fiber to the premises) networks, 339–340, 613
Full services access network (FSAN), 339
Fx VDSL1 frequency plan, *234*

G

G.983.1 standard, 345
G.983.3 standard, 345
G.9834 standard, 342
G.992.1 standard, 163
G.992.3 standard, 163–164
G.992.5 standard, 164
Galaxy satellite, 506
Gaming, 599
Gatekeepers, 589
General packet radio service (GPRS), 413
Geostationary orbit, 497
GFSK modulation, 441–442
Gigabit-capable passive optical networks, *see* GPONs
Gigabit Ethernet, 346–348
GI (Graded index), 374
Gilat Satellite Networks, 519
Global positioning system (GPS), 209, 567
global system for mobile communication (GSM), 417
Global system for mobile communications (GSM), 417
Gnutella, 59–60
Goodput, 14–15
GPC (Grant per connection) mode, 427
GPONs (Gigabit-capable passive optical networks), 351; *see also* PONs

encapsulation method, 351
standards, 342
GPRS (General packet radio service), 413
GPS (Global positioning system), 209, 567
Graded index (GI), 374
Grant per connection (GPC) mode, 427
Gratings, 367
Ground stations, 416
GSM (global system for mobile communication), 417
GSM (Global system for mobile communications), 417
GTSs (Guaranteed time slots), 466, 473
Guaranteed time slots (GTSs), 466, 473
Guard bands, 141

H

H.261 standard, *38*
H.263 standard, *38*
H.264/MPEG–4 Part 10 AVC standard, 37, *38*
H.323 standard, 596
HAN (Home area network), 392
Harmonically related carriers (HRCs), 267–268
HCF (Hybrid control function), 573
HCF (Hybrid coordination function), 14, 447
HC (Hybrid coordinator), 447
HDLC (High-level data link control), 461
HDSL (High-speed DSL), 165–166, 236
HDTV (High-definition television), 37
bandwidth, 353
growth of, 599
media compression, 17
Hermetic packaging, 365–367

HFC (Hybrid fiber-coaxial)
   networks, 255–259
   adaptation to upstream
      impairments, 319–325
   architecture of, 252–254
   cable plant architecture
      alternatives, 326–327
   DOCSIS downstream, 325–326
   DOCSIS standard, 110–111
   DOCSIS upstream, 326
   downstream signals, 257
   electrical delay, 264–265
   future technologies and
      architectures, 325–327
   growth of, 251–252
   hum modulation, 276–277
   impulse/burst noise, 275–276
   ingress, 272–274
   intermodulation distortion
      effects, 266–272
   laser clipping, 277
   MAC protocols, 300–318
   network management, 325
   passive channel effects, 259–266
   upstream signals, 257
Hidden node, 440
High-definition television, *see*
   HDTV
High-level data link control
   (HDLC), 461
High-pass filters, 172
High-performance LAN
   (HIPERLAN), 433–434
High-performance radio access
   (HIPERACCESS), 433
High-speed DSL (HDSL), 165–166,
   236
High-speed unlicensed metropolitan
   area networks (HUMANs),
   434
HIPERACCESS (High-performance
   radio access), 433
HIPERLAN/2, 448–452
   MAC layer, *440*, 449–450
   physical layer, 450–452

HIPERLAN (High-performance
   LAN), 433–434
HIPERLINK, 434
HOLD low-power state, 482
Home area network (HAN), 392
Home entertainment, 598–599
Home networks, 434–436
   c.LINK technology, *617*
   growth of, *585*
   next-generation, 616
HomePlug, 616, 618
Home Plug Power and Play (HPnP),
   619
Home Plug Power Line Alliance, 618
HomePNA, 616, 619
HomeRF Working Group (HRFWG),
   453, 462
Honeywell, 470
Hop, 123
Hot spots, 611
HPnP (Home Plug Power and Play),
   619
HRCs (Harmonically related
   carriers), 267–268
HRFWG (HomeRF Working Group),
   453, 462
Huffman codes, 74–75
Hughes Network Systems, 506,
   519–520
HUMANs (High-speed unlicensed
   metropolitan area networks),
   434
Hum modulation, 276–277
Hybrid control function (HCF), 573
Hybrid coordination function (HCF),
   14, 447
Hybrid coordinator (HC), 447
Hybrid fiber-coaxial networks, *see*
   HFC

I

IASs (Information access services),
   461
IBM, 455

IBSS (Independent basic service
set), 438
IDCT (Inverse discrete cosine
transform), 38
IDFT (Inverse discrete Fourier
transform), 215–216
IEEE 802.11 standard, 436–437
802.11a, *440*, 444–447, 611
802.11b, 432, 443–444, 446–447,
611, 618
802.11e, 447
802.11g, 442–443, 611
802.11i, 447–448, 574
as last mile alternative, 432
MAC layer, 438–441
management information base
(MIB), 572
network architecture, 437–438
network management, 571–576
physical layer, 441–447
quality of service management,
572–573
security management, 573–576
IEEE 802.15 standard, 465–475
802.15.3, 465–468
802.15.4, 468–475
task groups, 453–454
IEEE 802.16 standard, 419–420
802.16d, 563–564, 610
MAC layer, 420–428
network management, 563
physical layer, 428–432
working group, 434
IEEE 802.3 standard, 618
802.3ah, 576
802.3z, 375
IEEE P802.15 standard, 618
IETF (Internet Engineering Task
Force), 107, 523
IFFT (Inverse fast Fourier
transform), 402
I-frames, 28–29, 78
I-frames (Intracoded frames), 28–29
IGMP, 595–596
Image compression, 27–28, 34–36
Image files, download speed of, *8*

Impedance, 261
IMP (Instant messaging and
presence), 593
Impulse noise, 200, 275–276, 298
I-MSM (Inverted metal-
semiconductor-metal), 371
INCITS (International Committee
for Information Technology
Standards), 373
Inclination, 496
Incrementally related carriers
(IRCs), 267–268
Independent basic service set
(IBSS), 438
Industrial-scientific-medical (ISM)
band, 442
Information access services (IASs),
461
Information security, 7, 529–530,
607
Infrared Data Association, *see* IrDA
Infrared (IR), 395
Infrared LAN (IrLAN), 481
Infrastructure mode, 438
InGaAsN, 361–362
Ingress, 199, 272–274, 299–300
Initial maintenance IUC, 308
In-line filters, 173
I-nodes (Isochronous nodes), 463
Insertion gain transfer function, 194
Insertion loss, 194
Installation and maintenance, 530
Instant messaging and presence
(IMP), 593
Integrated circuits, 366
Integrated services digital network,
see ISDN
Intel, 455
Intelsat, 506
Interactive gaming, 599
Interactive media, 49–50
Interactive services, 251–252
Interactive television, 415
Interference, 161, 198–200, 612
Interleaving, 71–72
Intermodulation effects, 266–272

International Committee for Information Technology Standards (INCITS), 373

International Organization for Standardization (ISO), 30

International Telecommunications Union (ITU), 30–31

Internet, 416

Internet Engineering Task Force (IETF), 107, 523

Internet protocol (IP), *11*, 83–84

Internet protocol (IP)-based services, 587–598
  video over IP, 595–598
  voice over IP, 587–595

Internet service providers (ISPs), 534, 621

Internet streaming media alliance (ISMA), 86

Internet telephony, 56–57

Intersymbol interference (ISI), 219, 229, 377–378, 446

Interval usage codes (IUCs), 307–310

Intracoded frames (I-frames), 28–29

Intracoding, 77

IntServ, 107–109

Invensys Metering Systems, 470

Inverse discrete cosine transform (IDCT), 38

Inverse discrete Fourier transform (IDFT), 215–216

Inverse fast Fourier transform (IFFT), 402

Inverted metal-semiconductor-metal (I-MSM), 371

IP (Internet protocol), *11*, 83–84

IP (Internet protocol)-based services, 587–598
  video over IP, 595–598
  voice over IP, 587–595

IP-multicast, 48–49

IP telephony, 592

IRCs (Incrementally related carriers), 267–268

IrDA (Infrared Data Association), 459–462; *see also* WPANs
  advanced infrared (AIR) protocol, 461–462
  data protocol stack, *450*
  optional protocols, *460*
  physical layer, 460–461
  power management, 483
  range, 476–478
  serial infrared (SIR) data, 460–461
  support for LAN integration, 481
  support for voice, 479–480

IRIDIUM satellite system, 498

IR (Infrared), 395

IrLAN (Infrared LAN), 481

ISDN (Integrated services digital network), 160; *see also* DSL; Last mile copper access
  bit rates, 160–162
  data channels, 160
  vs. digital subscriber lines (DSL), 160–161
  service costs, 160
  splitters, 171–173
  static channels in, 50
  video coding standards, 36

ISI (Intersymbol interference), 219, 229, 377–378, 446

ISMA (Internet streaming media alliance), 86

ISM (Industrial-scientific-medical) band, 442

Isochronous nodes (I-nodes), 463

ISO (International Organization for Standardization), 30

Isolators, 368

ISPs (Internet service providers), 534, 621

ITU (International Telecommunications Union), 30–31

ITU JRG 8A–9B working group, 434

ITU-T Recommendation G.983, 342

ITU-T Recommendation G.983.1, 345, *577*

ITU-T Recommendation G.983.3, 345
ITU-T Recommendation G.984, 342
ITU-T Recommendation G.992.1, 163
ITU-T Recommendation G.992.3, 163–164
ITU-T Recommendation G.992.5, 164
IUCs (Interval usage codes), 307–310

J

J.83B standard, 277
Jitter, 71–73
Joint Photographic Experts Group (JPEG), 34
Joint Video Team (JVT), 37
JPEG–2000 standard, 35–36, 44, 82
JPEG images, 27–28, 34–36
JPEG (Joint Photographic Experts Group), 34
JPEG-LS standard, 35–36
JVT (Joint Video Team), 37

K

Ka-band, 502–503, *510*, 519
Kazaa, 59, 125
K-factor, 399–400
Killer applications, 586–587, 605
Kindred near-end crosstalk, 189–190
Ku-band, 501–503

L

L2CAP (Logical link control and adaptation protocol), 458
LAP (Link access protocol), 461
Laser clipping, 277
Laser diodes, 369

Last/first meter access networks, 578–579, *615*
Last/first mile access networks, 578–579, *615*
Last meter, 5
Last mile, 1
Last mile broadband, 4–5
   bit rate, 10
   competing technologies, 105–106
   next-generation, 607–619
   wireless access, *see* BWA
Last mile copper access, 159
   digital subscriber lines (DSL), 162–168
   spectral compatibility, 237–239
   twisted-pair lines, 178–188
   unbundling, 236–237
Last mile wireless access, 387–389
   3G systems, 417
   4G systems, 418
   home access architecture, 390–393
   IEEE 802.11 standard, 432
   IEEE 802.16 standard, 419–432
   local multipoint distribution service (LMDS), 410–415
   multichannel multipoint distribution system (MMDS), 408–410
   satellite communications, 415–416
Latency, 53–55
   media downloads, 53–54
   multihop wireless access, 126
   streaming, 54–55
LCH (Long transport channel), 450
LDAP (Lightweight directory access protocol), 110
LED (Light-emitting diode), 371
LEO (Low Earth orbiting) satellites, 498
Lifeline POTS, 176
Light-emitting diode (LED), 371
Lightning, 200
Lightweight directory access protocol (LDAP), 110

Linear equalizers, 378, 401
Linear predictive coding (LPC), 25
Line codes, 162
  in digital subscriber lines (DSL),
    214–215
  discrete multi-tone (DMT),
    215–222
  very high-bit-rate DSL (VDSL),
    225–227
Line-of-sight (LOS) wireless
  technology, 395
Link access protocol (LAP), 461
Link adaptation, 403–404
Link controller, 461
Link manager protocol (LMP), 458,
  461
LLC (Logical link control), 470–471
LMCS (Local multipoint
  communication systems), 433
LMDS (Local multipoint
  distribution service), 410–415;
  see also Last mile wireless
  access
  applications, 414–415
  bandwidth, 610
  deployment of, 562
  front-end technology, 412–413
  operating frequencies, 411–412
  radio-based solutions, 414
  transmission, 410–411
LMP (Link manager protocol), 458,
  461
Local area network, 335
Local loop, 168–169
  in broadband home access,
    391–392
  crosstalk, 188–198
  impulse noise, 200
  radio-frequency interference,
    198–200
Local multipoint communication
  systems (LMCS), 433
Local multipoint distribution
  service, *see* LMDS
Logical link control and adaptation
  protocol (L2CAP), 458

Logical link control (LLC), 470–471
Long grant IUC, 308
Long transport channel (LCH), 450
Loops, 159, 615–616
LOS (Line-of-sight) wireless
  technology, 395
Lossless compression, 21–22, 34
Lossy compression, 21–22
Low Earth orbiting (LEO) satellites,
  498
Low-latency media communication,
  55
Low-pass filters, 172
LPC (Linear predictive coding), 25

M

MAC (Media access control)
  broadband wireless access
    (BWA), 405–406
  channel acquisition, 427
  common part sublayer, 422–428
  Ethernet passive optical
    networks (EPONs), 342
  frame structure, 425
  HIPERLAN/2, *440*
  HIPERLAN/2 standard, 449–450
  IEEE 802.11a standard, *440*
  IEEE 802.11 standard, 14,
    438–441
  IEEE 802.15.3 standard,
    465–467
  IEEE 802.15.4 standard,
    470–473
  IEEE 802.16 standard, 420–422
  PDU formats, 423–424
  PDU transmission, 424–425
  PHY support, 425
  service-specific convergence
    sublayers, 422
  WDM passive optical networks,
    135–136
MAC protocols, 300–303
  burst profiles, 304

downstream description,
306–310
key elements, 305–306
model for, *303*
specifications, 303
upstream description, 310–316
Macrocells, 397
MAC scheduler, 316–317
Management database (MDB), 531
Management information base, *see*
MIB
MANs (Metropolitan area
networks), 132, 419, 610
Masquerade of information, 529
Maximum useful frequency,
181–182
MBone networks, 596–597
MC (Modulation coding), 12–13; *see
also* Channel coding; Network
coding (NC)
and broadband margin, 10
in IP protocol stack, *11*
and video compression, 29
MCUs (Multipoint control units),
588
MDB (Management database), 531
Mean opinion scores (MOSs), 39
Mean squared error (MSE), 41
Media access control, *see* MAC
Media access partitioning, 307
Media compression, 20–21
applications, 20
audio, 26–27
bit rate, 42–45
classification of, 21–22
defined, 17–18
dimensions of performance in,
23–24
error concealment in, 45–46
evaluation of algorithms, 39–42
image, 27–28
speech, 24–26
standards, 30–39
techniques, 22–23
video, 28–29
Media control, 84–85

Media delivery, 84–85
vs. data delivery, 18
file downloads, 53–54
streaming, 54–55
Media gateway controllers (MGCs),
589
Media gateway control protocol
(MGCP), 589
Media gateways (MGs), 589
Media rate, 69–71
Media streaming, *see* Streaming
ME (Motion estimation), 28
Message stream modification, 529
MetroCor fiber, *357*
Metropolitan area networks
(MANs), 132, 419, 610
MF-TDMA (Multifrequency TDMA),
514
MGCP (Media gateway control
protocol), 589
MGCs (Media gateway controllers),
589
MGs (Media gateways), 589
MIB (Management information
base), 532
broadband cable access network
(BCAN), 538–541
Ethernet passive optical
networks (EPONs), 577–578
IEEE 802.11, 572
mobile IP, *566*
Microcells, 397
Microfilters, 173
Microreflections, 261–263
Microsoft Corp., 455, 519
Microwave monolithic integrated
circuit (MMIC), 508
Microwave stations, 416
MIMO (Multiple input multiple
output), 13, 573
Minifiber nodes, 326–327
Minimal intermodal coupling, 375
Minimum shift keying (MSK), 474
MMDS (Multichannel multipoint
distribution system), 408–410,
561–562, 610

MMF (Multimode fiber), 358–359
  equalizing, 373–375
  spatially resolved equalization
    (SRE), 375–378
MMIC (Microwave monolithic
    integrated circuit), 508
Mobile communications, 416–418
Mobile IP, 564–567
Mobile networks, 559–560
Mobility management, 567–568
Modal noise, 359
Modems, 12–13
  central office (CO), 173–174
  connectivity, 257
  digital subscriber lines (DSL),
    171
  satellite broadband terminals,
    510–511
Mode selective loss (MSL), 359
Modification of information, 529
Modulation, 401
  single-carrier, 401
  ultra-wideband, 402
Modulation coding (MC), 12–13; *see
    also* Channel coding; Network
    coding (NC)
  and broadband margin, 10
  in IP protocol stack, *11*
  and video compression, 29
Molniya satellite, 498
MONET project, 141
Morse code communications, 273
MOSs (Mean opinion scores), 39
Motion-compensated prediction, 28
Motion estimation (ME), 28
Motion vector, 28
Motorola, 470
Moving Pictures Expert Group
    (MPEG), 36–37
MP3, 30, *33*
MPEG–1, 30
  applications, *33, 38*
  bit rate, *33*, 37
MPEG–2, *38*
MPEG–2 advanced audio coding
    (AAC), 30–34

MPEG–4, 33–34, 37, *38*
MPEG group of pictures, *29*
MPEG (Moving Pictures Expert
    Group), 36–37
MSE (Mean squared error), 41
MSK (Minimum shift keying), 474
MSL (Mode selective loss), 359
MSOs (Multiple system operators),
    325–326, 534, 605
Multibackbone networks, 596–597
Multibeam satellite system, 505
Multicast communications, 48–49
Multicast transmission, 604
Multichannel audio coding, 27
Multichannel multipoint
    distribution system (MMDS),
    408–410, 561–562, 610
Multifrequency TDMA (MF-TDMA),
    514
Multihop wireless access, 123–128
  advantages of, 123–125
  architecture of, *124*
  congestion, 125–126
  disadvantages of, 125–128
  latency, 126
  range of, 124–125
  security, 126–127
  service costs, 124–125
Multimedia, 435
Multimedia over Coax Alliance, 609
Multimode fiber, *see* MMF
Multiple antennas, 406–408
Multiple-file switching, 43–44,
    70–71
Multiple input multiple output
    (MIMO), 13, 573
Multiple system operators (MSOs),
    325–326, 534, 605
Multiplexing, 140–141, 546
Multipoint control units (MCUs),
    588
Multiservice operators, 338
Multistage banding, 140–141
Multitone modulation, 326
Music downloads, *8*

N

Napster, 60
National Research Council, 2
Natural audio, 33–34
NAV (Network allocation vector), 440
NC (Network coding), *11*, 13–15
Near-end crosstalk, *see* NEXT
Near-far effect, 240–241
Negative dispersion fiber, 360
Network allocation vector (NAV), 440
Network coding (NC), *11*, 13–15
Network file system (NFS), 154
Network interface unit (NIU), 534
Network management, 523–526
  asymmetric digital subscriber lines (ADSL), 549–555
  billing, 530
  cellular access networks, 564–571
  configuration management, 551–554
  fault management, 554–555
  heterogenous last/first mile/meter networks, 578–579
  IEEE 802.11 wireless LAN, 571–576
  IEEE 802.16d WirelessMAN/WiMax, 563–564
  IEEE 802.16 fixed wireless network, 563
  installation and maintenance, 530
  local multipoint distribution service (LMDS), 562
  Mobile IP, 564–567
  multichannel multipoint distribution services (MMDS), 561–562
  operations support systems (OSSs), 533
  optical access networks, 576–578
  performance management, 555
  protocols, 523–525
  reliability and dependability, 527
  resource and power management, 568
  security management, 528–530, 569–571, 573–576
  simple network management protocol (SNMP), 531–533, 555–558
  tiered service, 527–528
  types of services, 527
Network protocols, 83–86
  Internet protocol (IP), 83–84
  quality of service, 528
  real-time control protocol (RTCP), 84–85
  real-time streaming protocol (RTSP), 85
  real-time transport protocol (RTP), 84–85
  session announcement protocol (SAP), 86
  session description protocol (SDP), 86
  session initiation protocol (SIP), 85
  transmission control protocol (TCP), 83–84
  user datagram protocol (UDP), 83–84
Network security, 7
NEXT (Near-end crosstalk), 189–192; *see also* Crosstalk; FEXT
  alien, 190
  coupling, 191–192
  echo cancellation, 206–207
  and frequency-division duplexing, *210*
  kindred, 189–190
  symmetric and asymmetric services, *211*
NFS (Network file system), 154
NIU (Network interface unit), 534
Noise margins, 551–553
Nokia, 455

Non-interactive media, 49–50
Null information element, 309
Nyquist pulse-shaping factor, 326

O

OADMs (Optical add/drop
   multiplexers), 142
OAMP (operations, administration,
   maintenance, and
   provisioning), 576–577,
   619–621
ODN (Optical distribution network),
   121
OEICs (Optical electronic ICs), 366
OFDMA (Orthogonal frequency-
   division multiple access), 429
OFDM (Orthogonal frequency-
   division multiplexing), 402,
   429, 444–445, 618
OLS (Optical label switching),
   146–147
OLSR (Optical label switch router),
   147
OLT (Optical line terminator), 130,
   142, 336, 344–345
One-to-one communications, 48
ONT (Optical network termination),
   336, 344–345
ONU (Optical network unit), 130,
   336, 613
Open system interface (OSI), 523,
   567
operations, administration,
   maintenance, and
   provisioning (OAMP),
   576–577, 619–621
Operations support systems, *see*
   OSSs
Optical access, 129–130
   advanced technologies, 352–354
   ATM passive optical networks,
      343–346

Ethernet passive optical
   networks (EPONs), 346–350,
   576–578
fiber to the home (FTTH)
   networks, 335–343
Gigabit-capable passive optical
   networks (GPONs), 351
network management, 576–578
passive optical networks (PONs),
   130–132, 339–340
point-to-point topologies, *337*
Quality of Service (QoS), 139
spatially resolved equalization
   (SRE), 375–378
wavelength division
   multiplexing, 132–135
Optical add/drop multiplexers
   (OADMs), 142
Optical backbone networks,
   147–149
Optical components, 354–357
   for bidirectional transport,
      369–372
   coarse wavelength division
      multiplexing (CWDM),
      364–365
   fiber, 357–360
   interconnections, 365–367
   multimode fiber (MMF), 373–375
   packaging needs, 365–367
   passive optical devices, 367–368
   vertical cavity surface emitting
      laser (VCEL), 360–364
Optical cross-connects (OXCs), 142
Optical distribution network (ODN),
   121
Optical electronic ICs (OEICs), 366
Optical isolators, 368
Optical label switching (OLS),
   146–147
Optical label switch router (OLSR),
   147
Optical line terminator (OLT), 130,
   142, 336, 344–345
Optical network termination (ONT),
   336, 344–345

Optical network unit (ONU), 130, 336, 613
Orbits, 496–498
Orthogonal frequency-division multiple access (OFDMA), 429
Orthogonal frequency-division multiplexing (OFDM), 402, 429, 444–445, 618
OSI (Open system interface), 523, 567
OSSs (Operations support systems), 541–543
  in broadband services, 619–621
  DOCSIS 1.1, 544–545
  in video-on-demand (VoD) services, 530
OXCs (Optical cross-connects), 142

P

P2MP (Point-to-multipoint) network, 348, 577
P802.15 standard, 618
PacketCable, 108–109, 609
Packet data units, *see* PDUs
Packet error rate (PER), 284
Packetization, 75–76
Packets, 15–16
Packet-switched networks, 50, 417
Packet switching, 143–144
PAM (Pulse-amplitude modulation), 215
PanAmSat, 506, 519
Parabolic reflector antennas, 507
PARK low-power state, 482
Party lines, 183
Passive channel effects, 259–266
Passive optical devices, 367–368
Passive optical networks, *see* PONs
Path loss, 398–399
Payload data unit (PDU), 310
Payload header suppression (PHS), 315
PCF (Point control function), 573

PCF (Point coordination function), 14, 441
PDF (Probability density function), 64–65
PDU (Payload data unit), 310
PDUs (Packet data units), 406
  formats, 423–424
  transmission of, 424–425
Peak signal-to-noise ratio (PSNR), 41
Peer-to-peer file sharing, 149–150
Peer-to-peer networking, 57–60
Pegasus Development Corp., 506
Penultimate mile, 4–5
PEP (Performance-enhancing proxies), 510
Perceptual audio coders, 26
Perceptual evaluation of speech quality (PESQ), 40
Perceptual irrelevancy
  reduction of, 22–23
  removal of, 25
Performance-enhancing proxies (PEP), 510
Performance management, 555
PER (Packet error rate), 284
Personal computer network, 435–436
Personal video recorder (PVR), 599
Pervasive computing, 388, 418
PESQ (Perceptual evaluation of speech quality), 40
P-frames, 29, 78
P-frames (Predictively coded frames), 29
Philips, 470
Photodetectors, 376
Photonic integrated circuit (PIC), 370
Photoreceivers, *377*
PHS (Payload header suppression), 315
Physical carrier sense, 439
Physical layer
  10–66 GHz, 428
  2–11 GHz, 428–429

broadband wireless access (BWA), 401–405
convergence procedure, 440
HIPERLAN/2 standard, 450–452
IEEE 802.11 standard, 441–447
IEEE 802.15.3 standard, 467
IEEE 802.15.4 standard, 473–475
IEEE 802.16 standard, 429–432
IrDA, 460–461
protocol data unit, 474
service data unit, 474–475
Piconets, 457
PIC (Photonic integrated circuit), 370
Pitch predictor, 25
Plain old telephone service, *see* POTS
PLC (Power line communication), 618
p-n junctions, 370
Point control function (PCF), 573
Point coordination function (PCF), 14, 441
Point-to-multipoint (P2MP) network, 348, 577
Point-to-point communications, 48
deployment of, *131*
optical access networks, *337*
transmission losses in, 78
Point-to-point protocol (PPP), 461
Polar orbit, 497
Policies, 110
PONs (Passive optical networks), 130–132
architectures, 130–132
ATM, 343–346
deployment of, *131*, 339–340, 576
dual-fiber, *133*
Ethernet, 346–350
in Ethernet for the first mile (EFM), 120–121
Gigabit-capable, 351
next-generation, 613–614
power-splitting optimization, 341
single-fiber, *133*
splitters, 340–341
standards, 342–343
wavelength division multiplexing, 135–139
POTS (Plain old telephone service), 160
central office-based, 169–170
deploying digital subscriber lines (DSL) over, 171–178
lifeline, 176
splitters, 171–173
Power line communication (PLC), 618
Power line network, 614
Power lines, 200
Power management, 568
Power spectral density, *see* PSD
PPP (Point-to-point protocol), 461
Predictively coded frames (P-frames), 29
Pre-encoded media, 49
Privacy of information, 607
Probability density function (PDF), 64–65
Propagation velocity, 264–265
PSD (Power spectral density), 181
and crosstalk, 197–198
near-far effect, 240–241
and spectrum management classes, 239
upstream power backoff, 241–242
PSNR (Peak signal-to-noise ratio), 41
PSTN (Public switched telephone network), 169, 592
Public switched telephone network (PSTN), 169, 592
Pulse-amplitude modulation (PAM), 215
PVR (Personal video recorder), 599

Q

QAM (Quadrature amplitude modulation), 215

of downstream signals, 535
and forward error correction
  (FEC), 430
modems, 227
in multichannel multipoint
  distribution system (MMDS),
  409
in orthogonal frequency-division
  multiplexing (OFDM), 445
and upstream frequencies,
  263–264
QoS (Quality of Service), 52–53
  bandwidth, 121
  broadband services, 603–605, 607
  broadband wireless access
    (BWA), 396
  cellular access networks,
    568–569
  for CMTS scheduler, 116
  DiffServ, 109
  DOCSIS standard, 110–116, 544
  Ethernet for the first mile (EFM),
    121
  IEEE 802.11 standard, 572–573
  IntServ, 107–109
  for IP networks, 107
  in network management, 528,
    568–569
  optical access networks, 139
  policy support, 109–110
  stations, 447
  universal mobile
    telecommunications system
    (UMTS), *570*
QPSK (Quadrature phase shift
    keying)
  in digital video broadcasting
    (DVB) transmission, 413
  and forward error correction
    (FEC), 430
  and intersymbol interference,
    499–500
  and microreflections, 263–264
  of upstream signals, 535
  in very small aperture terminals
    (VSATs), 511

Quadrature amplitude modulation,
    see QAM
Quadrature phase shift keying, *see*
    QPSK
Quality of Service, *see* QoS
Quantization, 26–27

R

Radio-frequency amplifiers,
    252–254
Radio-frequency interference,
    198–200
  egress, 200
  ingress, 199
Radio frequency (RF), 395–396,
    534–535
Radio layer, 458
Radio link control (RLC), 425–427
RADIUS (Remote authentication
    dial-in user service), 575
Rain attenuation, 500–503
RAKE receiver, 401
RAM (Remote access multiplexer),
    176
Ranging request (RNG-REQ), 312,
    426–427
Ranging response (RNG-RSP), 312,
    426
Rate-compatible codes, 16
Rate control, 65–67
  and bandwidth, 67–69
  and media rate, 69–71
Rayleigh fading conditions, 400
RBOCs (Regional Bell operating
    companies), 236–237
Real-time control protocol (RTCP),
    84–85
Real-time encoding, 49
Real-time streaming protocol
    (RTSP), 85
Real-time transport protocol (RTP),
    84–85, 590–591
Reconfigurable OADMs (ROADMs),
    142

Reed-Solomon codes, 277–278, 403, 429
Regeneration, 504
Regional Bell operating companies (RBOCs), 236–237
Remote access multiplexer (RAM), 176
Remote authentication dial-in user service (RADIUS), 575
Remote terminal (RT), 169–170
  digital subscriber lines (DSL) deployment from, 175–177
  fiber to the home (FTTH) networks, 334
report (*Broadband: Bringing Home the Bits*), 2–3
Request/data IUC, 308
Request IUC, 308
Request to send (RTS) frames, 440
Reservation TDMA (R-TDMA), 512
Residential gateway (RG), 392, 616
Residential telephony, 160
Resource management, 568
Resource reservation protocol (RSVP), 108
Restricted mode launch (RML), 374–375
Retransmissions, 15, 71
Retrograde orbit, 497
Return link, *517*
Reversible variable length codewords (RVLCs), 75
RF (Radio frequency), 395–396, 534–535
RG (Residential gateway), 392, 616
RLC (Radio link control), 425–427
RML (Restricted mode launch), 374–375
RNG-REQ (Ranging request), 312, 426–427
RNG-RSP (Ranging response), 312, 426
ROADMs (Reconfigurable OADMs), 142
Robust audio tool, 598
Rogue access points, 126, 576

Routers, 143–144, 595–596
RSVP (Resource reservation protocol), 108
RTCP (Real-time control protocol), 84–85
R-TDMA (Reservation TDMA), 512
RTP (Real-time transport protocol), 84–85, 590–591
RT (Remote terminal), 169–170
  digital subscriber lines (DSL) deployment from, 175–177
  fiber to the home (FTTH) networks, 334
RTSP (Real-time streaming protocol), 85
RTS (Request to send) frames, 440
RVLCs (Reversible variable length codewords), 75

S

SAMA (Spread Aloha multiple access), 513
SAP (Session announcement protocol), 86
SA (Structured audio), 34
Satellite communications systems, 415–416; *see also* Last mile wireless access
  advantages of, 495–496
  antennas, *509*
  architecture, 503–506
  band selection, 500–503
  bandwidth allocations, *502*
  DirecWay, 519–520
  DOCSIS-S standard, 514–516
  DVB-RCS standard, 513–516
  earth stations, 506–511
  link budgets, 499–500
  modern architecture, 506
  modulation and coding, 511
  multi-access protocols, 512–513
  orbits, 496–498
  performance of, 505–506
  spot-beam antennas, 504–505

Starband, 519
switched-beam antennas,
    504–505
transponders, 503–504
very small aperture terminals
    (VSATs), 516–518
SBC Communications Inc., 339
SBS (Stimulated Brillouin
    scattering), 359
Scalable coding, 70–71
Scalable compression, 44–45
S-CDMA (Synchronous code
    division multiple access), 283;
    see also TDMA
    frame structures of, *297*
    frequency spectrum, *295*
    frequency-spreading property,
        292–293
    impulse noise, 298
    narrowband ingress, 299–300
    orthogonality property, 293–295
    power per code, 295–298
    synchronization sensitivity, 300
    vs. time division multiple access
        (TDMA), 298–300
    time-spreading property, 293
    transmit power dynamic range,
        298–299
SCH (Short transport channel), 450
SCM (Single-carrier modulation),
    235, 429
SCO (Synchronous connection
    oriented) link, 458–459
SDP (Session description protocol),
    86
SDR (Software defined radio), 419
SDUs (Service data units), 422
Search engines, 604
Second-generation (2G) networks,
    417
Second mile, 4–5
Secured socket shell (SSL), 569
Secure scalable streaming (SSS), 82
Secure transcoding, 81
Security, 7, 613

Security management, 528–530,
    569–571, 573–576
Self-crosstalk, 189–192
Sequence number (SN), 450
Serial infrared (SIR) data, 460
Service data units (SDUs), 422
Service discovery protocol, 461
Service ID (SID), 304
Service level agreements (SLAs),
    107
Service-specific convergence
    sublayer (SSCS), 471
SES Americom, 519
Session announcement protocol
    (SAP), 86
Session description protocol (SDP),
    86
Session initiation protocol (SIP), 85
    in voice over IP, 589–590
Shared wireless access protocol-
    cordless access, *see* SWAP-CA
SHDSL (Symmetric high-bit-rate
    DSL), 166
    bandwidth, 236
    bit rates, 236
    echo cancellation, 206
Short grant IUC, 308
Short interframe space (SIFS), 438
Short transport channel (SCH), 450
SID (Service ID), 304
SIDs (Station identifications), 312
SIFS (Short interframe space), 438
Signaling protocols, 588–590
Signal processing, 372–373
Signal-to-inference ratio (SIR),
    290–291
Signal-to-noise ratio (SNR), 41, 284,
    295–296
Simple network management
    protocol (SNMP), 523,
    531–533, 555–558
Single-beam satellite system, 505
Single-carrier modulation (SCM),
    235, 429
Single-channel audio coding, 27
Single-mode fiber (SMF), 335, 359

SIP (Session initiation protocol), 85
  in voice over IP, 589–590
SIR (Serial infrared) data, 460
SIR (Signal-to-inference ratio),
  290–291
Skype, 149
SLAs (Service level agreements),
  107
Small and medium enterprises
  (SMEs), 600
Small-office home-office (SOHO),
  527, 600
Smart Antennas Research Group,
  399–400
SmartBits 6000B, 111–113
SMEs (Small and medium
  enterprises), 600
SMF (Single-mode fiber), 335, 359
SMI (Structure of management
  information), 532
Smoothing, 72–73
SNIFF low-power state, 482
SNMP (Simple network
  management protocol), 523,
  531–533, 555–558
SNR (Signal-to-noise ratio), 41, 284,
  295–296
SN (Sequence number), 450
Soft access points, 126–127
Software defined radio (SDR), 419
SOHO (Small-office home-office),
  527, 600
SONET/SDH, 142–143
Source coding (SC), 17–18; *see also*
  Modulation coding (MC);
  Network coding (NC)
  and channel coding, 79–80
  in IP protocol stack, *11*
Spaceway, 506
Spatially resolved equalization
  (SRE), 375–378
Spectral compatibility, 237–239
Spectrum management classes,
  239–240
Speech coders, *31–32*
Speech compression, 24–26

evaluation of algorithms, 39–40
  standards, *31–32*
Speed, 5–6
Spirent SmartBits 6000B, 111–113
Split ratio, 341
Splitters, 171–173, 340–341, 367
Spot-beam antennas, 504–505
Spread Aloha multiple access
  (SAMA), 513
Spreading gain, 292
SRE (Spatially resolved
  equalization), 375–378
SSCS (Service-specific convergence
  sublayer), 471
SSL (Secured socket shell), 569
SSs, *see* Subscriber stations
SSS (Secure scalable streaming), 82
Standard carriers, 267–271
Star architecture, 335–336
Starband, 519
Static channels, 50–51
Station identifications (SIDs), 312
Station maintenance IUC, 308
Statistical redundancy, removal of,
  22–23
Stimulated Brillouin scattering
  (SBS), 359
Store-and-forward delays, 126
Stored media, 49
Streaming, 54–55
  bandwidth, 62–63
  buffering, 71–73
  end-to-end delay, 63
  error control, 73–80
  losses, 63–64
  rate control, 65–67
  video, *see* Video streaming
Structured audio (SA), 34
Structure of management
  information (SMI), 532
Subchannels, 215–218
Subscriber stations (SSs), 422–423
  channel acquisition, 427
  IP connectivity, 427–428
  MAC PDU formats, 423–424

MAC PDU transmissions,
424–425
radio link control, 425–427
Supercells, 397
Superframes, 208–209, 212–213,
472–473
SWAP-CA (Shared wireless access
protocol-cordless access),
462–465
data rate, 478
hybrid TDMA/CSMA frame, *464*
managed and ad hoc networks,
464
power management, 482–483
range, 476–478
support for LAN integration, 481
support for voice, 479–480
SWAP-lite, 465
SWAP-MM protocol, 464–465
Switched-beam antennas, 504–505
Symmetrical digital subscriber
lines, 165–166
Symmetric DSL, 235–236
Symmetric high-bit-rate DSL, *see*
SHDSL
Synchronous code division multiple
access, *see* S-CDMA
Synchronous connection oriented
(SCO) link, 458–459
Synthetic audio, 33–34

T

T1.413 standard, 163
T1.417 standard, 239
T1 lines, 165
Tag switching, 144–145
Taps, 255–256
TCM (Trellis-coded modulation),
283, 286, 291
TC-PAM (Trellis-coded pulse
amplitude modulation), 236
TCP (Transmission control
protocol), 83–84
goodput vs. throughput, 14–15

rate control, 67–69
in voice over IP, 590
TC (Transmission convergence)
sublayer, 430
TDD (Time-division duplexing),
207–209; *see also* Duplexing
vs. frequency-division
multiplexing, 404–405
hardware sharing, 213–214
power consumption, 214
superframes, 212–213
symmetric and asymmetric
services, 210–212
synchronization requirements,
214
TDMA (Time-division multiple
access), 136–139; *see also* S-
CDMA
broadband wireless access
(BWA), 404
DOCSIS 1.0, 280–283
in fixed wireless, 610–611
in hybrid fiber-coaxial (HFC)
networks, 257
impulse noise, 298
multifrequency, 514
narrowband ingress, 299–300
network management, 563
reservation, 512
synchronization sensitivity, 300
vs. synchronous code division
multiple access (S-CDMA),
298–300
transmit power dynamic range,
298–299
TDM (Time-division multiplexing),
175, 257
Telecommunications Act of 1996, 2
Telecommunications management
network (TMN), 536–538,
541–542, 621
Telephony network, 168–170
Teleteaching, 415
Telsat, 506
Temporal key integrity protocol
(TKIP), 574

Test-to-speech (TTS), 34
TFTP (Trivial file transfer protocol), 111
Third-generation (3G) networks, 417, 567–568
Third-Generation Partnership Project (3GPP), 568–569
Throughput, 14–15, 612
Tiered services, 527–528
Time-division duplexing, *see* TDD
Time-division multiple access, *see* TDMA
Time-division multiplexing (TDM), 175, 257
TKIP (Temporal key integrity protocol), 574
TLS (Transport layer security), 569
TMN (Telecommunications management network), 536–538, 541–542, 621
Toshiba, 455
Traffic analysis, 530
Transceivers, 366–367, 508–510
Transcoding, 42, 70, 80–83
Transfer function, 194
Transient hum modulation, 277
Transmission control protocol, *see* TCP
Transmission convergence (TC) sublayer, 430
Transmission lines, 259
Transponders, 503–504
Transport layer security (TLS), 569
Trellis-coded modulation (TCM), 283, 286, 291
Trellis-coded pulse amplitude modulation (TC-PAM), 236
Triple play, 527, 606
Trivial file transfer protocol (TFTP), 111
TTS (Test-to-speech), 34
Twisted-pair lines, 178–183
    attenuation, 180–181
    bridge taps, 183–188
    cable balance, 188
    maximum useful frequency, 181–182

U

UAs (User agents), 589
UCD (Upstream channel descriptor), 307
UDP (User datagram protocol), 68, 83–84, 590
UEP (Unequal error protection), 16–17
UGS (Unsolicited grant service), 314–315
UIUC (Uplink interval usage code), 426, 430
UL-MAP (Uplink MAP), 425, 430–431
Ultra-wideband modulation (UWBM), 402
UMTS (Universal mobile telecommunications system), 568–569
Unbundling, 236–237
Unequal error protection (UEP), 16–17
Unified messaging, 592–593
UNII (Unlicensed national information infrastructure), 444
Universal mobile telecommunications system (UMTS), 568–569
Universal user ID, 607
Unlicensed national information infrastructure (UNII), 444
Unshielded twisted pair (UTP), 179
Unsolicited grant service (UGS), 314–315
Uplink interval usage code (UIUC), 426, 430
Uplink MAP (UL-MAP), 425, 430–431

Upstream channel descriptor
(UCD), 307
Upstream modulation, 279
Upstream power backoff, 241–242
Upstream signals, 534
Usenet, 58
User agents (UAs), 589
User-based security model (USM),
545
User datagram protocol (UDP), 68,
83–84, 590
USM (User-based security model),
545
UTP (Unshielded twisted pair), 179
UWBM (Ultra-wideband
modulation), 402

V

Variable-bit-rate (VBR) channels,
51–52
Variable-bit-rate (VBR) coding, 51
Variable constellation multitone
(VCMT), 301
Variable-length codeword (VLC), 75
VBR (Variable-bit-rate) channels,
51–52
VBR (Variable-bit-rate) coding, 51
VCAT (Virtual concatenation), 143
VCEL (Vertical cavity surface
emitting laser), 360–364
line widths, 358
as photodetectors, 369
VCMT (Variable constellation
multitone), 301
VDSL1, 165, 199
frequency plans, *233–234*
VDSL2, 165, *199*
VDSL Olympics, 227
VDSL (Very high-bit-rate DSL),
164–165; *see also* ADSL; DSL
vs. asymmetric digital subscriber
lines (ADSL), 547
in broadband DSL access
network (BDAN), 546–547

cyclic suffix, 229–230
egress suppression, 232–233
frequency plans, 233–234
ingress, 199
line codes, 225–227
modems, 227
standards, 224–229
system parameters, 234–235
timing advance, 230–232
windowing, 232
Vectored transmissions, 243–244
Vegetation, 411
Verizon, 339
Vertical cavity surface emitting
laser, *see* VCEL
Very high-bit-rate DSL, *see* VDSL
Very small aperture terminals
(VSATs), 516–518
Video compression, 28–29
error-resilient, 74
scalable, 44–45
standards, 36–37
Video conferencing, 596–598
Video files, *8*
Video-on-demand (VoD), 530, 606
Video over IP, 595–598
Video streaming, 60–61
bandwidth, 62–63
delay jitter, 63
error propagation in, 76–77
incorrect state, 76
Internet protocols, 83–84
joint source/channel coding in,
79–80
loss rate, 63–64
media delivery and control
protocols, 84–85
media description and
announcement protocols, 86
standards and specifications,
86–87
Virtual carrier sense, 440
Virtual circuits, 144–145
Virtual concatenation (VCAT), 143
Visual audio tool, 598
Viterbi decoder, 452

VLC (Variable-length codeword), 75
VoD (Video-on-demand), 530, 606
Voice over IP (VoIP), 56–57
    call centers, 593–594
    cellular networks, 594–595
    data transport protocol, 590–591
    growth of, 587–595
    H.323b-based network, *589*
    instant messaging, 593
    IP telephony, 592
    public-switched telephone
        network (PSTN), 592
    signaling protocols, 588–590
    SIP-based network, *590*
    unified messaging, 592–593
Voice recognition, 592–593
Voice telephony, 592
Voltages, 200
VSATs (Very small aperture
    terminals), 516–518

W

WAP (Wireless application protocol),
    569–571
Waveform speech coders, *31*
WDM (Wavelength division
    multiplexing), 132–135; *see
    also* Frequency-division
    multiplexing (FDM)
    in passive optical networks, 338
    wavelength multiplexing,
        140–141
Web advertising, 604
Web content hosting, 604
WEP (Wired encryption privacy),
    448
WEP (Wired equivalent privacy),
    574–575
Wideband speech, 25–26
Wi-Fi, 123, 128
WiFi Alliance, 574
WiFi protected access (WPA), 574
Wildblue, 519
Wi-MAX, 128, 563–564, 610

Windowing, 232
Wired encryption privacy (WEP),
    448
Wired equivalent privacy (WEP),
    574–575
Wireless access, 121–128
    cost reduction in, 122–123
    long-range, 127–128
    multihop, 123–128
    reliability of, 122–123
Wireless application protocol (WAP),
    569–571
Wireless cable, *see* LMDS
Wireless DOCSIS, 128
Wireless local area networks, *see*
    WLANs
WirelessMAN, 419–420, 563–564
WirelessMAN-OFDM, 429
WirelessMAN-OFDMA, 429
WirelessMAN-SC2, 429
Wireless personal networks, *see*
    WPANs
Wireless transport layer security
    (WTLS), 569
WLANs (Wireless local area
    networks), 434–452
    HIPERLAN/2 standard, 448–452
    home networking applications,
        434–436
    IEEE 802.11 standard, 432,
        436–448
    and IEEE 802.15.3 standard, 468
    MAC layer, 438–441
    management information base
        (MIB), 572
    network architecture, 437–438
    network management, 571–576
    next-generation, 611–613
    quality of service management,
        572–573
    security management, 573–576
    vs. wireless personal networks
        (WPANs), 483–484
WPANs (Wireless personal
    networks), 452–454
    Bluetooth, 454–459

data rate, 478
IEEE 802.15 standard, 465–475
IrDA, 459–462
power management, 481–483
range, 476–478
shared wireless access protocol-
    cordless access (SWAP-CA),
    462–465
standards, 453–454
support for LAN integration,
    480–481
support for voice, 479–480
vs. wireless local area networks
    (WLANs), 483–484

WPA (WiFi protected access), 574
WTLS (Wireless transport layer
    security), 569

X

xTU-C, 174

Z

Zigbee Alliance, 470